ALGEBRA
FOR
COLLEGE
STUDENTS

ALGEBRA FOR COLLEGE STUDENTS

Michael G. Murphy   R.C. Pierce, Jr.

Kenneth E. Oberhoff   Dennis M. Rodriguez

University of Houston

HARCOURT BRACE JOVANOVICH, INC.

New York   San Diego   Chicago   San Francisco   Atlanta   London   Sydney   Toronto

To Cathy, Wanda, Linda, and Patricia

Cover Photograph: © 1982 Meryl Joseph

Technical Drawings by Jeanne Mulderig

ISBN: 0-15-502160-5
Library of Congress Catalog Card Number: 81-81895
Printed in the United States of America

Preface

Algebra for College Students is appropriate for courses that range in
length from one quarter to one semester. It provides a concise but
thorough and carefully written treatment of the basic concepts in
algebra needed for any further work in mathematics. Because students
come to this course with varying preparation, we have included a review of elementary ideas and techniques in Chapters 1 and 2. The
chapters that follow are structured for maximum independence, to
allow the instructor as much course flexibility as possible. Chapters 3
through 6 cover the standard material on first- and second-degree
equations and inequalities, higher-degree equations, and exponential
and logarithmic functions. In the final three chapters we discuss matrix algebra, sequences and series, and counting techniques and probability.

Our goal in writing this textbook is to present algebra in a direct,
readable, yet mathematically correct fashion that can be easily understood by the student. When a class is comfortable with the book it is
using, the instructor can devote lecture time to an exploration of mathematical ideas rather than an explication of the previous night's

reading assignment. In particular, *Algebra for College Students* offers the following important features:

· We provide a wealth of carefully chosen illustrative examples, often including explanatory running comments, which are set off with color. The end of an example is indicated by ■.

· Each section concludes with a generous set of exercises, graded and paired, at all levels.

· Throughout the text and exercises, we offer numerous applications to the physical and social sciences and to business. Particular areas of application are marketing, medicine, chemistry, demography, public health, education, automation, banking, manufacturing, nutrition, agriculture, ballistics, space exploration, voting, and games.

· Graphing techniques are emphasized in the sections on functions.

· *Notes* and *Warnings* for the student, clearly set off from the text with color, anticipate common errors or suggest alternative ways of working a problem.

· We approach new concepts intuitively before giving formal definitions.

· Important results and procedures are boxed for easy reference and review.

· A list of Key Terms and Formulas and a set of review exercises appear at the end of each chapter.

· Calculator exercises, identified by the symbol C, are included where appropriate.

Answers to odd-numbered exercises are given at the back of the book. The accompanying Instructor's Manual provides answers to the even-numbered exercises and additional test material.

We are indebted to reviewers Roger Breen (Florida Junior College), Philip Cheifetz (Nassau Community College), Nancy Jim Poxon (California State College, Sacramento), Ken Rager (Metropolitan State College), and Jack Twitchell (Mesa Community College) for their helpful comments and suggestions. We would also like to express our gratitude to the staff of Harcourt Brace Jovanovich: to Marilyn Davis for her guidance and encouragement, to Judy Burke for her meticulous editing of the manuscript, to designer Arlene Kosarin, to art editor Marian Griffith, and to production manager Robert Karpen. The expert typing of the manuscript was done by Jan Want.

MICHAEL G. MURPHY
KENNETH E. OBERHOFF
R. C. PIERCE, JR.
DENNIS M. RODRIGUEZ

Contents

Chapter 9   Counting Techniques and Probability

CHAPTER 1

Review
of
Basic
Algebra

This chapter provides a brief but thorough review of basic algebra. We will review the fundamental operations for the various types of numbers and related expressions. The chapter begins with a discussion of sets and then progresses through a review of the natural numbers, the integers, the rational numbers, the real numbers, and finally the entire complex number system. Operations with polynomials are also thoroughly reviewed.

1.1   Sets and Operations

A **set** is simply a collection of objects. These objects, called **elements** or **members,** may be numbers, people, marbles, events, or anything at all.

Sets are usually designated by capital letters A, B, C, and so on. The members of a set are listed or described within braces, { }. Thus the set A that contains only the numbers 2, 3, and 4 is written $A = \{2, 3, 4\}$. The symbol $\in$ means "is an element of" and is used to indicate set membership; for instance, $2 \in A$ means that 2 is an element of the set A. The symbol $\notin$ is the negation of $\in$, so that $5 \notin A$ means the element (number) 5 is not a member of the set A. The set of all counting numbers, called the **natural numbers,** is $N = \{1, 2, 3, 4, 5, \ldots\}$, where the three ellipsis dots mean "continue the pattern." Thus we are to understand that $6 \in N$, $7 \in N$, $8 \in N$, and so on.

We can also describe sets with **set-builder notation:** $\{x \mid x$ has property $P\}$. We read this as "the set of all elements x such that x has the property P."

Example 1   List the elements described by the set-builder notation

$$\{x \mid x \text{ is a natural number smaller than } 5\}$$

Solution   The elements satisfying the given property are the numbers 1, 2, 3, and 4. Thus the answer is $\{1, 2, 3, 4\}$. ■

Two sets are **equal** if they contain the same elements. Thus for the set in Example 1 we may write

$$\{x \mid x \text{ is a natural number smaller than } 5\} = \{1, 2, 3, 4\}$$

If $X = \{a, 2, 3\}$ and $Y = \{2, a, 3, 2\}$, then $X = Y$, since each set contains only the elements a, 2, and 3. The fact that the elements are not in the same order, or that the number 2 is listed twice in the set Y, does not affect equality.

A set R is a **subset** of another set S, written $R \subseteq S$, if each member of R is also a member of S.

Sets are often represented by geometric figures where the region within a figure represents the elements of a set. Such diagrams are called **Venn diagrams.** For example, to show that set R is a subset of set S, we would draw one geometric figure representing R that is completely contained inside another geometric figure representing S (Figure 1).

The symbol $\nsubseteq$ is the negation of $\subseteq$, so that $T \nsubseteq S$ means that T is not a subset of S.

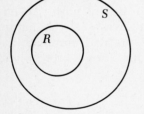

FIGURE 1

Example 2   If $A = \{0, 1, 2\}$, $B = \{2, 0, 1\}$, and $C = \{0, 2\}$, then $A \subseteq B$, $B \subseteq A$, $C \subseteq A$, and $C \subseteq B$, but $A \nsubseteq C$ and $B \nsubseteq C$. ■

In Example 2, $A \subseteq B$ and $B \subseteq A$ because $A = B$ (this is sometimes used as the definition of equality of sets). Moreover, $C \subseteq A$ but $C \neq A$, since $1 \in A$ but $1 \notin C$. Thus C is called a **proper subset** of A, and we write $C \subset A$ or $C \subsetneq A$.

The set operations of union and intersection generate new sets from given sets. The **union** of two sets A and B, written $A \cup B$, is the set of all elements that belong to either A or B or both.

$$A \cup B = \{x \mid x \in A \text{ or } x \in B\}$$

The **intersection** of two sets A and B, written $A \cap B$, is the set of all elements that belong to both A and B.

$$A \cap B = \{x \mid x \in A \text{ and } x \in B\}$$

Example 3 Let $A = \{1, 2, 3, x, y\}$ and $B = \{3, 4, 5, x, w\}$. Then

$$A \cup B = \{1, 2, 3, 4, 5, x, y, w\}$$

and

$$A \cap B = \{3, x\} \quad \blacksquare$$

The concepts of union and intersection can be extended to more than two sets—in fact, to any number of sets. The union is the set of all elements from all the sets. The intersection is the set of all elements common to every set.

If two sets such as $M = \{1, 2\}$ and $N = \{x, y\}$ have no common elements, then $M \cap N$ will be empty or void. We will use the symbol $\varnothing$ to denote this and refer to $\varnothing$ as the **empty set**. Thus $M \cap N = \varnothing$ means the two sets are **disjoint** or have no elements in common.

For a given situation or problem, we are usually concerned only with certain objects. The **universe** or **universal set** is the set of all objects under consideration.

Example 4 A die is a small cube having its sides numbered 1 through 6. When a die is thrown or rolled, one number will appear as the top number. Thus when throwing a die, our universal set U is

$$U = \{1, 2, 3, 4, 5, 6\} \quad \blacksquare$$

If U is the universal set under consideration and A is a subset of U, then the complement of A, written $\overline{A}$, is the set of all elements of U that are not in A.

$$\overline{A} = \{x \mid x \in U \text{ and } x \notin A\}$$

Example 5 Construct a Venn diagram showing the complement of a set A.

Solution See Figure 2. The rectangle represents the entire universe U. The circle represents the set A, and the shaded area outside the circle represents the solution $\overline{A}$. ∎

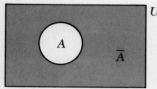

FIGURE 2

The **cross-product** of two sets A and B, written $A \times B$, is the set of all pairs of elements (a, b) whose first element is from A and whose second element is from B.

$$A \times B = \{(a, b) \mid a \in A \text{ and } b \in B\}$$

The order of the elements in the pair is important. Thus $(a, b) \neq (b, a)$ unless $a = b$.

Example 6 Construct $A \times B$ given $A = \{1, 2\}$ and $B = \{w, x, y\}$.

Solution $A \times B = \{(1, w), (1, x), (1, y), (2, w), (2, x), (2, y)\}$. ∎

The two sets do not have to be different to form the cross-product. We will see later, when we discuss the set of real numbers R, that $R \times R$ is a very important set in mathematics.

Our last concept in the study of sets is the power set of a universal set. The **power set** of U is the set of all subsets of U, including the empty set $\varnothing$ and U itself.

Example 7 Construct the power set of $U = \{a, b, c\}$.

Solution The power set of U is

$$P = \{\varnothing, \{a\}, \{b\}, \{c\}, \{a, b\}, \{a, c\}, \{b, c\}, U\}$$ ∎

In general, if U contains n elements, the number of sets in the power set of U is 2^n. In Example 7 U contains 3 elements, so there are $2^3 = 8$ elements in the power set.

If we consider the power set of U, as in Example 7, and the set operation $\cup$ (union) on these elements, we see that the union of any two subsets of U is again a subset of U. The same statement can be made for the set operation $\cap$ (intersection). This idea is important in the study of any operation on a set and is called **closure** with respect to that operation. The power set of a given set has closure with respect to union of its elements. It also has closure with respect to intersection of its elements. In Section 1.2 we will discuss the closure of certain sets of numbers with respect to familiar arithmetic operations.

Exercises 1.1

In Exercises 1–10 let $A = \{3, 4, 5, x, y\}$ and $B = \{4, x, y, z\}$. Insert $\in$ or $\notin$ in each blank to make a correct statement.

1. 3 ___∈___ A
2. x ___∈___ A
3. 3 ___∉___ B
4. x ___∈___ B
5. z ___∉___ A
6. z ___∈___ B
7. 1 ___∉___ A
8. 1 ___∉___ B
9. A ___∉___ B
10. B ___∉___ A

In Exercises 11–16 list the elements described by the set-builder notation. You may use ellipses to mean "continue the pattern" for infinite lists.

11. $\{x \mid x$ is a natural number smaller than 7$\}$ 1 2 3 4 5 6
12. $\{x \mid x$ is a natural number smaller than 10$\}$ 1 2 3 4 5 6 7 8 9
13. $\{y \mid y$ is a natural number larger than 6$\}$ 6, 7, 8, 9, - - - -
14. $\{y \mid y$ is an even (divisible by 2) natural number$\}$ 2, 4, 6, 8, 10 - - -
15. $\{x \mid x$ is a letter of the alphabet between C and H$\}$ D, E, F, G
16. $\{x \mid x$ is an odd (not divisible by 2) natural number$\}$ 1, 3, 5, 7, 9 - - - -

In Exercises 17–20 insert $=$ or $\neq$ in each blank to make a correct statement.

17. $\{a, b, c\}$ ___=___ $\{b, c, a\}$
18. $\{0, 1, 2, 3\}$ ___=___ $\{0, 1, 2, 3, 3\}$
19. $\{x, y, z\}$ ___≠___ $\{x, y, z, w\}$
20. $\{1, 2, 3, \ldots\}$ ___=___ $\{x \mid x$ is a natural number$\}$

In Exercises 21–30 let $X = \{1, 2, 3, 4\}$ and $Y = \{4, 6, 8, 10\}$. Insert $\subseteq$ or $\not\subseteq$ in each blank to make a correct statement.

21. X ___⊄___ Y
22. Y ___⊄___ X
23. X ___⊆___ $X \cup Y$
24. Y ___⊄___ $X \cup Y$
25. X ___⊄___ $X \cap Y$
26. $\varnothing$ ___⊆___ X
27. $\varnothing$ ___⊆___ Y
28. $\varnothing$ ___⊆___ $X \cap Y$
29. X ___⊆___ X
30. $X \cap Y$ ___⊄___ $X \cup Y$

In Exercises 31–40 the universal set U is the set of natural numbers smaller than 10. Let $A = \{1, 3, 5\}$ and $B = \{2, 4, 6\}$. Form the required set.

31. $A \cup B$
32. $A \cap B$
33. $\overline{A}$
34. $\overline{B}$

31. 1 2 3 4 5 6
32. ∅
33. 2, 4, 6, 7, 8, 9
34. 1, 3, 5, 7, 8, 9

35. 7, 8, 9
36
37 7,8,9
38
39 1,3,5
40 2,4,6

35. $\overline{A \cup B}$ 36. $\overline{A \cap B}$
37. $\overline{A} \cap \overline{B}$ 38. $A \cap U$
39. $A \cap A$ 40. $B \cup B$

In Exercises 41–45 construct a Venn diagram showing the required set or relationship.
41. Show the union of two disjoint sets.
42. Show the intersection of two sets one of which is a subset of the other.
43. Show the relationship of the set of natural numbers and $\{x \mid x$ is a natural number smaller than 5$\}$.
44. Show the union of three sets.
45. Show $\overline{A \cup B}$.

In Exercises 46–50 list the elements of $A \times B$.
46. $A = \{1, 2\}$, $B = \{3\}$ 47. $A = \{x, y, z\}$, $B = \{4, 5\}$
48. $A = \{4, 5\}$, $B = \{x, y, z\}$ 49. $A = \{2, 3\} = B$
50. $A = \{x \mid x$ is a natural number smaller than 5$\}$
    $B = \{y \mid y$ is a natural number smaller than 3$\}$
51. Is $A \times B$ equal to $B \times A$ always, sometimes, or never? Explain. *IF A=B*
52. Construct the power set of $U = \{1, 2\}$.
53. Construct the power set of $U = \{1, 2, 3, 4\}$. *U=2^N   U=2^4=16*

{∅}, {1}, {2}, {3}, {4}, {1,2}, {1,3}, {1,4}
{2,3}, {2,4}, {3,4}, {1,2,3}, {1,2,4} {1,3,4},
{2,3,4}, 4}

1.2 The Real Numbers

We begin our discussion with the set of natural numbers $N = \{1, 2, 3, 4, 5, \ldots\}$ and the basic operations addition ($+$), subtraction ($-$), multiplication ($\cdot$) and division ($\div$). The operations addition and multiplication are closed when applied to N. That is, the sum of any two natural numbers is a natural number and the product of any two natural numbers is a natural number. However, the operation subtraction is not closed on N. For example, if we subtract 5 from 3, the result is -2, which is not a natural number. A similar statement holds for division, since $3 \div 5$ is not a natural number.

The set of integers was developed in order to give closure to the operation subtraction. The set of integers is the set of all signed counting numbers and 0. Thus the set of integers is $Z = \{\ldots, -5, -4, -3, -2, -1, 0, 1, 2, 3, 4, 5, \ldots\}$. The set Z is closed under the operations addition, subtraction, and multiplication. We demonstrate this with a few examples, which will also serve to review the laws of signs.

Example 1   Add the following integers:

(a) 5   (b) -5 (c) -5 (d) 5
     15   -15 $+15$ -15

Solution   (a) The numbers have like signs (understood +), so we add to get 20 and use the common sign, obtaining $+20$.

(b) The numbers have like signs $(-)$, so we add the unsigned numbers to get 20 and use the common sign, obtaining -20 .

(c) The numbers have unlike signs (one +, one −), so we take the difference of the unsigned numbers to get 10 and use the sign of the larger unsigned number (+ from the 15), to obtain $+10$.

(d) As in part (c), the numbers have unlike signs, so we take the difference, which again is 10, and the sign of the larger unsigned number (− from the 15), to obtain -10 . ■

Note that each of the answers in Example 1 is an integer.

Example 2   In each case subtract the bottom number, called the **subtrahend**, from the top number:

(a)   5   (b)   −5   (c)   −5   (d)   5
     15   − 15   + 15   − 15

Solution   Note that each of these problems is written exactly as in Example 1. However, the operation is now subtraction. Recall the rule for subtraction: "Change the sign of the subtrahend and add."

(a) Change   5   to   5   and add to obtain -10 .
           15   − 15

(b) Change   −5   to   −5   and add to obtain $+10$.
           − 15   + 15

(c) Change   −5   to   −5   and add to obtain -20 .
           + 15   − 15

(d) Change   5   to   5   and add to obtain $+20$.
           − 15   + 15   ■

Each of the answers found in Example 2 is an integer. This demonstrates but does not prove that the integers are closed under subtraction. However, we will assume that the integers are closed under subtraction.

Finally, we review the rules of signs for multiplication of integers.

Example 3   Multiply the following:

(a) $(+2)(+3)$ (b) $(-2)(-3)$ (c) $(-2)(3)$ (d) $(2)(-3)$

Solution   (a, b) The answer to both parts is $+6$, since two like signs multiply to give "+".

(c, d) The answer to both parts is -6 , since two unlike signs multiply to give "−".   ∎

The integers are not closed under the fourth operation, division, since $3 \div 5$ is not an integer. For this reason the set of **rational numbers** was formed. The set of rational numbers Q is

$$Q = \left\{ \frac{x}{y} \,\middle|\, x \text{ and } y \text{ are integers and } y \neq 0 \right\}$$

Some examples of rational numbers are $\frac{2}{3}$, $5\frac{1}{4} = \frac{21}{4}$, $-6 = \frac{-6}{1}$, and $0 = \frac{0}{1}$. A rational number can be **reduced** if both the numerator and denominator are divisible by the same number. Thus

$$\frac{6}{4} = \frac{3 \cdot \cancel{2}}{2 \cdot \cancel{2}} = \frac{3}{2}$$

The rule for reducing a rational number is

$$\frac{a \cdot x}{b \cdot x} = \frac{a}{b}$$

The set of rational numbers is closed under the operations addition, subtraction, multiplication, and division (except for division by 0). The rules of signs for rational numbers are the same as those for integers.

The rule for multiplication of fractions is

$$\frac{a}{b} \cdot \frac{c}{d} = \frac{a \cdot c}{b \cdot d}$$

Example 4   Multiply the following rational numbers:

(a) $-5 \cdot \dfrac{2}{7}$ (b) $\dfrac{3}{4} \cdot \dfrac{-1}{3}$ (c) $\dfrac{-2}{9} \cdot \dfrac{-3}{4}$

Solution   (a) $-5 \cdot \dfrac{2}{7} = \dfrac{-5}{1} \cdot \dfrac{2}{7} = \dfrac{-5 \cdot 2}{1 \cdot 7} = \dfrac{-10}{7}$

(The integer 5 is written as $\frac{5}{1}$.)

(b) $\dfrac{3}{4} \cdot \dfrac{-1}{3} = \dfrac{\cancel{3} \cdot (-1)}{4 \cdot \cancel{3}} = \boxed{\dfrac{-1}{4}}$

(We can reduce by dividing equal factors.)

(c) $\dfrac{-2}{9} \cdot \dfrac{-3}{4} = \dfrac{\overset{-1}{\cancel{-2}} \cdot \overset{-1}{(\cancel{-3})}}{\underset{3}{\cancel{9}} \cdot \underset{2}{\cancel{4}}} = \dfrac{(-1) \cdot (-1)}{3 \cdot 2} = \boxed{\dfrac{1}{6}}$

$\left(\text{The same result is obtained by writing } \dfrac{-2 \cdot (-3)}{9 \cdot 4} = \dfrac{6}{36} = \dfrac{1}{6}.\right)$ ∎

The rule for adding two fractions that have the same denominator is

$$\boxed{\dfrac{a}{b} + \dfrac{c}{b} = \dfrac{a + c}{b}}$$

When the denominators are not the same, the fractions must be expressed in equivalent forms using common denominators, as will be shown in Example 5(b) and (c).

Example 5   Add the following rational numbers:

(a) $\dfrac{1}{7} + \dfrac{3}{7}$ (b) $\dfrac{1}{4} + \dfrac{1}{8}$ (c) $\dfrac{1}{6} + \dfrac{1}{4}$

Solution   (a) When the denominators are the same, combine the numerators. Our answer is

$$\dfrac{1}{7} + \dfrac{3}{7} = \dfrac{1 + 3}{7} = \boxed{\dfrac{4}{7}}$$

(b) When the denominators are not equal, find the lowest common denominator (L.C.D.), which is the smallest number into which each denominator divides evenly. Here both 4 and 8 divide into 8. Thus

$$\dfrac{1}{4} + \dfrac{1}{8} = \left(\dfrac{1}{4} \cdot \dfrac{2}{2}\right) + \dfrac{1}{8} = \dfrac{2}{8} + \dfrac{1}{8} = \dfrac{2 + 1}{8} = \boxed{\dfrac{3}{8}}$$

Note: In rewriting the fraction $\dfrac{1}{4}$ we used the property that $a \cdot 1 = a$. Multiplication by a form of 1, such as $\dfrac{2}{2}$, changes the appearance of a fraction but not its value.

(c) As in (b), we need the L.C.D. Since $6 = 2 \cdot 3$ and $4 = 2 \cdot 2$,

$$\text{L.C.D.} = \underline{2 \cdot 2 \cdot 3} = 12$$
$$\uparrow \quad \uparrow$$
$$4 \quad 6$$

Thus

$$\frac{1}{6} + \frac{1}{4} = \left(\frac{1}{6} \cdot \frac{2}{2}\right) + \left(\frac{1}{4} \cdot \frac{3}{3}\right) = \frac{2}{12} + \frac{3}{12} = \frac{2+3}{12} = \boxed{\frac{5}{12}} \quad \blacksquare$$

The rule for subtracting two fractions that have the same denominator is

$$\frac{a}{b} - \frac{c}{b} = \frac{a-c}{b}$$

As with addition, when the denominators are not the same, we must find a common denominator before subtracting. For example, $\frac{3}{5} - \frac{1}{4}$ must be rewritten with a common denominator of 20 to get

$$\frac{12}{20} - \frac{5}{20} = \frac{12-5}{20} = \frac{7}{20}$$

Recall from basic arithmetic that division of fractions was performed as follows:

$$\frac{2}{3} \div \frac{1}{5} = \frac{2}{3} \cdot \frac{5}{1} = \frac{10}{3}$$

Expressing this in a general fashion gives the rule for division:

$$\frac{a}{b} \div \frac{c}{d} = \frac{a}{b} \cdot \frac{d}{c} = \frac{a \cdot d}{b \cdot c}$$

The set of rational numbers satisfies the **field properties** for the operations addition and multiplication. If we let a, b, and c represent rational numbers, then these properties are as follows.

1. *Closure:* $a + b$ and $a \cdot b$ belong to the set for all elements a and b.
2. *Commutativity:* $a + b = b + a$ and $a \cdot b = b \cdot a$.
3. *Associativity:* $(a + b) + c = a + (b + c)$ and $(a \cdot b) \cdot c = a \cdot (b \cdot c)$.

4. *Identity elements:* $a + 0 = a$ (0 is the identity for addition).

$a \cdot 1 = a$ (1 is the identity for multiplication).

5. *Inverses:* $a + (-a) = 0$.

$$a \cdot \frac{1}{a} = 1. \left(\text{If } a = \frac{x}{y}, \text{ then } \frac{1}{a} = \frac{y}{x}. \right)$$

(There is no inverse for 0 under multiplication.)

6. *Distributivity:* $a \cdot (b + c) = a \cdot b + a \cdot c$.

Another way to describe the rational numbers is as the set of all numbers having a repeating decimal expansion (of fixed length). Thus, $5 = 5.00\overline{0}$ (repeating 0), $\frac{1}{3} = 0.33\overline{3}$ (repeating 3), $\frac{7}{30} = 0.23\overline{3}$ (repeating 3), and $\frac{8}{33} = 0.24\overline{24}$ (repeating 24), where the bar above the number indicates the repeating part, are all rational numbers.

There are many useful numbers that cannot be written as repeating decimals. Such numbers are called **irrational numbers.** Often we represent irrational numbers with special symbols such as $\sqrt{2}$, the square root of 2 (discussed in Section 1.8), which is approximately 1.4, and e (discussed in Chapter 6), which is approximately 2.71.

The set of **real numbers** is the set of all decimal numbers, both repeating and nonrepeating. The set of real numbers, which we call R, contains the rational numbers as a subset and satisfies the field properties.

Let us list the major subsets of R:

1. The natural numbers, $N = \{1, 2, 3, 4, \ldots\}$
2. The integers, $Z = \{\ldots -4, -3, -2, -1, 0, 1, 2, 3, 4, \ldots\}$

3. The rational numbers, $Q = \left\{ \dfrac{x}{y} \,\middle|\, x \in Z, y \in Z, y \neq 0 \right\}$

4. The irrational numbers, $\{x \in R \mid x \notin Q\}$

The real numbers can be plotted or graphed on a number line, as shown in Figure 3. The **positive** numbers are plotted in increasing order of size to the right of 0, and the **negative** numbers are plotted to the left of 0. A real number x is **less than** the real number y, written $x < y$ or $y > x$ (y is **greater than** x), if x is plotted to the left of y on the number line. We write $x \leq y$ if x is less than y or possibly equal to y and, similarly $y \geq x$ if $y > x$ or $y = x$.

We complete our discussion of real numbers in this section with the definition of absolute value and a discussion of order of operations.

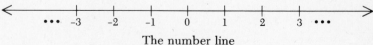

The number line

FIGURE 3

The **absolute value** of a number x, written $|x|$, is

$$|x| = \begin{cases} x & \text{if } x \geq 0 \\ -x & \text{if } x < 0 \end{cases}$$

Thus $|+5| = +5 = 5$ and $|-5| = -(-5) = +5 = 5$.

The **order of operations** is a rule that tells us how to evaluate an expression containing several operations.

ORDER OF OPERATIONS

1. Evaluate within parentheses
2. Exponents
3. Multiplication and division
4. Addition and subtraction

When multiplication and division appear in the same expression, we perform them from left to right. Thus $8 \div 4 \cdot 2 = (8 \div 4) \cdot 2 = 4$, and $6 \cdot 10 \div 5 = (6 \cdot 10) \div 5 = 12$. The same rule applies when addition and subtraction appear in the same expression.

Example 6   Evaluate:
(a) $3 - 3 \cdot 5 + 2$ (b) $(3 - 3) \cdot 5 + 2$
(c) $3 - 3 \cdot (5 + 2)$

Solution   (a) $3 - 3 \cdot 5 + 2 = 3 - (3 \cdot 5) + 2$

(multiply before adding or subtracting)

$$= 3 - 15 + 2$$
$$= -12 + 2 \quad \text{(add and subtract from left to right)}$$

$$= \boxed{-10}$$

(b) $(3 - 3) \cdot 5 + 2 = (0 \cdot 5) + 2$ (evaluate within parentheses first)
$$= 0 + 2$$

$$= \boxed{2}$$

(c) $3 - 3 \cdot (5 + 2) = 3 - (3 \cdot 7)$ (evaluate within parentheses first)
$$= 3 - 21$$

$$= \boxed{-18} \quad \blacksquare$$

Exercises 1.2

handwritten:
a. 8, 2, -8, -2, -5
b. 2, 8, -2, -8, 5
c 16, -16, 15, -15, 0

In Exercises 1–5
(a) add the two numbers;
(b) subtract the bottom number from the top number;
(c) multiply the two numbers.

1.	$+5$	2.	$+5$	3.	-5
	$+3$		-3		-3

4.	-5	5.	0
	$+3$		-5

In Exercises 6–20 perform the indicated operation.

6. $\dfrac{1}{2} + \dfrac{1}{3}$ $5/6$ 7. $\dfrac{1}{2} - \dfrac{1}{3}$ $1/6$

8. $\dfrac{1}{2} \cdot \dfrac{1}{3}$ $1/6$ 9. $\dfrac{1}{2} \div \dfrac{1}{3}$ $3/2$

10. $\dfrac{3}{4} + \dfrac{3}{8}$ $1\frac{1}{8}$ 11. $\dfrac{3}{4} - \dfrac{3}{8}$ $3/8$

12. $\dfrac{3}{4} \cdot \dfrac{3}{8}$ $9/32$ 13. $\dfrac{3}{4} \div \dfrac{3}{8}$ 2

14. $5 + \dfrac{1}{10}$ $5\frac{1}{10}$ 15. $5 \cdot \dfrac{1}{10}$ $\frac{1}{2}$

16. $\dfrac{1}{10} \div 5 = 1/50$ 17. $\dfrac{1}{10} - 5$ $-4\frac{9}{10}$

18. $\dfrac{2}{9} \cdot \dfrac{3}{4} \cdot \dfrac{8}{7}$ $\dfrac{4}{21}$ 19. $\dfrac{4}{7} + \dfrac{3}{14} - \dfrac{5}{6}$

handwritten: $24 + 9 - 35 = \dfrac{-2}{42} = \left\{\dfrac{1}{21}\right\}$

20. $\dfrac{1}{5}\left(\dfrac{1}{4} + \dfrac{1}{6}\right) = 1/12$

In Exercises 21–30 identify the field property illustrated.
21. $2 + 3 = 3 + 2$ *Com*
22. $(2 + 3) + 4 = 2 + (3 + 4)$
23. $2 \cdot 3 = 3 \cdot 2$ *com*
24. $2 \cdot (3 \cdot 4) = (2 \cdot 3) \cdot 4$
25. $0 + 5 = 5$ *Id el*
26. $1 \cdot 6 = 6$

27. $2(6 + 2) = 12 + 4$ *distributivity*
28. $2 \cdot \dfrac{1}{2} = 1$

29. $3 + (-3) = 0$ *INVERSE*
30. $\dfrac{3}{4} + \dfrac{9}{37}$ is a rational number

handwritten number line: ← -3 0 .5 1.45 √5 →

31. Construct a number line and plot the points 0, 5, -3, $\frac{1}{2}$, $\frac{15}{2}$, and $\sqrt{2}$.
    (*Hint:* For $\sqrt{2}$, use the approximate value 1.4.)
32. Calculate the decimal expansion of the following rational numbers by dividing the denominator into the numerator. Identify the repeating part of the decimal expansion.

(a) $\dfrac{2}{3}$ $.\overline{66}$ (b) $\dfrac{1}{4}$ $.25$

(c) $\dfrac{5}{6}$ $.30$ (d) $\dfrac{2}{5}$ $.4$

33. Calculate the following absolute values:

   (a) $\left|-\dfrac{2}{3}\right|$ 2/3

   (b) $|6 - 3|$ 3

   (c) $\left|\dfrac{1}{4} - \dfrac{1}{3}\right|$ 1/12

   (d) $\left|\dfrac{1}{4}\right| - \left|\dfrac{1}{3}\right|$ − 1/12

In Exercises 34–40 use the order of operations to evaluate.

34. $4 - 4 \div 2$ 2

35. $3 \div 6 \cdot 4$ 2

36. $6 \cdot 4 \div 3$ 8

37. $2 - 2(3 - 6 \cdot 5)$ 56

38. $\dfrac{1}{2} + \dfrac{2}{3} \div \dfrac{4}{9}$

39. $\dfrac{1}{2} \cdot \dfrac{2}{3} \div \dfrac{4}{9}$ 1/3 × 9/4   9/12   3/4

40. $\left(\dfrac{1}{2} + \dfrac{2}{3}\right) \div \dfrac{4}{9}$

1.3   Addition and Subtraction of Polynomials

A **variable** is a symbol, such as x, which is used to denote one member of a given set. In this section we will use variables to represent real numbers and will discuss some of the general properties of combining variables. If n is any natural number and x is any real number, then we define x^n by

$$x^n = \underbrace{x \cdot x \cdot x \cdots x}_{n \text{ times}}$$

Thus x^n means "take the product of n x's," so that $5^4 = 5 \cdot 5 \cdot 5 \cdot 5 = 625$ and $(\frac{2}{3})^2 = \frac{2}{3} \cdot \frac{2}{3} = \frac{4}{9}$. In the expression x^n, the variable n is called the **exponent** and the variable x is called the **base**. Any product of a fixed real number and a finite number of variables with exponents is called a **term**. The **degree of the term** is the sum of the exponents of its variables. The fixed real number is called the **constant** or the **coefficient,** or more precisely, the **numerical coefficient**. A sum of a finite number of terms, with positive integer exponents on all the variables, is called a **polynomial**. The **degree of the polynomial** is the maximum of the degrees of its terms.

Example 1   Find the degree of the polynomial $2xy^3 - 3w + \frac{1}{2}$. State the number of terms and identify the numerical coefficient of the second term.

Solution The degree of the first term is 4 (exponent of x plus exponent of y). The degree of the second term is 1 (exponent of w). The degree of the third term is 0 (no variables). The degree of the polynomial is the largest of these three numbers, 4.

The polynomial has three terms: $2xy^3$, $-3w$, and $\frac{1}{2}$. The numerical coefficient of the second term is -3. (the sign of the term is part of the coefficient.) ■

For contrast, the next example lists several expressions that are not polynomials.

Example 2 The following are not polynomials:

(a) $5 + \dfrac{1}{x}$ (no variable is allowed in the denominator).

(b) $3x^2y + 7y^{-3} - 6$ (a negative exponent such as -3 is not allowed).

(c) $2xy^4 + 3\sqrt{x}$ (a variable under a radical sign is not allowed). ■

Our objective in this section is to state the rules for addition and subtraction of polynomials. Remember that the variables used in each of our polynomials represent real numbers, so that each of the rules must hold when the variables are replaced by fixed constants.

Example 3 Demonstrate that the value of $x + x$ is the same as the value of $2x$ when

(a) x is replaced by 5.   (b) x is replaced by -3.
(c) x is replaced by 0.

Solution (a) $x + x = 5 + 5 = 10$
$2x = 2 \cdot 5 = 10$
Result: The values obtained are equal.
(b) $x + x = (-3) + (-3) = -6$
$2x = 2 \cdot (-3) = -6$
Result: The values obtained are equal.
(c) $x + x = 0 + 0 = 0$
$2x = 2 \cdot 0 = 0$
Result: The values obtained are equal.

We could continue this process and check other real values for x. In each case we would find that the two expressions give equal results. ■

For any natural number n and any variable x we have

$$n \cdot x = \underbrace{x + x + \cdots + x}_{n \text{ times}}$$

This property allows us to combine $2x + 3x$, since $2x + 3x = (x + x) + (x + x + x) = 5x$. This also fits the distributive property (Section 1.2): $2x + 3x = (2 + 3)x = 5x$. However, this approach does not enable us to combine $2x + 3y$, since $2x + 3y = (x + x) + (y + y + y)$, which cannot be simplified any further. This case also does not fit the distributive property. Thus we have arrived at our rule for combining terms of a polynomial: The variables (including the exponents) in each term must be the same so that we can apply the distributive property to combine the numerical coefficients. The variables and their exponents will not change.

Example 4   Combine, if possible:
(a) $2xy^2 + 3xy^2$ (b) $2xy^2 - 3xy^2$
(c) $3x^2y^2 - 5xy^2$

Solution   (a) $2xy^2 + 3xy^2 = (2 + 3)xy^2$ (distributive property)

$$= 5 \cdot xy^2 = \boxed{5xy^2}$$

(b) $2xy^2 - 3xy^2 = (2 - 3)xy^2$ (distributive property)

$$= (-1) \cdot xy^2 = \boxed{-1xy^2} = \boxed{-xy^2}$$

(c) Since the exponents of the variables are not the same, (x has exponent 2 in the first term and exponent 1 in the second term), the terms cannot be combined. ∎

Two polynomials are added by rearranging their terms and combining the like terms. Two polynomials are subtracted by changing the sign of each term in the subtrahend and then adding.

Example 5   Combine the polynomials:
(a) $(2x - 3x^2 + 4y) + (5x^2 - 6y + 7)$.
(b) $(2x - 3x^2 + 4y) - (5x^2 - 6y + 7)$

Solution (a) $(2x - 3x^2 + 4y) + (5x^2 - 6y + 7) = \underline{2x} - \underline{3x^2 + 5x^2} + \underline{4y - 6y}$

$\underline{+ 7}$ (rearrange terms)

$= 2x + (-3 + 5)x^2 + (4 - 6)y$

$+ 7$ (distributive property)

$= \boxed{2x + 2x^2 - 2y + 7}$

(combine)

(b) $(2x - 3x^2 + 4y) - (5x^2 - 6y + 7) = 2x - 3x^2 + 4y - 5x^2 + 6y$

$- 7$ (change signs)

$= \underline{2x} - \underline{3x^2 - 5x^2} + \underline{4y + 6y}$

$\underline{- 7}$ (rearrange terms)

$= 2x + (-3 - 5)x^2 + (4 + 6)y$

$- 7$ (distributive property)

$= \boxed{2x - 8x^2 + 10y - 7}$

(combine) ∎

Exercises 1.3

In Exercises 1–5 give
(a) the number of terms in the expression;
(b) the degree of the polynomial;
(c) the additional information requested.

1. $2x^2y - 3xy^3z^4 + 5$; numerical coefficient of the second term

2. $\frac{1}{2}x - 3y$; exponent of x in the first term

3. $x^2 - y^3 + z - 3$; numerical coefficient of the second term
4. $3xy^2z$; exponent of z in the first term

5. $\frac{7}{8}y + \frac{4}{5}z^2 - 7a^2x + 14y^4 + 15$; degree of the third term

In Exercises 6–10 explain why the expression is not a polynomial.

6. $x^2 - 3\sqrt{y} + 2$

7. $x^2 - \sqrt{2}y + \frac{1}{z}$

8. $x^{1/2} + 3$

9. $\frac{x}{y} + 7z - \frac{2}{3}$

10. $x^2 - \frac{7}{8}y^{-4} + 6x^4y$

In Exercises 11–28 combine the polynomials.

11. $2x + 5x$

12. $3y + 7y$

13. $3x^2y + 14x^2y$

14. $15xy^3z + 16xy^3z$

15. $\frac{2}{3}x + \frac{3}{4}x$

16. $\frac{7}{8}y^2 + \frac{5}{12}y^2$

17. $2x - 3x$

18. $10y - 4y$

19. $15x^2y^3 - 10x^2y^3 + x^2y^3$

20. $3xy - 7x^2y + 6xy$

21. $(3x + 2y) + (5x - 7y)$

22. $(5x^2y - 4y) + (7z - 2x^2y)$

23. $(4x - 5y) - (2x - 4y)$ 24. $(3x^2y + 6z) - (x^2y - 5z + w)$

25. $(3x^2y - 7xy^3) + (x^3y - xy^3) - (2xy - 7xy^3)$

26. $\left(\dfrac{2}{3} x^2y + \dfrac{3}{4} x - 5\right) + \left(\dfrac{5}{3} x^2y - \dfrac{1}{2} x + 3\right)$

27. $\left(\dfrac{2}{3} x^2y + \dfrac{3}{4} x - 5\right) - \left(\dfrac{5}{3} x^2y - \dfrac{1}{2} x + 3\right)$

28. $(x^2 + y^2 + z^2) - (a^2 - b^2 + c^2)$

1.4   Multiplication and Factoring of Polynomials

In this section we will be primarily concerned with polynomials of one, two, or three terms. We will refer to each of these with the following terminology:

A **monomial** is a polynomial with one term.
A **binomial** is a polynomial with two terms.
A **trinomial** is a polynomial with three terms.

We previously defined x^n as the product of n x's. From this definition it follows that

$$x^n \cdot x^m = x^{n+m}$$

since $x^n \cdot x^m = \underbrace{\underbrace{x \cdot x \cdot x \cdots x}_{n \text{ times}} \cdot \underbrace{x \cdot x \cdot x \cdots x}_{m \text{ times}}}_{n + m \text{ times}} = x^{n+m}$

Using this formula, we can easily multiply two or more monomials as shown in Example 1.

Example 1   Multiply the following monomials:

$$(2x^2y)(3x)$$

Solution   First we rearrange:

$(2x^2y)(3x) = 2 \cdot 3 \cdot x^2 \cdot x \cdot y$ (commutative property of multiplication)

Next we simplify:

$$(2 \cdot 3)(x^2 \cdot x)(y) = \boxed{6x^3y} \quad \blacksquare$$

The process used in Example 1 can be applied to any (finite) number of products of monomials. However, for polynomials with two or more terms, we will multiply only two terms at a time by using the distributive property. We will see that, in every situation, the process simplifies to multiplying monomials.

Example 2 Multiply $3x^2y(2x - 7y^2)$.

Solution
$$3x^2y(2x - 7y^2) = 3x^2y(2x) + 3x^2y(-7y^2) \qquad \text{(distributive property)}$$
$$= \underline{3 \cdot 2 \cdot x^2 \cdot x \cdot y} + \underline{3 \cdot (-7) \cdot x^2 \cdot y \cdot y^2}$$
$$\text{(rearrange each monomial)}$$
$$= 6x^3y - 21x^2y^3 \qquad \text{(simplify)} \quad \blacksquare$$

The solution $6x^3y - 21x^2y^3$ was found by applying the distributive property to obtain two products of monomials. Then the monomial multiplication was used to finish the problem.

The distributive property may be used to multiply any two polynomials. In each case we apply the distributive property until we have only products of monomials, as Example 3 shows.

Example 3 Multiply $(2x^2y - 3x)(5x - 3y + 7z)$.

Solution
$$(2x^2y - 3x)(5x - 3y + 7z) = \underline{(2x^2y - 3x)(5x)} + \underline{(2x^2y - 3x)(-3y)}$$
$$+ \underline{(2x^2y - 3x)(+7z)}$$
$$= 5x(2x^2y - 3x) - 3y(2x^2y - 3x)$$
$$+ 7z(2x^2y - 3x)$$
$$= \underline{5x(2x^2y)} + \underline{5x(-3x)}$$
$$- \underline{3y(2x^2y)} - \underline{3y(-3x)} + \underline{7z(2x^2y)}$$
$$+ \underline{7z(-3x)}$$
$$= \underline{5 \cdot 2 \cdot x \cdot x^2 \cdot y} + \underline{5(-3) \cdot x \cdot x}$$
$$\underline{- 3 \cdot 2 \cdot x^2 \cdot y \cdot y} - \underline{3(-3) \cdot y \cdot x}$$
$$\underline{+ 7 \cdot 2 \cdot x^2 \cdot y \cdot z} + \underline{7(-3) \cdot x \cdot z}$$
$$= 10x^3y - 15x^2 - 6x^2y^2 + 9xy$$
$$+ 14x^2yz - 21xz \quad \blacksquare$$

The product of three polynomials is found by a two-step process:

Step 1: Multiply any two of the polynomials together.
Step 2: Multiply the remaining polynomial by the result of Step 1.

This procedure is demonstrated in Example 4.

Example 4 Multiply $(2x + y)(x - 3y)(3x + 2y)$.

Solution *Step 1:* Multiply any two of the polynomials:

$$(2x + y)(x - 3y) = (2x + y)(x) + (2x + y)(-3y)$$
$$= 2x \cdot x + y \cdot x + 2x(-3y) + y(-3y)$$
$$= 2x^2 + xy - 6xy - 3y^2$$
$$= 2x^2 - 5xy - 3y^2 \quad \text{(result of Step 1)}$$

Step 2: Multiply the remaining polynomial, $3x + 2y$, by the result of Step 1, which is $2x^2 - 5xy - 3y^2$:

$$(2x^2 - 5xy - 3y^2)(3x + 2y) = (2x^2 - 5xy - 3y^2) \cdot (3x)$$
$$+ (2x^2 - 5xy - 3y^2)(+2y)$$
$$= 2x^2(3x) - 5xy(3x) - 3y^2(3x) + 2x^2(2y)$$
$$- 5xy(2y) - 3y^2(2y)$$
$$= 6x^3 - 15x^2y - 9xy^2 + 4x^2y$$
$$- 10xy^2 - 6y^3$$
$$= 6x^3 - 11x^2y - 19xy^2 - 6y^3$$
$$\text{(final result)} \quad \blacksquare$$

Example 5 gives some "special" products. These are important because the results are used in both directions. At times we are asked to multiply. The reverse process of writing a polynomial as a product of two or more polynomials is called **factoring.**

Example 5 Multiply these "special" products:
(a) $(x + y)(x - y)$
(b) $(x + y)(x + y)$ [this is the same as $(x + y)^2$]
(c) $(x + b)(x + d)$ where b and d are real numbers
(d) $(ax + b)(cx + d)$ where a, b, c, and d are real numbers

Solution (a) $(x + y)(x - y) = (x + y)(x) + (x + y)(-y)$

$$= x^2 + yx - xy - y^2 = \boxed{x^2 - y^2}$$

(b) $(x + y)(x + y) = x(x + y) + y(x + y)$

$$= x^2 + xy + yx + y^2 = \boxed{x^2 + 2xy + y^2}$$

(c) $(x + b)(x + d) = (x + b)x + (x + b)d$

$$= x^2 + bx + dx + bd = \boxed{x^2 + (b + d)x + bd}$$

(d) As in part (c), we multiply to get

$$\boxed{(ac)x^2 + (bc + ad)x + (bd)} \quad . \quad \blacksquare$$

Thus, using Example 5(a) to factor $x^2 - y^2$, we write

$$x^2 - y^2 = (x - y)(x + y)$$

This is called the **difference of squares** formula.

Example 6 Factor each difference of squares:
(a) $x^2 - 9$ (b) $4x^2 - 49$
(c) $(x + y)^2 - 1$

Solution (a) $x^2 - 9 = x^2 - 3^2$
$$= (x - 3)(x + 3)$$
(b) $4x^2 - 49 = (2x)^2 - 7^2$
$$= (2x - 7)(2x + 7)$$
(c) $(x + y)^2 - 1 = [(x + y) - 1][(x + y) + 1]$
$$= (x + y - 1)(x + y + 1)$$ ■

In order to use Example 5(b), we write

$$x^2 + 2xy + y^2 = (x + y)^2$$

This is called recognizing a **perfect square**. A second form, which can be verified by multiplication, is

$$x^2 - 2xy + y^2 = (x - y)^2$$

Each of these requires the end terms to be perfect squares and the middle term to be twice the product of the square roots of these end terms.

Example 7 Factor the following perfect squares:
(a) $x^2 + 6x + 9$ (b) $4x^2 - 20xy + 25y^2$

Solution (a) $x^2 + 6x + 9 = x^2 + 6x + 3^2$
$$= (x)^2 + 2(x)(3) + (3)^2$$

$$= (x + 3)^2$$
(b) $4x^2 - 20xy + 25y^2 = (2x)^2 - 20xy + (5y)^2$
$$= (2x)^2 - 2(2x)(5y) + (5y)^2$$

$$= (2x - 5y)^2$$ ■

Our third method, called the **trial and error** process for factoring trinominals, uses the results of Example 5(c) and (d). Restating the result of part (c), we have

$$x^2 + (b + d)x + \overline{bd} = (x + b)(x + d)$$

The result of Example 5(d) is similar. We will not use these as formulas. In most cases that would be too difficult. However, the process above shows that the end terms of the trinomial must be factored and these factors must be used to build the terms of the two binomials in the final result. Thus we will determine the factors of the first and last terms and check all the possibilities. This is why the process is called trial and error.

Example 8 Factor
(a) $x^2 + 5x + 6$ (b) $x^2 - 5x + 6$

Solution (a) Factoring x^2 gives $x \cdot x$, and factoring 6 gives $1 \cdot 6$ or $2 \cdot 3$. Thus we check

$$(x + 1)(x + 6) = x^2 + 7x + 6 \quad \text{and} \quad (x + 2)(x + 3) = x^2 + 5x + 6$$

The correct factoring is $(x + 2)(x + 3)$.
(b) Again, 6 can be factored as $1 \cdot 6$ or $2 \cdot 3$. We check

$$(x - 1)(x - 6) = x^2 - 7x + 6 \quad \text{and} \quad (x - 2)(x - 3) = x^2 - 5x + 6$$

The correct factoring is $(x - 2)(x - 3)$. ■

Note that when the trinomial ends with a positive term, such as $+6$ in Example 8, the signs in the binomial are both plus or both minus. In fact, they are the same as the middle term of the trinomial. When the last sign in the trinomial is negative, as in $x^2 + 5x - 6$, the signs in the binomials will be opposites.

Example 9 Factor:
(a) $x^2 + 5x - 6$ (b) $x^2 - 5x - 6$

Solution Factoring -6 gives $-1 \cdot 6$, $1 \cdot (-6)$, $-2 \cdot 3$, or $2 \cdot (-3)$. Thus we check all four possibilities:

$$(x - 1)(x + 6) = x^2 + 5x - 6 \quad \text{and} \quad (x + 1)(x - 6) = x^2 - 5x - 6$$

and

$$(x - 2)(x + 3) = x^2 + x - 6 \quad \text{and} \quad (x + 2)(x - 3) = x^2 - x - 6$$

(a) The solution is $(x - 1)(x + 6)$.
(b) The solution is $(x + 1)(x - 6)$. ■

Now we consider a case where the coefficient of the first term is not 1.

Example 10   Factor $4x^2 - 13xy + 3y^2$.

Solution   Write parentheses for the two binomials:

$$(\quad)(\quad)$$

Determine the signs to put in the binomials. In this case the last sign of the trinomial is plus, so the sign in each binomial is the same as the sign of the middle term, which is minus:

$$(\quad - \quad)(\quad - \quad)$$

Factor the first and last terms:

$$4x^2 = (1x)(4x) \quad \text{or} \quad (2x)(2x)$$

$$3y^2 = (1y)(3y)$$

Write all possibilities, using factors of $4x^2$ as first terms of the binomials and factors of $3y^2$ as second terms:

$$(1x - 1y)(4x - 3y)$$

$$(1x - 3y)(4x - 1y)$$

$$(2x - 1y)(2x - 3y)$$

Check the products of these binomials by multiplying:

$$(1x - 1y)(4x - 3y) = 4x^2 - 7xy + 3y^2$$

$$(1x - 3y)(4x - 1y) = 4x^2 - 13xy + 3y^2$$

$$(2x - 1y)(2x - 3y) = 4x^2 - 8xy + 3y^2$$

The correct factorization is

$$(x - 3y)(4x - y) \quad ■$$

The distributive property also gives us a factoring method. If a factor is common to all terms of a polynomial, then the factor may be taken out:

$$ax + b\underline{x} = \underline{x}(a + b)$$

This is called **taking out a common factor.** When factoring, we always remove common factors first and then try to apply our other methods.

Example 11   Factor completely: $2ax^4 + 2ax^2 - 4a$.

Solution   First we take out the common factor, $2a$:

$$2ax^4 + 2ax^2 - 4a = 2a(x^4 + x^2 - 2)$$

Next we factor the trinomial:

$$x^4 + x^2 - 2 = (x^2 - 1)(x^2 + 2)$$

Now try to factor the binomials:

$$x^2 - 1 = (x - 1)(x + 1)$$

$$x^2 + 2 \quad \text{does not factor}$$

The complete process of factoring looks as follows:

$$\begin{aligned}
2ax^4 + 2ax^2 - 4a &= 2a(x^4 + x^2 - 2) \\
&= 2a(x^2 - 1)(x^2 + 2) \\
&= 2a(x - 1)(x + 1)(x^2 + 2) \quad \blacksquare
\end{aligned}$$

Formulas for the sum and difference of cubes are

$$x^3 + y^3 = (x + y)(x^2 - xy + y^2)$$

$$x^3 - y^3 = (x - y)(x^2 + xy + y^2)$$

These formulas can be verified by multiplication.

Example 12   Factor:
(a) $x^3 + 8$ (b) $x^3 - 64$

Solution   (a) $x^3 + 8 = x^3 + 2^3$. Thus

$$\begin{aligned}
x^3 + 8 &= (x + 2)(x^2 - x \cdot 2 + 2^2) \\
&= (x + 2)(x^2 - 2x + 4)
\end{aligned}$$

(b) $x^3 - 64 = x^3 - 4^3$. Thus

$$\begin{aligned}
x^3 - 64 &= (x - 4)(x^2 + x \cdot 4 + 4^2) \\
&= (x - 4)(x^2 + 4x + 16) \quad \blacksquare
\end{aligned}$$

Exercises 1.4   In Exercises 1–30 multiply.

1. $(2x)(3x^2y)$

2. $(4x^2y)(-5x^3y^2)$

3. $(\frac{2}{3}xy)(\frac{9}{8}y^2z)$

4. $(\frac{3}{4}x^2y^3)(\frac{8}{27}xy^3z^2)$

5. $(-2x^2y)(-3xy^3)(5yz)$

6. $(-6xy^3)(4x^3z)(xyz^2)$

7. $2xy^2(3x - 4y)$

8. $3x^2y^3(2xy - 3z + 2)$

9. $-5x^2y(3y + 2x - 7x^2y^3)$

10. $\frac{2}{3}xy\left(\frac{1}{2}x^2 - \frac{3}{4}y\right)$

11. $(x + y)(x - y)$

12. $(x + y)(x + y)$

13. $(x - y)(x - y)$

14. $(x - 2)(x - 3)$

15. $(x - 2)(x + 3)$

16. $(x + 2)(x + 3)$

17. $(2x - 3)(5x - 7)$

18. $(2x + 3)(5x + 7)$

19. $(2x - 3)(5x + 7)$

20. $(2x + 3)(5x - 7)$

21. $(x + y - 2)(x + y - 3)$

22. $(x + y)(x^2 - xy + y^2)$

23. $(x - y)(x^2 + xy + y^2)$

24. $(x + 3)(x^2 - 3x + 9)$

25. $(x - 5)(x^2 + 5x + 25)$

26. $(2x + 3y)(2x + 3y)$

27. $(2x - 3y)(2x - 3y)$

28. $(2x - 3y)(2x + 3y)$

29. $(x + 2)(x - 2)(x^2 + 4)$

30. $(x + 3)(x - 3)(x - 1)(x + 1)$

In Exercises 31–70 factor completely.

31. $x^2 - 9$

32. $x^2 - 4$

33. $x^2 - 36$

34. $4x^2 - 49$

35. $9 - 64x^2$

36. $x^2 + 6x + 9$

37. $x^2 + 10x + 25$

38. $4x^2 + 28x + 49$

39. $16x^2 + 8xy + y^2$

40. $x^2 + 6x + 5$

41. $x^2 + 6x + 8$

42. $x^2 + 8x + 12$

43. $x^2 - 7x + 10$

44. $x^2 + 7x + 10$

45. $x^2 - 6x + 8$

46. $x^2 - 6x + 5$

47. $x^2 + 6x - 7$

48. $x^2 - 6x - 7$

49. $x^2 - 3x - 10$

50. $x^2 + 3x - 10$

51. $3x^2 + 5x - 2$

52. $3x^2 - 5x - 2$

53. $3x^2 - x - 2$

54. $3x^2 + x - 2$

55. $6x^2 + 25x + 4$

56. $6x^2 - 10x + 4$

57. $5x^2 + 2xy - 3y^2$

58. $5x^2 - 14xy - 3y^2$

59. $x^4 - 5x^2 + 4$

60. $8x + 6xy$

61. $2ax + 4ay + 2a$

62. $x^2y - x^2z + 2x^2$

63. $3x^2y + 6xy + 3y$

64. $x^3 - 9x$

65. $x^3 - 1$

66. $8x^3 + 27$

67. $x^{16} - 1$

68. $x^6 - 7x^3 - 8$

69. $3x^2 - 50x - 10,000$

70. $18x^2 + 29x - 40$

1.5   Fractions and Division

An **algebraic fraction** is the quotient of two polynomials. Thus

$$\frac{x + y}{x - y}, \quad \frac{x - 3}{y}, \quad \text{and} \quad \frac{x^2 + 2xy - y^2}{a - b}$$

are algebraic fractions. The rules for working with these fractions are very similar to those for numerical fractions (Section 1.2). Thus we will want to discuss reduction, multiplication, and division. In Section 1.6 we will also consider addition, subtraction, and more about fractions.

The rule for reduction of algebraic fractions looks the same as in Section 1.2.

$$\frac{P \cdot Q}{P \cdot R} = \frac{Q}{R}$$

However, here P, Q, and R represent factors of the polynomials.

Example 1   Reduce $\dfrac{x + y}{x^2 + 5xy + 4y^2}$.

Solution   First we factor completely:

$$\frac{(x + y)}{(x + y)(x + 4y)}$$

Next we divide out the common factor, $(x + y)$:

$$\frac{\overset{1}{\cancel{(x + y)}}}{\underset{1}{\cancel{(x + y)}(x + 4y)}}$$

The solution is

$$\frac{1}{x + 4y} \qquad \blacksquare$$

The rules for multiplication and division also look the same as the rules in Section 1.2.

$$\frac{P}{Q} \cdot \frac{R}{S} = \frac{P \cdot R}{Q \cdot S} \quad \text{and} \quad \frac{P}{Q} \div \frac{R}{S} = \frac{P}{Q} \cdot \frac{S}{R}$$

While the process remains the same, P, Q, R, and S now represent polynomials.

Example 2   Multiply $\dfrac{x + y}{a^2 - ab} \cdot \dfrac{ax + ay}{x^2 + 2xy + y^2}$.

Solution First we factor completely:

$$\frac{(x + y)}{a(a - b)} \cdot \frac{a(x + y)}{(x + y)(x + y)}$$

Next we divide out the common factors (reduce):

$$\frac{\overset{1}{\cancel{(x + y)}}}{\cancel{a}(a - b)} \cdot \frac{\overset{1}{\cancel{a}}\overset{1}{\cancel{(x + y)}}}{\cancel{(x + y)}(x + y)}$$

Finally, we multiply to obtain the solution:

$$\boxed{\frac{1}{a - b}} \quad \blacksquare$$

The division process is completed by multiplying by the reciprocal of the divisor.

Example 3 Divide: $(x + 2) \div \dfrac{x^2 + 3x + 2}{x^2 - 3x}$

Solution First we apply the definition of division:

$$(x + 2) \div \frac{x^2 + 3x + 2}{x^2 - 3x} = \frac{x + 2}{1} \cdot \frac{x^2 - 3x}{x^2 + 3x + 2}$$

Next we factor completely, divide out common factors, and multiply, as in Example 2.

$$\frac{\overset{1}{\cancel{(x + 2)}}}{1} \cdot \frac{x(x - 3)}{\cancel{(x + 2)}(x + 1)}$$

The solution is

$$\boxed{\frac{x(x - 3)}{x + 1}} \quad \blacksquare$$

We will consider two special cases of division: when the denominator of a fraction is a monomial, and when the denominator is a binomial. We refer to these processes as **short division** (the monomial case) and **long division** (the binomial case). (Note, however, that the long division process is not limited to binomials. It can be used with any polynomial of two or more terms.)

Example 4 Divide: $\dfrac{x^2 + 3x + 1}{3x}$.

Solution Since the denominator is a monomial, we use short division. Divide each term of the numerator by the denominator:

$$\frac{x^2}{3x} + \frac{3x}{3x} + \frac{1}{3x}$$

Next, we simplify (reduce each term):

$$\frac{x \cdot \overset{1}{\cancel{x}}}{3 \cdot \cancel{x}} + \frac{\overset{1}{\cancel{3}} \cdot \cancel{x}}{\cancel{3} \cdot \cancel{x}} + \frac{1}{3x}$$

The solution is

$$\boxed{\frac{x}{3} + 1 + \frac{1}{3x}} \quad \blacksquare$$

The process in Example 4 will be used only when the denominator is a monomial and when we want the final answer to be a collection of simplified terms. If the denominator is a binomial, then long division is required.

Example 5 Divide: $\dfrac{x^3 + 3x^2 - 2x + 1}{x + 2}$.

Solution Since the denominator is a binomial, we use long division.
Step 1: Rewrite the division as

$$x + 2 \overline{) x^3 + 3x^2 - 2x + 1}$$

Step 2: Divide the first term of the dividend, x^3, by the first term of the divisor, x, to obtain $x^3/x = x^2$. Place the answer above the division box.

$$\begin{array}{r} x^2 \\ x + 2 \overline{) x^3 + 3x^2 - 2x + 1} \end{array}$$

Step 3: Multiply the result, x^2, by the divisor, $x + 2$, to obtain $x^3 + 2x^2$, and place this result under the dividend:

$$\begin{array}{r} x^2 \\ x + 2 \overline{) x^3 + 3x^2 - 2x + 1} \\ x^3 + 2x^2 \leftarrow x^2(x + 2) \end{array}$$

Step 4: Subtract (change signs and add):

$$\begin{array}{r} x^2 \\ x + 2 \overline{)\,x^3 + 3x^2 - 2x + 1} \\ +\ x^3 + 2x^2 \\ \hline x^2 - 2x + 1 \end{array}$$

Step 5: Repeat Steps 2, 3, and 4 on this result, $x^2 - 2x + 1$:

$$\frac{x^2}{x} = x$$

$$\begin{array}{r} x^2 + x \\ x + 2 \overline{)\,x^3 + 3x^2 - 2x + 1} \\ +\ x^3 + 2x^2 \\ \hline x^2 - 2x + 1 \\ +\ x^2 + 2x \\ \hline -\ 4x + 1 \end{array}$$

Step 6: Repeat Steps 2, 3, and 4 on the result, $-4x + 1$:

$$\frac{-4x}{x} = -4$$

$$\begin{array}{r} x^2 + x - 4 \\ x + 2 \overline{)\,x^3 + 3x^2 - 2x + 1} \\ -\ x^3 + 2x^2 \\ \hline x^2 - 2x + 1 \\ -\ x^2 + 2x \\ \hline -\ 4x + 1 \\ + + \\ \hline -\ 4x - 8 \\ \hline +\ 9 \end{array}$$

Repeat Steps 2, 3, and 4 until the division in Step 2 is not possible. Here we stop, since 9 is not divisible by x.
The result is

$$x^2 + x - 4, \quad \text{with remainder} +9$$

or

$$x^2 + x - 4 + \frac{+9}{x + 2} \qquad \blacksquare$$

Exercises 1.5

In Exercises 1–10 reduce.

1. $\dfrac{x}{x^2 + 2x}$

2. $\dfrac{x + 3}{x^2 + 4x + 3}$

3. $\dfrac{x - 2}{x^2 - 2x}$

4. $\dfrac{2x - 1}{2x^2 + 5x - 3}$

5. $\dfrac{x - y}{x^2 - y^2}$

6. $\dfrac{x^2 - xy}{x^3 - y^3}$

7. $\dfrac{x^2 - 2x + 1}{x^4 - 1}$

8. $\dfrac{x^3 - 9x}{x^3 + 6x^2 + 9x}$

9. $\dfrac{3x^2 + 16x + 5}{3x^2 + x}$

10. $\dfrac{y - x}{x^2 + 4xy - 5y^2}$

In Exercises 11–20 perform the indicated operation and simplify.

11. $\dfrac{x}{x + 1} \cdot \dfrac{x + 2}{x^2 + 2x}$

12. $\dfrac{x}{3y} \cdot \dfrac{-6y^2}{x^3}$

13. $\dfrac{-2}{5y} \cdot \dfrac{y}{4}$

14. $\dfrac{2}{x - 3} \cdot \dfrac{x}{2}$

15. $\dfrac{x^2}{-3y} \div \dfrac{2x}{x - 3}$

16. $\dfrac{x + 1}{x + 2} \div \dfrac{x + 1}{x - 2}$

17. $\dfrac{x^2 + 2x + 1}{x^2 + 3x + 2} \div \dfrac{2x + 2}{x + 2}$

18. $\dfrac{x^3 + y^3}{x^2 - y^2} \cdot \dfrac{x^2 - xy}{x^2 + xy + y^2}$

19. $\dfrac{x^3 - 8}{x^2 - 4} \div \dfrac{2x^2 + 4x + 8}{x^2 + 2x}$

20. $\dfrac{xy - 2x^2}{y - 2} \cdot \dfrac{2y - 4}{2x^2 + xy - y^2}$

In Exercises 21–30 divide.

21. $\dfrac{xy^2 - z}{3yz}$

22. $\dfrac{2y + 3x + 5}{10xy}$

23. $\dfrac{x^2 + 3x - 1}{x + 1}$

24. $\dfrac{x^2 - 2x + 3}{x - 3}$

25. $\dfrac{x^2 - 2x + 3}{-3x}$

26. $\dfrac{x^3 - 3x^2 + 2x - 5}{x + 2}$

27. $\dfrac{2x^4 + 3x^2 - 5x - 3}{x - 3}$

28. $\dfrac{6x^3 - x^2 + 5}{2x + 1}$

29. $\dfrac{6x^2 - 3x^2y + 3xy}{-3xy}$

30. $\dfrac{3x^3 - 2x + 4 - 10x^2}{3x + 2}$

(*Hint:* Rearrange the terms in the numerator so that the exponents are in descending order of size.)

1.6   More About Fractions

We complete our discussion of fractions by showing how to add and subtract them. Fractions are added or subtracted only if their denominators are the same.

The rule for addition is

$$\frac{P}{Q} + \frac{R}{Q} = \frac{P + R}{Q}$$

The rule for subtraction is

$$\frac{P}{Q} - \frac{R}{Q} = \frac{P - R}{Q}$$

Example 1   Combine $\dfrac{3x}{x + 2} - \dfrac{x + 1}{x + 2}$.

Solution   The denominators are the same, so we subtract by combining the numerators.

$$\frac{3x}{x + 2} - \frac{x + 1}{x + 2} = \frac{3x - (x + 1)}{x + 2}$$

$$= \frac{3x - x - 1}{x + 2} = \boxed{\frac{2x - 1}{x + 2}}$$

Note that the subtraction changes the sign of both terms in the numerator of the fraction being subtracted. ∎

Before we add or subtract fractions with different denominators, let us review the concept of lowest (or least) common denominator (L.C.D.). In Section 1.2 we saw that any two rational numbers could be added by calculating their L.C.D., rewriting each with the L.C.D. as denominator, and then adding. Let us review this process with an example.

Example 2   Combine $\frac{3}{40} + \frac{2}{15}$.

Solution   *Step 1:* Factor each denominator:

$$40 = 2 \cdot 2 \cdot 2 \cdot 5 \quad \text{and} \quad 15 = 3 \cdot 5$$

Step 2: List the different factors that occur:

$$2, 3, 5$$

Step 3: For each factor, list the *maximum* number of times it occurs *in any single denominator:*

Factor 2 from 40 occurs *3* times
Factor 3 from 15 occurs *1* time
Factor 5 from either occurs *1* time

Step 4: The L.C.D. is the product of the factors (from Step 2) with the exponents found in Step 3:

$$\text{L.C.D.} = 2^3 \cdot 3^1 \cdot 5^1$$

This is the same as

$$\underbrace{2 \cdot 2 \cdot 2 \cdot 5}_{40} \cdot \underbrace{5 \cdot 3}_{15}$$

Step 5: To rewrite each fraction with the L.C.D. as the denominator, multiply the numerator and the denominator by the missing factors:

$$\frac{3}{40} = \frac{3}{2 \cdot 2 \cdot 2 \cdot 5}\left(\frac{\cdot\, 3}{\cdot\, 3}\right) \qquad \frac{2}{15} = \frac{2}{3 \cdot 5}\left(\frac{\cdot\, 2 \cdot 2 \cdot 2}{\cdot\, 2 \cdot 2 \cdot 2}\right)$$

$$= \frac{9}{\text{L.C.D.}} \qquad\qquad\qquad = \frac{16}{\text{L.C.D.}}$$

Step 6: Combine:

$$\frac{9}{\text{L.C.D.}} + \frac{16}{\text{L.C.D.}} = \frac{9 + 16}{\text{L.C.D.}} = \frac{25}{\text{L.C.D.}}$$

Step 7: Replace the L.C.D. with its value and reduce (factor and divide out):

$$\frac{25}{2 \cdot 2 \cdot 2 \cdot 3 \cdot 5} = \frac{5 \cdot \overset{1}{\cancel{5}}}{2 \cdot 2 \cdot 2 \cdot 3 \cdot \cancel{5}} = \frac{5}{2 \cdot 2 \cdot 2 \cdot 3} = \boxed{\frac{5}{24}} \quad \blacksquare$$

We will follow the process used in Example 2 for any addition or subtraction problem involving fractions. In each case we will factor the denominators, build their L.C.D., rewrite the fractions so that each has the L.C.D. as its denominator, and then combine and reduce.

Example 3 Combine

$$\frac{3}{2x + 4} + \frac{1}{x^2 + x - 2} - \frac{1}{x^3 + x^2 - 2x}$$

Solution First we factor the denominators:

$$\frac{1}{2(x + 2)} + \frac{1}{(x + 2)(x - 1)} - \frac{1}{x(x + 2)(x - 1)}$$

Next we build the L.C.D.:

Factors	Maximum number of times appearing
2	1
x	1
$x + 2$	1
$x - 1$	1

The L.C.D. is $2^1 \cdot x^1 \cdot (x + 2)^1 \cdot (x - 1)^1$.
Rewrite each fraction:

$$\left[\frac{3}{2(x + 2)} \cdot \left(\frac{x(x - 1)}{x(x - 1)}\right)\right] + \left[\frac{1}{(x + 2)(x - 1)} \cdot \left(\frac{2x}{2x}\right)\right]$$

$$-\left[\frac{1}{x(x + 2)(x - 1)} \cdot \left(\frac{2}{2}\right)\right] = \frac{3x^2 - 3x}{\text{L.C.D.}} + \frac{2x}{\text{L.C.D.}} - \frac{2}{\text{L.C.D.}}$$

Combine:

$$\frac{3x^2 - 3x + 2x - 2}{\text{L.C.D.}} = \frac{3x^2 - x - 2}{\text{L.C.D.}}$$

Finally, we replace the L.C.D. with its value and reduce:

$$\frac{3x^2 - x - 2}{2x(x + 2)(x - 1)} = \frac{(3x + 2)(x - 1)}{2x(x + 2)(x - 1)} = \frac{3x + 2}{2x(x + 2)} \quad \blacksquare$$

Example 3 demonstrates how to add or subtract fractions. In the previous section we saw how to multiply or divide fractions. When we have combinations of these operations, we rely on the order or operations (as discussed in Section 1.2). Let us recall this order:

1. evaluate within parentheses
2. exponents
3. multiply and divide (from left to right)
4. add and subtract (from left to right)

Let us apply this to the next example.

Example 4   Combine:

(a) $\dfrac{3}{x} + \dfrac{2}{x} \cdot \dfrac{5}{y}$

(b) $\left(\dfrac{3}{x} + \dfrac{2}{x}\right) \cdot \dfrac{5}{y}$

Solution   (a) $\dfrac{3}{x} + \dfrac{2}{x} \cdot \dfrac{5}{y} = \dfrac{3}{x} + \dfrac{10}{xy}$ (multiply)

$$= \dfrac{3}{x} \cdot \dfrac{y}{y} + \dfrac{10}{xy} \qquad \text{(L.C.D.} = xy\text{)}$$

$$= \dfrac{3y}{xy} + \dfrac{10}{xy} = \boxed{\dfrac{3y + 10}{xy}}$$

(b) $\left(\dfrac{3}{x} + \dfrac{2}{x}\right) \cdot \dfrac{5}{y} = \left(\dfrac{3 + 2}{x}\right) \cdot \dfrac{5}{y}$ (evaluate within parentheses)

$$= \dfrac{5}{x} \cdot \dfrac{5}{y} \qquad \text{(multiply)}$$

$$= \boxed{\dfrac{25}{xy}} \quad \blacksquare$$

We will use the order of operations to simplify complex fractions. A **complex fraction** is a fraction whose numerator or denominator (or both) contain a fraction. For example,

$$\dfrac{x + \dfrac{1}{x}}{y}$$

is a complex fraction, since the fraction $1/x$ appears in its numerator. In order to simplify such fractions, we will consider the numerator and denominator to be in parentheses and rewrite

$$\dfrac{\left(x + \dfrac{1}{x}\right)}{(y)} \quad \text{as} \quad \left(x + \dfrac{1}{x}\right) \div (y)$$

We will work a similar problem as our next example.

Example 5   Simplify the complex fraction

$$\dfrac{x + \dfrac{1}{x}}{\dfrac{y}{x}}$$

Solution

$$\frac{x + \dfrac{1}{x}}{\dfrac{y}{x}} = \left(x + \frac{1}{x}\right) \div \left(\frac{y}{x}\right) \qquad \text{(rewrite)}$$

$$= \left(\frac{x}{1} \cdot \frac{x}{x} + \frac{1}{x}\right) \div \left(\frac{y}{x}\right) \qquad \text{(evaluate within parentheses)}$$

$$= \left(\frac{x^2 + 1}{x}\right) \div \frac{y}{x}$$

$$= \frac{x^2 + 1}{\cancel{x}} \cdot \frac{\overset{1}{\cancel{x}}}{y} \qquad \text{(change division to multiplication and reduce)}$$

$$= \boxed{\frac{x^2 + 1}{y}} \quad \blacksquare$$

Computational Note: Another way to work the problem is to calculate the L.C.D. for the numerator and denominator of the entire complex fraction. Then multiply numerator and denominator by this L.C.D. In Example 5 this L.C.D. is x, and we multiply to get

$$\frac{x + \dfrac{1}{x}}{\dfrac{y}{x}} \cdot \frac{x}{x} = \frac{\left(x + \dfrac{1}{x}\right) \cdot x}{\dfrac{y}{x} \cdot x} = \frac{x^2 + \dfrac{x}{x}}{\dfrac{xy}{x}} = \boxed{\frac{x^2 + 1}{y}}$$

Exercises 1.6

In Exercises 1–30 combine and simplify.

1. $\dfrac{2x}{x + 2} + \dfrac{4}{x + 2}$

2. $\dfrac{3x - 2}{x + 5} + \dfrac{7 - 2x}{x + 5}$

3. $\dfrac{x^2}{x - 3} - \dfrac{x + 6}{x - 3}$

4. $\dfrac{3x}{x - 2} - \dfrac{10 - x^2}{x - 2}$

5. $\dfrac{x - 8}{x^2 - 4x} + \dfrac{1}{x - 4}$

6. $\dfrac{2x - 9}{3x - 9} + \dfrac{1}{x - 3}$

7. $\dfrac{5x - 1}{2x^2 - 2x} - \dfrac{2}{x - 1}$

8. $\dfrac{4x - 2}{3x^2 - 6x} - \dfrac{1}{3x}$

9. $\dfrac{1}{x - 3} + \dfrac{1}{x + 2}$

10. $\dfrac{2}{x + 1} - \dfrac{3}{x - 1}$

11. $\dfrac{x + 1}{x^3 - x^2} - \dfrac{1}{x^2 - 2x + 1}$

12. $\dfrac{2}{x^2 - 4} + \dfrac{x}{x^2 + 5x + 6}$

13. $\dfrac{2}{x} + \dfrac{3}{x - 1} - \dfrac{x - 2}{x^2 - x}$

14. $\dfrac{x - 1}{x + 2} + \dfrac{2x + 12}{x^2 - 4} - \dfrac{4}{x - 2}$

15. $\dfrac{x}{3x^2 + x - 2} + \dfrac{x + 10}{3x^2 + 4x - 4} - \dfrac{1}{x^2 + 3x + 2}$

16. $1 - \dfrac{1}{x + 1} - \dfrac{x}{(x + 1)^2}$

17. $\dfrac{2}{y} - \dfrac{3}{y} \cdot \dfrac{1}{x}$

18. $\left(\dfrac{2}{y} - \dfrac{3}{y}\right) \cdot \dfrac{1}{x}$

19. $\dfrac{3}{x^2} - \dfrac{1}{xy} \div \dfrac{x}{y}$

20. $\left(\dfrac{3}{x^2} - \dfrac{1}{xy}\right) \div \dfrac{x}{y}$

21. $\dfrac{x}{x + 1} - \dfrac{x}{x + 1} \cdot \dfrac{1}{x + 2}$

22. $\dfrac{2x}{x - 3} - \dfrac{6}{x - 1} \div \dfrac{x - 3}{x - 1}$

23. $\dfrac{\dfrac{1}{xy}}{\dfrac{y^2}{x}}$

24. $\dfrac{\dfrac{x + 2}{x - 2}}{\dfrac{x - 2}{x}}$

25. $\dfrac{x - \dfrac{1}{x}}{x + 1}$

26. $\dfrac{2x - \dfrac{8}{x}}{\dfrac{2x - 4}{x^2 + x}}$

27. $\dfrac{\dfrac{x}{y} + \dfrac{1}{3}}{\dfrac{2}{y^2} - \dfrac{1}{x}}$

28. $\dfrac{\dfrac{x}{3} - \dfrac{3}{x}}{\dfrac{x - 2}{x} - \dfrac{1}{x}}$

29. $\dfrac{\dfrac{x^2}{x - 2} + \dfrac{3x + 2}{x - 2}}{\dfrac{x}{x - 2} + \dfrac{1}{x - 2}}$

30. $\dfrac{\dfrac{1}{x - 1} - \dfrac{1}{x + 1}}{\dfrac{1}{x - 1} - \dfrac{1}{x - 3}}$

1.7 Exponents

In Section 1.3 we defined variable and exponent and discussed the meaning of a variable x with the exponent n, where n is a natural number. Recall that x^n means the product of n x's. In Section 1.4 we showed that $x^n \cdot x^m = x^{n+m}$, for n and m natural numbers. We now wish to extend our discussion to the rules for exponents, where the exponents may be integers or rational numbers. These rules can be established algebraically in much the same way that Rule 1 was shown in Section 1.4. The rules for evaluating a variable raised to a rational number exponent (other than an integer) will be explained in Section 1.8.

RULES FOR EXPONENTS

1. $x^m \cdot x^n = x^{m+n}$

2. $\dfrac{x^m}{x^n} = x^{m-n}$ $(x \neq 0)$

3. $(x^m)^n = x^{m \cdot n}$

4. $(x \cdot y)^n = x^n \cdot y^n$

5. $\left(\dfrac{x}{y}\right)^n = \dfrac{x^n}{y^n}$ $(y \neq 0)$

In these rules x and y are any real numbers (except where certain values are excluded) and m and n are rational numbers.

Two important results are consistent with our five rules of exponents:

$$x^0 = 1 \qquad (\text{if } x \neq 0)$$

$$x^{-n} = \frac{1}{x^n} \qquad (\text{if } x \neq 0)$$

We define $x^0 = 1$ because, on the one hand, $x^n/x^n = 1$, and on the other hand, by Rule 2 of exponents $x^n/x^n = x^{n-n} = x^0$. To see why we define $x^{-n} = 1/x^n$, notice that $x^2/x^3 = 1/x$. But by Rule 2 of exponents,

$$\frac{x^2}{x^3} = x^{2-3} = x^{-1}$$

Thus $1/x = x^{-1}$ must be our definition.

The expression 0^0 is undefined. A common mistake is to try to apply the rule of exponents to sums, but $(x + y)^n \neq x^n + y^n$ (except for special cases such as $x = 0$ or $y = 0$).

Warning:

1. 0^0 is not defined
2. $(x + y)^n \neq x^n + y^n$

Example 1 Simplify the following expressions. The results should not contain zero exponents or negative exponents.

(a) $x^5 \cdot x^2$ (b) $\dfrac{x^5}{x^2}$ (c) $(x^2)^5$ (d) $(2x)^5$

(e) $\left(\dfrac{2}{x}\right)^5$ (f) $(x + y)^0$ (g) $3x^{-5}$

Solution   (a) $x^5 \cdot x^2 = x^{5+2} = x^7$ (Rule 1)

(b) $\dfrac{x^5}{x^2} = x^{5-2} = x^3$ (Rule 2)

(c) $(x^2)^5 = x^{2 \cdot 5} = x^{10}$ (Rule 3)
(d) $(2x)^5 = 2^5 \cdot x^5 = 32x^5$ (Rule 4)

(e) $\left(\dfrac{2}{x}\right)^5 = \dfrac{2^5}{x^5} = \dfrac{32}{x^5}$ (Rule 5)

(f) $(x + y)^0 = 1$ if $x + y \neq 0$ (by the definition of x^0) and is undefined if $x + y = 0$.

(g) $3x^{-5} = 3 \cdot \dfrac{1}{x^5}$ (definition of x^{-n})

$$= \dfrac{3}{1} \cdot \dfrac{1}{x^5} = \dfrac{3}{x^5} \quad \blacksquare$$

Example 2 Simplify $\dfrac{2x^2y^{-3}z^0}{5^{-2}x^3y}$

Solution We must perform the division

$$(2x^2y^{-3}z^0) \div (5^{-2}x^3y)$$

Using the order of operations (Section 1.2), we evaluate within the parentheses to obtain

$$\left(\dfrac{2x^2}{1} \cdot \dfrac{1}{y^3} \cdot 1\right) \div \left(\dfrac{1}{5^2} \cdot \dfrac{x^3y}{1}\right) = \left(\dfrac{2x^2}{y^3}\right) \div \left(\dfrac{x^3y}{5^2}\right)$$

Division is the next operation to be performed:

$$\dfrac{2x^2}{y^3} \div \dfrac{x^3y}{25} = \dfrac{2x^2}{y^3} \cdot \dfrac{25}{x^3y} = \dfrac{2 \cdot 25 \cdot \overset{1}{\cancel{x^2}}}{\underset{x}{\cancel{x^3}} \cdot y^3 \cdot y} = \boxed{\dfrac{50}{xy^4}}$$

A second way to work the problem is to group the like factors of the problem and use the rules for exponents. (This works only when dividing a monomial by a monomial.)

$$\left(\frac{2}{5^{-2}}\right)\left(\frac{x^2}{x^3}\right)\left(\frac{y^{-3}}{y}\right)\left(\frac{z^0}{1}\right) = 2 \cdot 5^2 \cdot x^{2-3} \cdot y^{-3-1} \cdot z^0$$

$$= 50 \cdot x^{-1} \cdot y^{-4} \cdot 1$$

$$= 50 \cdot \frac{1}{x} \cdot \frac{1}{y^4} \cdot 1$$

$$= \frac{50}{xy^4} \quad \blacksquare$$

In Example 2 we divided a monomial by a monomial. However, the first method used in Example 2 will also work for the more complicated multiterm problem, as Example 3 demonstrates.

Example 3 Simplify $\dfrac{x^{-1}}{y^{-2} + x}$

Solution First we rewrite the problem:

$$\frac{x^{-1}}{y^{-2} + x} = (x^{-1}) \div (y^{-2} + x)$$

Then we proceed as in Example 2.

$$(x^{-1}) \div (y^{-2} + x) = \left(\frac{1}{x}\right) \div \left(\frac{1}{y^2} + \frac{x}{1}\right) \qquad \text{(rules of exponents)}$$

$$= \frac{1}{x} \div \frac{1 + xy^2}{y^2} \qquad \text{(add within parentheses)}$$

$$= \frac{1}{x} \cdot \frac{y^2}{1 + xy^2} \qquad \text{(invert to multiply)}$$

$$= \frac{y^2}{x(1 + xy^2)} \quad \blacksquare$$

Finally, we define $x^{1/q}$ where $x \geq 0$ and q is any natural number:

$$x^{1/q} = y \quad \text{where} \quad y^q = x \text{ and } y \geq 0$$

Example 4 Simplify the following:
(a) $9^{1/2}$ (b) $8^{1/3}$ (c) $64^{-1/3}$

Solution (a) $9^{1/2} = \boxed{3}$, since $3^2 = 9$

(b) $8^{1/3} = \boxed{2}$, since $2^3 = 8$

(c) By rule 4, $64^{-1/3} = 1/64^{1/3}$. Now $64^{1/3} = 4$, since $4^3 = 64$. Thus

$$64^{-1/3} = \boxed{\dfrac{1}{4}} \quad \blacksquare$$

Note that in Example 4(a), we also have $(-3)^2 = 9$. However, our definition requires us to take the *positive* value 3. For the negative number whose square is 9, we will write $-9^{1/2} = -3$.

Next we extend our definition of $x^{1/q}$ to all values of x when q is an *odd* number (that is, when q is not divisible by 2). In this case there will be a unique real solution y to the problem, but y will be negative when x is negative.

Example 5 Simplify:
(a) $(-8)^{1/3}$
(b) $-16^{1/2}$

Solution (a) $(-8)^{1/3} = \boxed{-2}$, since $(-2)^3 = -8$. Note that we may work with a negative base, -8, since $q = 3$ is odd.

(b) $-16^{1/2}$ means $-(16^{1/2})$, and we know that $16^{1/2} = 4$, since $4^2 = 16$. Thus

$$-(16^{1/2}) = \boxed{-4} \quad \blacksquare$$

In Example 5(b) the exponent $\frac{1}{2}$ applies only to the base 16. Thus $-16^{1/2}$ is not the same as $(-16)^{1/2}$. In fact, $(-16)^{1/2}$ is not defined, since there is no real number y such that $y^2 = -16$. (However, this problem will have a *complex* number solution, as will be shown in Section 1.9.)

Finally, for a real number x and natural numbers p and q, we define

$$\boxed{x^{p/q} = (x^{1/q})^p}$$

where $x^{1/q}$ is defined as in our previous discussion.

Example 6 Simplify (if possible):
(a) $9^{3/2}$
(b) $9^{-3/2}$
(c) $(-8)^{2/3}$
(d) $(-9)^{3/2}$

Solution (a) $9^{3/2} = (9^{1/2})^3 = 3^3 = \boxed{27}$

(b) $9^{-3/2} = \dfrac{1}{9^{3/2}} = \boxed{\dfrac{1}{27}}$

(c) $(-8)^{2/3} = ((-8)^{1/3})^2 = (-2)^2 = \boxed{4}$

(d) $(-9)^{3/2}$ means $[(-9)^{1/2}]^3$. But $(-9)^{1/2}$ is not a real number, since there is no real number whose square is -9. Thus $[(-9)^{1/2}]^3$ is not a real number. ■

Exercises 1.7

In Exercises 1–64 simplify the expression using the laws of exponents. Each answer should contain only positive exponents; reduce all fractions to lowest terms.

1. $y^4 \cdot y^5$

2. $x^2 \cdot x^{-1}$

3. $z^3 \cdot z^5 \cdot z^{-4}$

4. $a^5 \cdot a^{3/2} \cdot a^{-1/2}$

5. $\dfrac{x^6}{x^3}$

6. $\dfrac{y^{-3}}{y^4}$

7. $\dfrac{y^{-5}}{y^{-2}}$

8. $\dfrac{a^{6/5}}{a^{1/5}}$

9. $(x^3)^2$

10. $(x^2)^3$

11. $(y^{-2})^{-3}$

12. $(a^{-3})^5$

13. $(3x)^3$

14. $(4y)^2$

15. $(3y)^3$

16. $(5a^{-1})^2$

17. $\left(\dfrac{3}{x}\right)^2$

18. $\left(\dfrac{x^2}{y}\right)^4$

19. $\left(\dfrac{x^{-2}}{y^3}\right)^{-2}$

20. $\left(\dfrac{2a}{b^{-1}}\right)^3$

21. $(a + b)^0$

22. $(2a)(3b)^0$

23. $\dfrac{3a^2b^0}{6ab^3}$

24. $(a + b)^4(a + b)^{-4}$

25. $\dfrac{(2x + y)^3}{(2x + y)^3}$

26. $\dfrac{3xy^{-2}z^0}{2^2x^3y}$

27. $\dfrac{4^{-1}x^{-2}y^3}{2^{-2}x^{-2}y^5}$

28. $\dfrac{2^0ab^2}{-3a^3b^{-3}}$

29. $\dfrac{5x^0y^{-3}}{5^{-2}x^3y^4}$

30. $\dfrac{4^2a^{-3}b^2c^0}{8a^4b^{-3}c^2}$

31. $\dfrac{x^2}{y^{-3} + x^2}$

32. $\dfrac{a^{-1} + b^{-1}}{ab}$

33. $\dfrac{y^{-1} + y^2}{x^{-2}}$

34. $\dfrac{x^{-2} + y^{-2}}{xy^{-1}}$

35. $\dfrac{a^{-2}}{a^{-1} + a^{-2}}$

36. $\dfrac{x + y^{-1}}{x - y^{-1}}$

37.	$25^{1/2}$	38.	$36^{1/2}$
39.	$16^{1/4}$	40.	$32^{1/5}$
41.	$(-27)^{1/3}$	42.	$(-125)^{1/3}$
43.	$(-32)^{1/2}$ — NO REAL SOLUTION	44.	$(-64)^{1/3}$
45.	$-(4^{1/2})$	46.	$-36^{1/2}$
47.	$-16^{1/4}$	48.	$-64^{1/2}$
49.	$16^{3/4}$	50.	$64^{2/3}$
51.	$25^{3/2}$ $(25^{\frac{1}{2}})^3$ $5 \times 5 \times 5 = 125$	52.	$81^{1/2}$
53.	$81^{-1/2}$	54.	$25^{-3/2}$
55.	$64^{-2/3}$	56.	$-64^{-2/3}$
57.	$(-25)^{1/2}$	58.	$-25^{1/2}$
59.	$(-16)^{1/2}$	60.	$49^{1/2}$
61.	$\dfrac{5^{-2}}{25^{-1}}$	62.	$(2^{-3})(64^{1/2})$
63.	$(81^{1/2})(9^{-2})(3^2)$	64.	$\dfrac{(100^{1/2})(5^2)}{50}$

1.8   Radicals

In the previous section we defined $x^{1/q}$ for two cases:

1. $x \geq 0$ and q any natural number.
2. x any real number and q an odd natural number.

We call $x^{1/q}$ the **qth root** of x or the **qth radical.** The symbol for writing this in radical form is $\sqrt[q]{x}$. That is,

$$\sqrt[q]{x} = x^{1/q}$$

Note that $x^{p/q} = (x^{1/q})^p = (\sqrt[q]{x})^p$. Also, $x^{p/q} = (x^p)^{1/q} = \sqrt[q]{x^p}$. Thus a radical can always be written as an expression with a fractional exponent, and conversely, an expression with a fractional (rational number) exponent can always be written as a radical.

Example 1   Write the following in radical form:
(a) $x^{1/2}$ (b) $y^{2/3}$
(c) $(x^2 + 2x + 1)^{3/2}$

Solution   (a) $x^{1/2} = \sqrt[2]{x}$
(b) $y^{2/3} = \sqrt[3]{y^2}$ or $(\sqrt[3]{y})^2$
(c) $(x^2 + 2x + 1)^{3/2} = (\sqrt[2]{x^2 + 2x + 1})^3$ ∎

In radicals of the form $\sqrt[q]{x}$, we will be working most often with $q = 2$ and $q = 3$. We will refer to $\sqrt[2]{x}$ as the **square root** of x and simply write $\sqrt{x}$. That is, when no value of q is written, we mean $q = 2$. We will refer to $\sqrt[3]{x}$ as the **cube root** of x.

Example 2 Write the following in exponential form:

(a) $\sqrt{x + 1}$ (b) $\sqrt[3]{x^2}$

(c) $(\sqrt[4]{x + y})^5$

Solution (a) $\sqrt{x + 1} = \sqrt[2]{x + 1} = (x + 1)^{1/2}$

(b) $\sqrt[3]{x^2} = (x^2)^{1/3} = x^{2/3}$

(c) $(\sqrt[4]{x + y})^5 = ((x + y)^{1/4})^5 = (x + y)^{5/4}$ ∎

When working a problem, we can change from radical form to exponential form or from exponential form to radical form depending on which is more convenient. In the next example we use this process to verify a product rule for radicals by using one of our exponential rules.

Example 3 Verify that $(\sqrt[q]{A})(\sqrt[q]{B}) = \sqrt[q]{A \cdot B}$.

Solution By definition, $\sqrt[q]{A} = A^{1/q}$ and $\sqrt[q]{B} = B^{1/q}$. Thus

$$\sqrt[q]{A}\sqrt[q]{B} = A^{1/q} \cdot B^{1/q}$$
$$= (A \cdot B)^{1/q} \qquad \text{(Rule 4 for exponents)}$$
$$= \sqrt[q]{A \cdot B} \qquad \text{(definition of radical)}$$

Conclusion: $\boxed{\sqrt[q]{A}\sqrt[q]{B} = \sqrt[q]{A \cdot B}}$ ∎

Let us use the result of Example 3 to simplify some radicals.

Example 4 Simplify:

(a) $\sqrt{2x}\sqrt{8x}$ (b) $\sqrt[3]{3xy^2}\sqrt[3]{9y}$

Solution (a) $\sqrt{2x}\sqrt{8x} = \sqrt{2x \cdot 8x}$
$$= \sqrt{16x^2}$$
$$= \boxed{4x} \qquad \text{(since } 4x \cdot 4x = 16x^2\text{)}$$

(b) $\sqrt[3]{3xy^2}\sqrt[3]{9y} = \sqrt[3]{27xy^3}$

$\qquad\qquad\quad = \sqrt[3]{27y^3}\sqrt[3]{x}$

$\qquad\qquad\quad = \boxed{3y\sqrt[3]{x}} \qquad$ (since $3y \cdot 3y \cdot 3y = 27y^3$) ■

Since radical factors of an expression are equivalent to exponential factors, we can combine terms of an expression if the radicals are the same.

Example 5 Combine:

(a) $3\sqrt{x} + 4\sqrt{x}$

(b) $\sqrt{x^2y} + 2x\sqrt{y}$

(c) $4x\sqrt[3]{xy} - 5\sqrt[3]{x^4y}$

Solution (a) $3\sqrt{x} + 4\sqrt{x} = \sqrt{x}(3 + 4)$

$\qquad\qquad\qquad\qquad = \sqrt{x}(7) = \boxed{7\sqrt{x}}$

(b) $\sqrt{x^2y} + 2x\sqrt{y} = \sqrt{x^2}\sqrt{y} + 2x\sqrt{y}$

$\qquad\qquad\qquad\quad = x\sqrt{y} + 2x\sqrt{y}$

$\qquad\qquad\qquad\quad = x\sqrt{y}(1 + 2)$

$\qquad\qquad\qquad\quad = x\sqrt{y}(3) = \boxed{3x\sqrt{y}}$

(c) $4x\sqrt[3]{xy} - 5\sqrt[3]{x^4y} = 4x\sqrt[3]{xy} - 5\sqrt[3]{x^3}\sqrt[3]{xy}$

$\qquad\qquad\qquad\qquad = 4x\sqrt[3]{xy} - 5x\sqrt[3]{xy}$

$\qquad\qquad\qquad\qquad = x\sqrt[3]{xy}(4 - 5)$

$\qquad\qquad\qquad\qquad = -1x\sqrt[3]{xy} = \boxed{-x\sqrt[3]{xy}} \qquad$ ■

Combining the ideas of Examples 4 and 5, we can expand to multiplying expressions involving radicals.

Example 6 Multiply:

(a) $(\sqrt{2} + \sqrt{3})(\sqrt{2} - 5\sqrt{3})$

(b) $(\sqrt{x} + \sqrt{y})(\sqrt{x} - \sqrt{y})$

(c) $(\sqrt{x} - 5\sqrt{2})^2$

Solution   (a) $(\sqrt{2} + \sqrt{3})(\sqrt{2} - 5\sqrt{3}) = \sqrt{2}(\sqrt{2} - 5\sqrt{3}) + \sqrt{3}(\sqrt{2} - 5\sqrt{3})$
$$= \sqrt{2}\sqrt{2} - 5\sqrt{2}\sqrt{3} + \sqrt{3}\sqrt{2}$$
$$- 5\sqrt{3}\sqrt{3}$$
$$= \sqrt{4} - 5\sqrt{6} + \sqrt{6} - 5\sqrt{9}$$
$$= 2 - 4\sqrt{6} - 5 \cdot 3$$
$$= 2 - 15 - 4\sqrt{6}$$
$$= \boxed{-13 - 4\sqrt{6}}$$

(b) $(\sqrt{x} + \sqrt{y})(\sqrt{x} - \sqrt{y}) = \sqrt{x}(\sqrt{x} - \sqrt{y}) + \sqrt{y}(\sqrt{x} - \sqrt{y})$
$$= \sqrt{x}\sqrt{x} - \sqrt{x}\sqrt{y} + \sqrt{y}\sqrt{x} - \sqrt{y}\sqrt{y}$$
$$= \sqrt{x^2} - \sqrt{y^2} = \boxed{x - y}$$

Another way to perform this multiplication is to write

$(\sqrt{x} + \sqrt{y})(\sqrt{x} - \sqrt{y}) = (\sqrt{x})^2 - (\sqrt{y})^2$ (by the difference of squares
formula from Section 1.4)
$$= x - y$$

(c) $(\sqrt{x} - 5\sqrt{2})^2 = (\sqrt{x} - 5\sqrt{2})(\sqrt{x} - 5\sqrt{2})$
$$= \sqrt{x}(\sqrt{x} - 5\sqrt{2}) - 5\sqrt{2}(\sqrt{x} - 5\sqrt{2})$$
$$= \sqrt{x}\sqrt{x} - 5\sqrt{x}\sqrt{2} - 5\sqrt{2}\sqrt{x} + 25\sqrt{2}\sqrt{2}$$
$$= x - 10\sqrt{2x} + 25 \cdot 2$$
$$= \boxed{x - 10\sqrt{2x} + 50} \quad \blacksquare$$

Now we consider fractions that involve a radical. We have a quotient rule that is similar to the product rule verified in Example 3. The rule is

$$\sqrt[q]{\frac{A}{B}} = \frac{\sqrt[q]{A}}{\sqrt[q]{B}}$$

The proof is also similar to that given in Example 3 and is left for the reader. When simplifying fractions involving radicals, we will

1. Simplify the radical itself as much as possible and combine terms, as in Examples 4 and 5.
2. Rationalize the denominator, if possible; that is, rewrite the fraction so that its denominator does not contain any radicals (see Example 7).

In order to rationalize the denominator of a fraction, we multiply the numerator and denominator of the fraction by an expression chosen so that the denominator of the resulting fraction will be free of radicals.

For the monomial case of square roots, this necessary factor is generally the radical itself that appears in the denominator. For $x/\sqrt{2}$ we multiply numerator and denominator by $\sqrt{2}$ to obtain

$$\frac{x \cdot \sqrt{2}}{\sqrt{2} \cdot \sqrt{2}} = \frac{x \cdot \sqrt{2}}{\sqrt{4}} = \frac{x\sqrt{2}}{2}$$

For the binomial case of square roots, such as $x/(\sqrt{x} - 2)$, we multiply by the conjugate of the denominator, $\sqrt{x} + 2$, to obtain

$$\frac{x}{\sqrt{x} - 2} \cdot \frac{\sqrt{x} + 2}{\sqrt{x} + 2} = \frac{x\sqrt{x} + 2x}{\sqrt{x^2} - 4} = \frac{x\sqrt{x} + 2x}{x - 4}$$

Recalling that $a^2 - b^2 = (a - b)(a + b)$, we can see that multiplying a binomial expression $(a + b)$ containing a square root by its conjugate $(a - b)$ will produce $a^2 - b^2$ and will always eliminate the (square root) radical. The next example shows the steps involved in rationalizing fractions.

Example 7   Simplify as much as possible. Rationalize the denominators:

(a) $\sqrt{\dfrac{1}{8}}$

(b) $\dfrac{\sqrt{6}}{\sqrt{3} + 7}$

(c) $\dfrac{\sqrt{x}}{\sqrt{x} - \sqrt{y}}$

Solution   (a) $\sqrt{\dfrac{1}{8}} = \dfrac{\sqrt{1}}{\sqrt{8}}$ (quotient rule)

$= \dfrac{1}{\sqrt{8}} \cdot \dfrac{\sqrt{8}}{\sqrt{8}}$ (rationalize)

$= \dfrac{\sqrt{8}}{\sqrt{64}}$ (multiply)

$= \dfrac{\sqrt{4}\sqrt{2}}{8}$ (product rule, simplify)

$= \dfrac{2\sqrt{2}}{8} = \boxed{\dfrac{\sqrt{2}}{4}}$

(b) $\dfrac{\sqrt{6}}{\sqrt{3} + 7} = \dfrac{\sqrt{6}}{\sqrt{3} + 7} \cdot \dfrac{\sqrt{3} - 7}{\sqrt{3} - 7}$ (rationalize, using the conjugate)

$= \dfrac{\sqrt{18} - 7\sqrt{6}}{\sqrt{9} - 49}$ (multiply)

$= \dfrac{\sqrt{9}\sqrt{2} - 7\sqrt{6}}{3 - 49}$ (simplify)

$= \boxed{\dfrac{3\sqrt{2} - 7\sqrt{6}}{-46}}$

(c) $\dfrac{\sqrt{x}}{\sqrt{x} - \sqrt{y}} = \dfrac{\sqrt{x}}{\sqrt{x} - \sqrt{y}} \cdot \dfrac{\sqrt{x} + \sqrt{y}}{\sqrt{x} + \sqrt{y}}$ (multiply by the conjugate)

$\qquad = \dfrac{\sqrt{x^2} + \sqrt{xy}}{\sqrt{x^2} - \sqrt{y^2}}$ (simplify)

$\qquad = \dfrac{x + \sqrt{xy}}{x - y}$ ∎

Computational Note: In Example 7(a) we could have simplified the denominator $\sqrt{8}$ by first writing $\sqrt{8} = \sqrt{4} \cdot \sqrt{2} = 2\sqrt{2}$. Then the denominator could be simplified by only multiplying by $\sqrt{2}$. This would have given the solution immediately as

$$\frac{1}{\sqrt{8}} = \frac{1}{2\sqrt{2}} \cdot \frac{\sqrt{2}}{\sqrt{2}} = \frac{\sqrt{2}}{4}$$

In Example 8 we will consider fractional expressions containing a radical, where the radical contains a sum instead of a product. There is no sum rule for radicals. The expression under the radical should be combined and then factored if possible.

Example 8 Simplify:

(a) $\dfrac{6 - \sqrt{25 - 16}}{3}$

(b) $\dfrac{x + \sqrt{x^2 + 3x^2}}{x}$

(c) $\dfrac{x - \sqrt{x^2 + 2x + 1}}{x}$

Solution (a) $\dfrac{6 - \sqrt{25 - 16}}{3} = \dfrac{6 - \sqrt{9}}{3}$ (simplify within the radical)

$\qquad = \dfrac{6 - 3}{3}$

$\qquad = \dfrac{3}{3} = \boxed{1}$

(b) $\dfrac{x + \sqrt{x^2 + 3x^2}}{x} = \dfrac{x + \sqrt{4x^2}}{x}$ (simplify within the radical)

$\qquad = \dfrac{x + 2x}{x}$

$\qquad = \dfrac{3x}{x} = \boxed{3}$

(c) $\dfrac{x - \sqrt{x^2 + 2x + 1}}{x} = \dfrac{x - \sqrt{(x + 1)^2}}{x}$ (write as a product by factoring within the radical)

$$= \dfrac{x - (x + 1)}{x}$$

$$= \dfrac{x - x - 1}{x} = \boxed{\dfrac{-1}{x}} \quad \blacksquare$$

Exercises 1.8

In Exercises 1–10 write the expression in radical form.

1. $y^{1/5}$
2. $z^{2/3}$
3. $(x + y)^{3/4}$
4. $(x - y)^{1/2}$
5. $(xy)^{2/3}$
6. $(x^2 + 2x + 1)^{3/5}$
7. $(ax + b)^{1/3}$
8. $(x^2 + 3x + 1)^{1/2}$
9. $(x + y)^{5/2}$
10. $(2x + 4y)^{2/7}$

In Exercises 11–20 write the expression in exponential form.

11. $\sqrt{x}$
12. $\sqrt{x + y}$
13. $\sqrt[3]{x}$
14. $\sqrt[3]{y^4}$
15. $(\sqrt[3]{y})^4$
16. $\sqrt[4]{3xy^3}$
17. $(\sqrt[4]{2x + 1})^3$
18. $\sqrt[3]{(2x + 1)^2}$
19. $\sqrt[3]{x^2 y}$
20. $\sqrt{x^2 + y}$

In Exercises 21–35 perform the indicated operation and simplify.

21. $\sqrt{5x}\sqrt{20x}$
22. $\sqrt[3]{2x^2 y}\sqrt[3]{4xy}$
23. $\sqrt{8x}\sqrt{2xy}$
24. $\sqrt{2x}\sqrt{2xy}\sqrt{3y}$
25. $\sqrt[3]{25a^2 b}\sqrt[3]{5b^2}\sqrt[3]{a}$
26. $4\sqrt{y} + 2\sqrt{y} - 3\sqrt{y}$
27. $\sqrt{4x^2 y} + 4x\sqrt{y} - \sqrt{9x^2 y}$
28. $3a\sqrt[3]{ab} + 4\sqrt[3]{a^4 b} - \sqrt[3]{8a^4 b}$
29. $(2\sqrt{x})(3\sqrt{x}) + 2x$
30. $(3\sqrt[3]{x})(4\sqrt[3]{x})(\sqrt[3]{x})$
31. $(\sqrt{2} + \sqrt{5})(\sqrt{2} - 3\sqrt{5})$
32. $(\sqrt{3} - 2\sqrt{2})(3\sqrt{3} + 4\sqrt{2})$
33. $(\sqrt{x} + 2\sqrt{y})(3\sqrt{x} - 4\sqrt{y})$
34. $(\sqrt{xy} + 2\sqrt{y})^2$
35. $(x\sqrt{y} + y\sqrt{x})^2$

In Exercises 36–46 simplify the expression.

36. $\sqrt{\dfrac{1}{3}}$
37. $\sqrt{\dfrac{2}{5}}$
38. $\dfrac{\sqrt{3}}{\sqrt{2} - \sqrt{3}}$
39. $\dfrac{\sqrt{5}}{\sqrt{2} + 5}$
40. $\sqrt{\dfrac{x}{x + y}}$
41. $\dfrac{8 - 3\sqrt{26} - 10}{3}$
42. $\dfrac{15 + 2\sqrt{52} - 2}{5}$
43. $\dfrac{\sqrt{x}}{\sqrt{x} + 2\sqrt{y}}$

44. $\dfrac{\sqrt{2} + \sqrt{3}}{\sqrt{2} - 2\sqrt{3}}$

45. $\dfrac{x + \sqrt{5x^2 + 4x^2}}{12x}$

46. $\dfrac{x - \sqrt{x^2 + 6x + 9}}{3x}$

1.9 The Complex Number System

In the previous section we were able to simplify square roots of positive numbers but not square roots of negative numbers. Thus we could compute $\sqrt{9} = 3$, but $\sqrt{-9}$ was undefined. We now define the symbol i as

$$i = \sqrt{-1}$$

With this definition we can simplify $\sqrt{-9}$ by writing $\sqrt{-9} = \sqrt{-1 \cdot 9} = \sqrt{-1}\sqrt{9} = i \cdot 3 = 3i$. From our definition it must also be true that

$$i^2 = -1$$

Using this, we can check that $\sqrt{-9} = 3i$, since $(3i)(3i) = 9i^2 = 9(-1) = -9$.

The set of all numbers of the form $x + yi$ where $i = \sqrt{-1}$ and x and y are real numbers is called the set of **complex numbers.** The number x is called the **real part** of the complex number $x + yi$, and the number yi is called the **imaginary part** of the complex number.

Example 1 Identify the real part and the imaginary part of each of the following complex numbers:
(a) $2 - 3i$ (b) -3
(c) $\frac{2}{3}i$

Solution (a) The real part is 2, and the imaginary part is $-3i$
(b) $-3 = -3 + 0i$, so the real part is -3, and the imaginary part is $0i$.
(c) $\frac{2}{3}i = 0 + \frac{2}{3}i$, so the real part is 0 and the imaginary part is $\frac{2}{3}i$. ∎

Note: The final form of a complex number should be $x + yi$. In particular, fractions such as $\dfrac{1 + 2i}{3}$ should be rewritten as $\frac{1}{3} + \frac{2}{3}i$.

Since $i = \sqrt{-1}$ is a radical, we can use the definitions of Section 1.8 to simplify, add, subtract, multiply, or divide.

Example 2   Simplify:

(a) $\sqrt{-25x^2}$

(b) $\sqrt{\dfrac{-18x}{y}}$

Solution (a) $\sqrt{-25x^2} = \sqrt{-1 \cdot 25x^2}$ (factor)

$\qquad\qquad\qquad = \sqrt{-1} \cdot \sqrt{25x^2}$ (product rule)

$\qquad\qquad\qquad = i \cdot 5x$ (definition of i; simplify)

$\qquad\qquad\qquad = \boxed{5xi}$

(b) $\sqrt{\dfrac{-18x}{x}} = \sqrt{-1 \cdot \dfrac{18x}{y}}$ (factor)

$\qquad\qquad\quad = \sqrt{-1}\,\sqrt{\dfrac{18x}{y}}$ (product rule)

$\qquad\qquad\quad = i\,\dfrac{\sqrt{18x}}{\sqrt{y}}\left(\dfrac{\sqrt{y}}{\sqrt{y}}\right)$ (rationalize)

$\qquad\qquad\quad = i\,\dfrac{\sqrt{9 \cdot 2 \cdot xy}}{\sqrt{y^2}}$

$\qquad\qquad\quad = i \cdot \dfrac{3\sqrt{2xy}}{y} = \boxed{\dfrac{3\sqrt{2xy}}{y}\,i}$ ∎

Example 3 illustrates addition, subtraction, and multiplication with complex numbers.

Example 3   Perform the indicated operations and simplify:
(a) $(2 + i) + (3 - 9i)$ (b) $(2 + i) - (3 - 9i)$
(c) $(2 + i)(3 - 9i)$ (d) $(-2i)(3ix)(y)$
(e) $(5 - 2i)^2$ (f) $(5 - 2i)(5 + 2i)$

Solution (a) $(2 + i) + (3 - 9i) = 2 + i + 3 - 9i$

$\qquad\qquad\qquad\qquad\quad = \underline{2 + 3} + \underline{i - 9i}$ (rearrange terms)

$\qquad\qquad\qquad\qquad\quad = \boxed{5 \quad - \quad 8i}$ (combine)

(b) $(2 + i) - (3 - 9i) = 2 + i - 3 + 9i$

$\qquad\qquad\qquad\qquad\quad = 2 - 3 + i + 9i$ (rearrange terms)

$\qquad\qquad\qquad\qquad\quad = \boxed{-1 + 10i}$ (combine)

(c) $(2 + i)(3 - 9i) = 6 - 18i + 3i - 9i^2$
$\qquad = 6 - 15i - 9(-1) \qquad$ (combine; $i^2 = -1$)
$\qquad = 6 - 15i + 9$

$\qquad = \boxed{15 - 15i} \qquad$ (combine)

(d) $(-2i)(3ix)(y) = -6i^2xy$
$\qquad = -6(-1)xy \qquad (i^2 = -1)$

$\qquad = \boxed{6xy}$

(e) $(5 - 2i)(5 - 2i) = 25 - 10i - 10i + 4i^2$
$\qquad = 25 - 20i - 4 \qquad$ (combine; $i^2 = -1$)

$\qquad = \boxed{21 - 20i} \qquad$ (combine)

(f) $(5 - 2i)(5 + 2i) = 25 + 10i - 10i - 4i^2$
$\qquad = 25 - 4(-1) \qquad$ (combine; $i^2 = -1$)
$\qquad = 25 + 4$

$\qquad = \boxed{29} \quad \blacksquare$

The conjugate of $x + yi$ is $x - yi$. In Example 3(f) we see that multiplying a complex number by its conjugate results in a real number. This fact can be used in simplifying fractions, as it was in the section on ordinary radicals.

Simplify as much as possible. Write the answer in the form $x + yi$.

Example 4

(a) $\dfrac{5}{i}$ $\qquad$ (b) $\dfrac{1 + 2i}{3 + i}$ $\qquad$ (c) $\dfrac{9 - \sqrt{8 - 26}}{6}$

Solution (a) $\dfrac{5}{i} \cdot \dfrac{i}{i} = \dfrac{5i}{i^2} = \dfrac{5i}{-1} = \boxed{-5i}$

(b) $\dfrac{1 + 2i}{3 + i} \cdot \dfrac{3 - i}{3 - i} = \dfrac{3 - i + 6i - 2i^2}{9 - i^2}$ $\qquad$ (multiply by conjugate of denominator)

$\qquad = \dfrac{3 + 5i + 2}{9 + 1} \qquad (i^2 = -1)$

$\qquad = \dfrac{5 + 5i}{10}$

$\qquad = \dfrac{\not5(1 + i)}{\not5 \cdot 2}$

$\qquad = \dfrac{1 + i}{2} = \boxed{\dfrac{1}{2} + \dfrac{1}{2}i}$

(c) $\dfrac{9 - \sqrt{8 - 26}}{6} = \dfrac{9 - \sqrt{-18}}{6}$

$= \dfrac{9 - i\sqrt{9 \cdot 2}}{6}$ (bring out i; $18 = 9 \cdot 2$)

$= \dfrac{9 - 3i\sqrt{2}}{6}$

$= \dfrac{\cancel{3}(3 - i\sqrt{2})}{\cancel{3} \cdot 2}$

$= \dfrac{3 - i\sqrt{2}}{2} = \dfrac{3}{2} - \dfrac{\sqrt{2}}{2} i$ ∎

On occasion we may encounter higher exponents of i than 2. The following example shows how to handle such exponents.

Example 5   Simplify i^6 and i^7.

$i^6 = i^{2 \cdot 3} = (i^2)^3 = (-1)^3 = \boxed{-1}$

$i^7 = i^6 \cdot i = (-1)i = \boxed{-i}$ ∎

↑
just computed

Solution   The values of i^n form a repeating pattern:

$$i = \sqrt{-1} = i$$
$$i^2 = -1$$
$$i^3 = -i$$
$$i^4 = 1$$
$$i^5 = i$$
$$i^6 = -1$$
$$\vdots$$

or

$$i^n = \begin{cases} i \text{ for } n = 1, 5, 9, 13, \ldots \\ -1 \text{ for } n = 2, 6, 10, 14, \ldots \\ -i \text{ for } n = 3, 7, 11, 15, \ldots \\ 1 \text{ for } n = 4, 8, 12, 16, \ldots \end{cases}$$

As a practical matter, it is usually more convenient to compute the value of i^n by rewriting with respect to i^2, since $i^2 = -1$ and $(-1)^m$ is 1 when m is even and is -1 when m is odd.

Example 6 Substitute $1 + i$ for x in the expression $x^2 - 2x + 2$ and simplify.

Solution
$$
\begin{aligned}
(1 + i)^2 - 2(1 + i) + 2 &= 1 + 2i + i^2 - 2 - 2i + 2 \quad \text{(multiply)} \\
&= 1 + (-1) \quad \text{(combine } 2i, -2i, -2, 2; i^2 = -1) \\
&= 0 \quad \blacksquare
\end{aligned}
$$

We will encounter complex numbers again in Chapters 4 and 5.

Exercises 1.9
In Exercises 1–10 identify the real part and the imaginary part for the complex number.

1. $1 + i$
2. $2 - 3i$
3. 5
4. $6i$
5. $\dfrac{5 + 2i}{3}$
6. $\sqrt{2} - \sqrt{3}i$
7. $x + yi$
8. $2a + 3bi$
9. $\sqrt{7} - wi$
10. $5i - \frac{1}{2}$

In Exercises 11–20 simplify the expression.

11. $\sqrt{-49}$
12. $\sqrt{-100}$
13. $\sqrt{-\dfrac{1}{4}}$
14. $(-25)^{1/2}$
15. $\sqrt{-24}$
16. $\sqrt{-16xy^3}$
17. $\sqrt{-45c^2d^5}$
18. $\sqrt{\dfrac{-2x}{9y^2}}$
19. $-\sqrt{-4x^8}$
20. $\sqrt[3]{-8x^6}$

In Exercises 21–36 perform the indicated operations and simplify. Write the answer in the form $x + yi$.

21. $(2 + 3i) + (5 - i)$
22. $(2 + 3i) - (5 - i)$
23. $(2 + 3i)(5 - i)$
24. $\dfrac{2 + 3i}{5 - i}$
25. $(2 - 3i)^2$
26. $(2 - 3i)(2 + 3i)$
27. $(6 - i) - (1 - 9i)$
28. $(2i)(-3i)$
29. $\dfrac{2 - \sqrt{4 - 20}}{4}$
30. $(4 - 3i) + (-2 + 6i)$

31. $\dfrac{7 + 2i}{i}$

32. $(\sqrt{6} + 2i)^2$

33. $\dfrac{5 - 3i}{1 + i}$

34. $\dfrac{5 + \sqrt{9 - 84}}{15}$

35. $(\sqrt{2} + xi)(\sqrt{2} - xi)$

36. $\dfrac{1 + i}{1 - i}$

37. Simplify i^3, i^4, and i^5.
38. Simplify i^8, i^{10}, and i^{99}.
39. Substitute $-1 + 2i$ for x in the expression $x^2 + 2x + 5$ and simplify.
40. Substitute $\dfrac{1}{2} + \dfrac{\sqrt{3}}{2} i$ for x in the expression $x^2 - x + 1$ and simplify.

Key Terms and Formulas

Set:
  element
  subset
  union
  intersection
  complement
  universal set
  cross-product
  power set
  set-builder notation
  Venn diagram
Number:
  natural number
  integer
  rational number
  irrational number
  real number
  complex number
Number line
Field properties:
  closure
  commutativity
  associativity
  identity
  inverse
  distributivity
Absolute value
Order of operations

Polynomial:
  variable
  exponent
  base
  numerical coefficient
  degree
Monomial
Binomial
Trinomial
Fraction:
  algebraic fraction
  complex fraction
Long division
Radical:
  square root
  cube root
Formulas for fractions:

$$\frac{a \cdot x}{b \cdot x} = \frac{a}{b} \quad \text{(reduction)}$$

$$\frac{a}{b} \cdot \frac{c}{d} = \frac{ac}{bd} \quad \text{(multiplication)}$$

$$\frac{a}{b} + \frac{c}{b} = \frac{a + c}{b} \quad \text{(addition)}$$

$$\frac{a}{b} - \frac{c}{b} = \frac{a - c}{b} \quad \text{(subtraction)}$$

$$\frac{a}{b} \div \frac{c}{d} = \frac{a}{b} \cdot \frac{d}{c} \quad \text{(division)}$$

$$n \cdot x = \underbrace{x + x + \cdots + x}_{n \text{ times}}$$

(n is a natural number)

$$x^n = \underbrace{x \cdot x \cdots x}_{n \text{ times}}$$

(n is a natural number)

Formulas for factoring:

$x^2 - y^2 = (x - y)(x + y)$ (difference of squares)

$x^2 + 2xy + y^2 = (x + y)^2$ (perfect square)

$x^2 - 2xy + y^2 = (x - y)^2$ (perfect square)

$x^2 + (b + d)x + bd = (x + b)(x + d)$ (trial and error)

$ax + bx = x(a + b)$ or $(a + b)x$ (common factor)

$x^3 + y^3 = (x + y)(x^2 - xy + y^2)$ (sum of cubes)

$x^3 - y^3 = (x - y)(x^2 + xy + y^2)$ (difference of cubes)

Formulas for exponents and radicals:

$$x^m \cdot x^n = x^{m+n}$$

$$\frac{x^m}{x^n} = x^{m-n}$$

$$(x^m)^n = x^{mn}$$
$$(xy)^n = x^n \cdot y^n$$

$$\left(\frac{x}{y}\right)^n = \frac{x^n}{y^n}$$

$$x^0 = 1$$

$$x^{-n} = \frac{1}{x^n}$$

$$x^{p/q} = (x^{1/q})^p$$

$$\sqrt[q]{x} = x^{1/q}$$

$$\sqrt[q]{xy} = \sqrt[q]{x} \cdot \sqrt[q]{y}$$

$$\sqrt[q]{\frac{x}{y}} = \frac{\sqrt[q]{x}}{\sqrt[q]{y}}$$

$$i = \sqrt{-1}$$

$$i^2 = -1$$

Review Exercises

1. If $X = \{2, 3, a, b\}$ and $Y = \{3, a\}$, put the correct symbol ($\in, \notin, \subseteq,$ or $\nsubseteq$) in each blank.
    (a) 2 _____ X
    (b) c _____ X
    (c) Y _____ X
    (d) X _____ Y

2. List the elements described by the set-builder notation

    $$\{m \mid m \text{ is a natural number smaller than 8}\}$$

3. Let the universal set U be the set of natural numbers smaller than 6, the set $A = \{2, 4\}$, and the set $B = \{4, 5\}$. Form the following sets:
    (a) $A \cup B$
    (b) $A \cap B$
    (c) $\overline{A}$
    (d) $\overline{A} \cup \overline{B}$
    (e) $\overline{\overline{B}}$

4. List the elements of $\{a, b\} \times \{c, d, e\}$.

5. Construct the power set of $\{x, y\}$.

In Exercises 6–11 perform the indicated operations.

6. $\dfrac{3}{5} + \dfrac{5}{6} - \dfrac{1}{2}$

7. $\dfrac{3}{5} \cdot \dfrac{5}{6} \div \dfrac{1}{2}$

8. $5 - 5(2 + 3)$

9. $(5 - 5)(2 + 3)$

10. $\dfrac{2}{3}\left(\dfrac{1}{5} + \dfrac{3}{4}\right)$

11. $\dfrac{2}{3} - \left(\dfrac{1}{5} + \dfrac{3}{4}\right)$

In Exercises 12–14 calculate the absolute value.

12. $|5 - 7|$

13. $\left|\dfrac{1}{2} - \dfrac{3}{4}\right|$

14. $\left|\dfrac{1}{2}\right| - \left|\dfrac{3}{4}\right|$

15. For the expression $x^2 - \frac{2}{3}xy^2z + 4y^3 - 10$:
    (a) identify the number of terms.
    (b) state the degree of the polynomial.
    (c) give the numerical coefficient of the second term.
    (d) give the exponent of z in the second term.

In Exercises 16–24 perform the indicated operations.

16. $(2xy^2)(-3x^3y)(xy^2)$

17. $2xy^2 - 3x^3y + xy^2$

18. $2xy^2(-3x^3y + xy^2)$

19. $(2x + y^2)(-3x^3y + xy^2)$

20. $(x - 3y)(2x + 3y)$

21. $(x - 3y) + (2x + 3y)$

22. $(x - 3y) - (2x + 3y)$

23. $\left(\dfrac{3}{4}xy^2 + \dfrac{1}{2}x^2y\right) - \left(\dfrac{5}{6}xy^2 - \dfrac{2}{7}x^2y\right)$

24. $(x + 1)(x - 1)(3x^2 + 3)$

In Exercises 25–35 factor the expression completely.

25. $x^2 + 7x + 10$

26. $x^2 - 2x - 15$

27. $x^2 + 15x$

28. $2x^3 - 250$

29. $6x^2 + 13x - 5$

30. $8x^2 + 34x + 21$

31. $x^5 + 64x^2$

32. $16x^3 - 44x^2 - 42x$

33. $81x^2 - 72xy + 16y^2$

34. $x^6 - y^6$

35. $3x^5 - 30x^3 + 27x$

In Exercises 36–45 combine the expressions and simplify.

36. $\dfrac{x^2 - 8}{x - 2} + \dfrac{2x}{x - 2}$

37. $\dfrac{x + 1}{x - 3} - \dfrac{12}{x^2 - 3x}$

38. $\dfrac{x}{x - 2} + \dfrac{3}{x + 2}$

39. $\dfrac{6}{3x - 5} + \dfrac{x}{2x + 1} - \dfrac{26}{6x^2 - 7x - 5}$

40. $\dfrac{3}{3x - 1} + \dfrac{1}{9x^2 - 1} + \dfrac{2}{3x + 1}$

41. $\dfrac{2}{x^2 - 3x + 2} + \dfrac{x}{x^2 - 1} - \dfrac{1}{x^2 - x - 2}$

42. $\dfrac{x}{z^2} - \dfrac{1}{y} \cdot \dfrac{x + 2}{y}$

43. $\dfrac{2}{x - 2} - \dfrac{3}{x - 2} \div \dfrac{x + 3}{x - 2}$

44. $\dfrac{\dfrac{3}{xy^2}}{\dfrac{2}{x} + \dfrac{x}{y}}$

45. $\dfrac{\dfrac{1}{x+1} - \dfrac{2}{x-1}}{\dfrac{3}{x+2} + \dfrac{x}{x+1}}$

In Exercises 46–60 simplify the expression.

46. $x^2 \cdot x^{3/2} \cdot x^{1/2}$

47. $\dfrac{xy^{-2}}{x^{-2}y^3}$

48. $\left(\dfrac{x^{-3}y}{z^0}\right)^{-2}$

49. $\dfrac{x^{-1}}{y^{-2} + x}$

50. $(-8x^3)^{1/3}$

51. $(25y^6)^{-1/2}$

52. $\dfrac{3^{-1}x^{-2}y^3}{6x^{-3}y^{-1}}$

53. $-(36x^2)^{1/2}$

54. $\dfrac{x^{-2} + y^{-3}}{x^{-3}y}$

55. $\sqrt{\dfrac{2}{3x}}$

56. $\dfrac{\sqrt{2}}{\sqrt{2} + 1}$

57. $\sqrt{\dfrac{x+1}{x-1}}$

58. $\dfrac{3y - \sqrt{4y^2}}{y}$

59. $\dfrac{\sqrt{2} + \sqrt{3}}{\sqrt{2} - \sqrt{3}}$

60. $\dfrac{x + \sqrt{3x^2 + x^2}}{3x^2}$

In Exercises 61–70 perform the indicated operations and simplify.

61. $\sqrt{3x^3}\sqrt{27x}$

62. $\sqrt{16x}\sqrt{xy^3}$

63. $\sqrt{3x}\sqrt{3xy}\sqrt{3y}$

64. $\sqrt[3]{2x^2y}\sqrt[3]{4xy^3}$

65. $\sqrt[3]{16x^4y}\sqrt[3]{2xy^2}\sqrt[3]{2xy}$

66. $3y\sqrt{x} + y\sqrt{16x} - \sqrt{25xy^2}$

67. $(\sqrt{7} + \sqrt{2})(\sqrt{14} - \sqrt{2})$

68. $(\sqrt{x} + 2\sqrt{y})^2$

69. $2\sqrt[3]{x}(3\sqrt[3]{x^2} + \sqrt[3]{y})$

70. $\sqrt{18x^2y} + \sqrt[3]{54x} - x\sqrt{50y}$

In Exercises 71–75 simplify the expression.

71. $\sqrt{-25x^2y}$

72. $\sqrt{\dfrac{-36x}{9y^2}}$

73. $-\sqrt{-16x^{16}}$

74. $\sqrt[3]{-27x^9}$

75. $\sqrt{-\dfrac{2}{9}x^4}$

In Exercises 76–85 perform the indicated operations and simplify.

76. $(2 + 4i) + (6 - 5i)$

77. $\dfrac{1 - 3i}{2 - i}$

78. $(3 + 4i)(3 - 4i)$

79. $(3 + 4i) - (3 - 4i)$

80. $(3 - 4i)^2$

81. $(\sqrt{2} + 3i)^2$

82. $\dfrac{6 - 3i}{2i}$

83. $\dfrac{6 + \sqrt{16 - 25}}{4}$

84. $(x + 7i)(x - 7i)$

85. $(x + \sqrt{-7})(x - \sqrt{-7})$

CHAPTER 2

First-Degree Equations and Inequalities

2.1   Statements in One Variable

Solving equations is one of our major objectives in college algebra and a cornerstone in the practical application of algebra.

An **equation** is a mathematical statement that one algebraic expression is equal to another.

Example 1 $\dfrac{4}{2} = 2$ (This is an **identity**—it is always true.)

$5x = 30$ (This is a **conditional equation**—its truth depends on the value substituted for the variable, x.)

$\dfrac{1}{c} = \dfrac{1}{a} + \dfrac{1}{b}$ (This is a **literal equation**—it involves more than one variable.)   ■

If a variable, x, in an equation is replaced by a value or expression that makes the equation a true statement (identity), then the value or expression is called a **solution** of the equation for x.

Example 2   2 is a solution for x in the equation $4x = 8$, since $4 \cdot 2 = 8$ is an identity.   ■

To **solve** an equation for x means to find algebraically all the solutions of the equation for x. We will build the skills needed to solve equations step by step.

First, we need to make some natural observations regarding equality. If we perform the same operation on equal expressions, then the resulting expressions are equal.

PROPERTIES OF EQUALITY

Addition: If $s = t$, then $s + c = t + c$.
Multiplication: If $s = t$, then $cs = ct$.

The two statements above actually address the four basic operations, since division is a special form of multiplication (remember: do not divide by 0) and subtraction is a special form of addition.

To solve an equation of the form

$$ax = b \quad \text{with } a \neq 0$$

simply divide both sides of the equation by a, which gives the solution

$$x = \frac{b}{a}$$

If $ax = b$ with $a \neq 0$,   then $x = \dfrac{b}{a}$

Example 3   Solve $4x = 8$, $-3y = 15$, and $-10z = -5$ for their respective unknowns.

Solution

$$4x = 8 \qquad -3y = 15 \qquad -10z = -5$$

$$\frac{\cancel{4}x}{\cancel{4}} = \frac{8}{4} \qquad \frac{-\cancel{3}y}{-\cancel{3}} = \frac{15}{-3} \qquad \frac{-\cancel{10}z}{-\cancel{10}} = \frac{-5}{-10}$$

$$x = 2 \qquad\qquad y = -5 \qquad\qquad z = \frac{1}{2} \qquad ■$$

Example 4   Solve $2x + 3 = 4x - 5$ for x.

Solution

$$2x + 3 = 4x - 5 \quad \text{(statement of problem)}$$

$$2x + \cancel{3} - \cancel{3} = 4x - 5 - 3 \quad \text{(subtract 3 from each side)}$$

$$2x = 4x - 8 \quad \text{(combine terms)}$$

$$2x - 4x = \cancel{4x} - \cancel{4x} - 8 \quad \text{(subtract 4x from each side)}$$

$$-2x = -8 \quad \text{(combine terms)}$$

$$\frac{-\cancel{2}x}{-\cancel{2}} = \frac{-8}{-2} \quad \text{(divide by } -2 \text{ on each side)}$$

$$\boxed{x = 4} \quad \text{(simplify)}$$

Check: $2 \cdot 4 + 3 \stackrel{?}{=} 4 \cdot 4 - 5$

$$8 + 3 \stackrel{?}{=} 16 - 5$$

$$11 \stackrel{\checkmark}{=} 11 \quad \blacksquare$$

Example 4 gives us a useful pattern for solving this type of equation. The terms not involving x are moved to the right side of the equation, and the terms involving x are moved to the left side. This gives an equation of the form $ax = b$, which was solved earlier.

Example 5   Solve $2x - (x - 4) = 5(x + 2)$ for x.

Solution

$$2x - (x - 4) = 5(x + 2) \quad \text{(statement of problem)}$$

$$2x - x + 4 = 5x + 10 \quad \text{(remove grouping symbols)}$$

$$x + 4 = 5x + 10 \quad \text{(combine like terms)}$$

$$x = 5x + 6 \quad \text{(subtract 4 from both sides)}$$

$$-4x = 6 \quad \text{(subtract 5x from both sides)}$$

$$x = -\frac{6}{4} \quad \text{(divide both sides by } -4)$$

$$\boxed{x = -\frac{3}{2}} \quad \text{(reduce fraction)}$$

Check: $2\left(-\dfrac{3}{2}\right) - \left(-\dfrac{3}{2} - 4\right) \overset{?}{=} 5\left(-\dfrac{3}{2} + 2\right)$

$-\dfrac{6}{2} - \left(-\dfrac{3}{2} - \dfrac{8}{2}\right) \overset{?}{=} 5\left(-\dfrac{3}{2} + \dfrac{4}{2}\right)$

$-\dfrac{6}{2} - \left(-\dfrac{11}{2}\right) \overset{?}{=} 5\left(\dfrac{1}{2}\right)$

$-\dfrac{6}{2} + \dfrac{11}{2} \overset{?}{=} \dfrac{5}{2}$

$\dfrac{5}{2} \overset{\checkmark}{=} \dfrac{5}{2}$ ∎

A natural extension of our work is the solution of equations that involve fractions, such as solving

$$\frac{7}{x} + 4 = \frac{1}{2} \quad \text{for } x$$

As a general principle we will convert the fractional problem into the type that we have already studied. This is done by multiplying both sides of the equation by the least common denominator (L.C.D.) of all the fractions in the equation. In the example above the L.C.D. is $2x$. The following step-by-step solution shows how the method works.

Example 6   Solve $\dfrac{7}{x} + 4 = \dfrac{1}{2}$ for x.

Solution

$\dfrac{7}{x} + 4 = \dfrac{1}{2}$ (L.C.D. $= 2x$)

$\left(2x \cdot \dfrac{7}{x}\right) + (2x \cdot 4) = \left(2x \cdot \dfrac{1}{2}\right)$ (multiply both sides by L.C.D.)

$14 + 8x = x$ (simplify)

$8x = x - 14$ (subtract 14 from both sides)

$7x = -14$ (subtract x from both sides)

$x = -2$ (divide by 7)

Check: $\dfrac{7}{-2} + 4 \stackrel{?}{=} \dfrac{1}{2}$

$-\dfrac{7}{2} + \dfrac{8}{2} \stackrel{?}{=} \dfrac{1}{2}$

$\dfrac{1}{2} \stackrel{\checkmark}{=} \dfrac{1}{2}$ ■

The next example shows the method again but also points out the need for reasonable caution in all uses of mathematical techniques.

Example 7 Solve $2 - \dfrac{1}{w + 1} = \dfrac{w}{w + 1}$ for w.

Solution $2 - \dfrac{1}{w + 1} = \dfrac{w}{w + 1}$ (L.C.D. $= w + 1$)

$[2 \cdot (w + 1)] - \left[\dfrac{1}{(w + 1)} \cdot (w + 1)\right] = \left[\dfrac{w}{(w + 1)} \cdot (w + 1)\right]$

(multiply both sides by the L.C.D.)

$2w + 2 - 1 = w$ (simplify)

$2w + 1 = w$ (combine terms)

$2w = w - 1$ (subtract 1 from both sides)

$w = -1$ (subtract w from both sides)

Check: $2 - \dfrac{1}{-1 + 1} \stackrel{?}{=} \dfrac{-1}{-1 + 1}$

$2 - \dfrac{1}{0} \stackrel{?}{=} \dfrac{-1}{0}$

But division by 0 is not defined; thus there is no solution . ■

The result of "no solution" may seem quite mysterious, since we followed the usual steps for solving a fractional equation. The difficulty is that the indicated value $w = -1$ gives undefined expressions (0 denominators) when the problem is checked. *To be a correct answer, the apparent solution must satisfy the original equation!* To prevent our being satisfied with incorrect solutions, we should make a mental note that certain values of the unknown will result in undefined expressions. A glance at the denominators tells us which values would result in a denominator being 0.

Example 8 Consider $\dfrac{2}{x-1} + \dfrac{5}{x} = \dfrac{1}{3}$. A value of $x = 1$ makes the first denominator zero, a value of $x = 0$ makes the second denominator zero, and the third denominator is never zero. We conclude that $x = 1$ and $x = 0$ *cannot* be acceptable solutions. These are called **exclusions** or **restrictions** on x. ■

As we mentioned in Example 1, a literal equation is a single equation that involves more than one variable. Our objective is to solve for a specified variable in terms of the others.

Example 9 Solve $2x - 3y = 5$ for y.

Solution
$$2x - 3y = 5 \qquad \text{(original equation)}$$
$$-3y = 5 - 2x \qquad \text{(subtract } 2x \text{ from both sides to isolate } y)$$
$$y = \frac{5 - 2x}{-3} \qquad \text{(divide both sides by } -3)$$
or
$$y = \frac{2x - 5}{3} \qquad \text{(multiply numerator and denominator by } -1)\quad ■$$

The technique is to isolate the desired variable on the left side of the equation and then divide by the coefficient of this variable. Example 9 comprises one step of a process that will be discussed in Chapter 3. The next example is an application that involves rewriting a formula.

Example 10 Solve $1/R = 1/S + 1/T$ for S. This formula arises in the study of electricity. The variables R, S, and T are the resistances, measured in ohms, of resistors in an electrical circuit. The formula relates the resistance of two parallel resistors to a single equivalent resistor as depicted in Figure 1. In effect the problem is to rewrite the formula so that the value of S can be determined if the values of R and T are known.

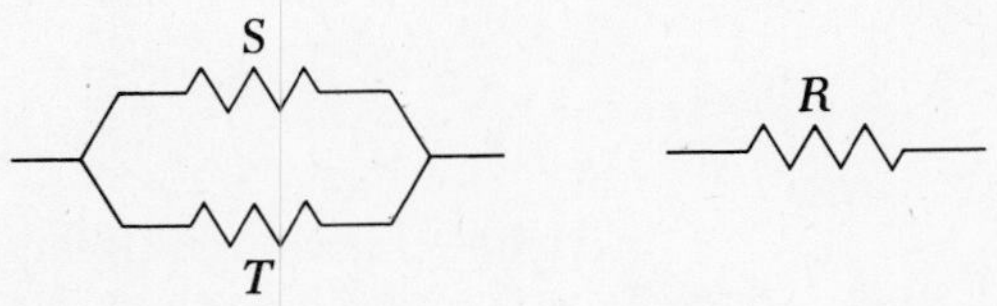

FIGURE 1

Solution

$$\frac{1}{R} = \frac{1}{S} + \frac{1}{T} \qquad \text{(L.C.D. = } RST\text{)}$$

$$\cancel{R}ST \cdot \frac{1}{\cancel{R}} = R\cancel{S}T \cdot \frac{1}{\cancel{S}} + RS\cancel{T} \cdot \frac{1}{\cancel{T}} \qquad \text{(multiply both sides by L.C.D.)}$$

$$ST = RT + RS \qquad \text{(simplify)}$$

$$ST - RS = RT \qquad \text{(subtract } RS \text{ from both sides to isolate } S\text{)}$$

$$(T - R)S = RT \qquad \text{(factor out } S\text{)}$$

$$\frac{\cancel{(T - R)}S}{\cancel{(T - R)}} = \frac{RT}{T - R} \qquad \text{(divide both sides by } T - R\text{)}$$

$$\boxed{S = \frac{RT}{T - R}} \qquad \text{(simplify)} \quad \blacksquare$$

Equations use the notion of two expressions being the same. Another basic notion is that of order. We will consider expressions related by "less than" ($<$), "less than or equal to" ($\leq$), "greater than" ($>$), and "greater than or equal to" ($\geq$). Such statements are called **inequalities.**

Example 11 Some examples of inequalities are

$$2 \leq 3, \qquad 3x + 1 > 2(x - 4), \quad \text{and} \quad \frac{x}{2} \geq 3 \quad \blacksquare$$

The need to solve inequalities often arises in applications of algebra. A solution of an inequality in one variable is a value that results in a true statement when substituted for the variable in the inequality. Most inequalities have infinitely many solutions.

Example 12 The values $x = 1$ and $x = 4$ are solutions of $2x - 7 < 5$, since

$$2 \cdot 1 - 7 < 5 \quad \text{or} \quad -5 < 5$$

and

$$2 \cdot 4 - 7 < 5 \quad \text{or} \quad 1 < 5$$

are true statements. In fact, any real number less than 6 is a solution of $2x - 7 < 5$. Thus the set of solutions or the **solution set** of $2x - 7 < 5$ is $\{x \mid x < 6\}$. $\quad \blacksquare$

The following properties are needed to solve inequalities. The properties are stated for one type of inequality but hold in like manner for the other types.

PROPERTIES OF INEQUALITY

Addition: If $s < t$, then $s + c < t + c$.

Multiplication: $\begin{cases} \text{If } s < t \text{ and } c > 0, \text{ then } sc < tc. \\ \text{If } s < t \text{ and } c < 0, \text{ then } sc > tc. \end{cases}$

To paraphrase these properties, adding the same quantity to both sides preserves the type of inequality, multiplying both sides by the same *positive* quantity preserves the type of inequality, but multiplying both sides by the same *negative* quantity reverses the type of inequality.

Example 13   (a) Since $-1 < 5$, it follows that $2 < 8$ (add 3).
(b) Since $4 \geq 2$, it follows that $3 \geq 1$ (add -1).
(c) Since $-6 > -7$, it follows that $-12 > -14$ (multiply by 2).
(d) Since $-4 \leq +10$, it follows that $+2 \geq -5$ (multiply by $-\frac{1}{2}$).
Note in (b) that adding a negative covers subtraction and in (d) that multiplying by $-\frac{1}{2}$ is like dividing by -2.   ■

Inequalities are solved in the same manner as equations.

Example 14   Solve $2x - 3 < 5(x + 3)$ for x.

Solution $2x - 3 < 5(x + 3)$ (statement of problem)

$2x - 3 < 5x + 15$ (remove grouping symbols)

$2x < 5x + 18$ (add 3 to both sides)

$-3x < 18$ (subtract $5x$ from both sides)

$x > -6$ (divide both sides by -3)

Note that dividing by -3 reverses the type of inequality.   ■

The solution found in Example 14 is the set of all real numbers greater than -6. Another way of describing this is as the set of all

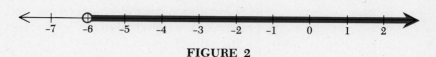

FIGURE 2

numbers on the number line to the right of -6. Graphically, this is represented in Figure 2. The open circle at -6 indicates that -6 itself is *not* a valid solution. If -6 were included, we would use a solid circle. The thick line and arrow represent all the solutions of the example inequality.

Example 15   Graph each inequality:
(a) $x < 0$ (b) $x \geq 2$ (c) $x \leq -\frac{3}{4}$

Solution   (a) $x < 0$:

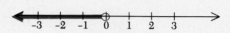

(b) $x \geq 2$:

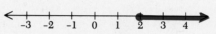

(c) $x \leq -\frac{3}{4}$:

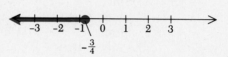

　　　　　　Equations and inequalities that involve absolute values are the final topics in this section.

Example 16   Solve $|x - 1| = 5$ for x.

Solution   Since the absolute value of either 5 or -5 is 5, the equation will be true if $x - 1 = 5$ or if $x - 1 = -5$.

$$x - 1 = 5 \qquad\qquad x - 1 = -5$$

$$\boxed{x = 6} \quad \text{or} \quad \boxed{x = -4}$$

$$\textit{Check:} \qquad x = 6 \qquad\qquad x = -4$$

$$|6 - 1| \overset{?}{=} 5 \qquad |-4 - 1| \overset{?}{=} 5$$

$$|5| \overset{?}{=} 5 \qquad\qquad |-5| \overset{?}{=} 5$$

$$5 \overset{\checkmark}{=} 5 \qquad\qquad 5 \overset{\checkmark}{=} 5 \quad\blacksquare$$

Example 17   Solve and graph $|2x - 3| \leq 7$.

Solution   The absolute value of $2x - 3$ will be less than or equal to 7 provided that $2x - 3$ falls in the range -7 to 7.

$$-7 \le 2x - 3 \le 7 \qquad (2x - 3 \text{ is between } -7 \text{ and } 7)$$

$$-4 \le 2x \le 10 \qquad (\text{add } 3)$$

$$\boxed{-2 \le x \le 5} \qquad (\text{divide by } 2)$$

The graph appears in Figure 3. ■

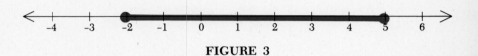

FIGURE 3

Example 18   Solve and graph $\left| \dfrac{x - 7}{2} \right| > 3$.

Solution   The absolute value of $(x - 7)/2$ will be greater than 3 provided that $(x - 7)/2$ is greater than 3 *or* less than -3.

$$\frac{x - 7}{2} < -3 \qquad\qquad \frac{x - 7}{2} > 3$$

$$x - 7 < -6 \qquad\qquad x - 7 > 6$$

$$\boxed{x < 1} \qquad \text{or} \qquad \boxed{x > 13}$$

The graph appears in Figure 4. ■

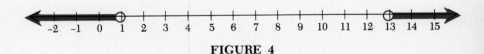

FIGURE 4

In this section we have learned how to solve equations in one variable using the properties of equality. This included simple, fractional, literal, and absolute value equations. We have also learned how to solve and graph a variety of inequalities in one variable by using the properties of inequality. The next section shows the use of equations to solve a variety of practical problems that are stated in everyday language.

Exercises 2.1   1. Show that $x = 3$ is a solution of $x + 7 = 4x - 2$.

2. Show that $x = 2$ is a solution of $\dfrac{3}{x - 4} + 5 = \dfrac{7}{2}$.

In Exercises 3–16 solve the equation.

3. $5x = 20$
4. $-3y = 9$
5. $2x = -14$
6. $-4x = -16$
7. $3z = z - 6$
8. $6x - 3 = -21$
9. $2x + 5 = 4x + 9$
10. $11x - 6 = 3x + 24$
11. $8(x - 2) = 4x$
12. $4(1 - c) = 2$
13. $2x - 2 = 6(x - 3)$
14. $5(x + 3) = 3(x - 1)$
15. $w - (2w + 7) = 5(w + 1)$
16. $-2(3x - 1) = 3(x + 2) - 4$

In Exercises 17–19 identify the values of the unknown that cannot be acceptable solutions.

17. $\dfrac{2}{x} + 3 = \dfrac{5}{2}$

18. $\dfrac{1}{x - 1} - \dfrac{2}{x + 2} = 4$

19. $\dfrac{7}{2y - 3} + 1 = \dfrac{2}{2y - 3}$

In Exercises 20–25 solve the equation.

20. $\dfrac{4}{x} = \dfrac{1}{2}$

21. $2 - \dfrac{1}{x} = \dfrac{3}{x}$

22. $\dfrac{8}{x + 1} - 3 = \dfrac{5}{x + 1}$

23. $\dfrac{7}{2y - 3} + 1 = \dfrac{2}{2y - 3}$

24. $\dfrac{4x + 3}{2x + 5} = \dfrac{-7}{2x + 5}$

25. $\dfrac{x}{x - 1} = \dfrac{3x}{1 - x} + 2$

[*Hint:* $1 - x = (-1)(x - 1)$.]

26. Solve for y: $3x + y = 1$.
27. Solve for x: $2x + 3y = 7$.
28. Solve for y: $5x - 2y = -6$.

Exercises 29–31 involve formulas from business and science that will be used in Section 2.2.

29. Solve for T: $I = PRT$ (simple interest)
30. Solve for w: $A = 2l + 2w$ (perimeter of rectangle)
31. Solve for r: $d = rt$ (distance-rate-time formula)
32. Solve for T: $\dfrac{1}{R} = \dfrac{1}{S} + \dfrac{1}{T} + \dfrac{1}{U}$ (three resistors in parallel)
33. Solve for x: $S = 2xy + 2xz + 2yz$ (surface area of box)
34. Solve for C: $F = \dfrac{9}{5} C + 32$ (temperature conversion from Celsius to Fahrenheit)
35. Graph on the number line: $\{x \mid x > -2\}$
36. Graph on the number line: $\{x \mid x < 7\}$
37. Graph on the number line: $\{x \mid x \geq 1\}$

In Exercises 38–45 solve the inequality and graph the solutions on the number line.

38. $2x + 3 < 7$
39. $x - 2 \geq 2x + 5$
40. $6x - (x + 6) > 2(x - 4) - 1$
41. $\dfrac{x}{3} \leq -8$

42. $\dfrac{2 - 3x}{4} > 5$ 43. $-3 < x + 1 < 3$

44. $-1 < 2x - 3 \leq 1$ 45. $-8 \leq -2x \leq 8$

In Exercises 46–50 solve the equation.

46. $|x + 3| = 1$ 47. $|2x - 1| = 5$

48. $\left|\dfrac{3x - 2}{4}\right| = \dfrac{1}{2}$ 49. $|x - 4| = 0$

50. $|3x + 7| = -1$

In Exercises 51–53, solve the inequality and graph the solutions on the number line.

51. $|4x - 6| > 2$ 52. $|2x + 1| \leq 5$
53. $|x - 6| \geq 10$

Exercises 54–55 are challenging problems designed to test your depth of understanding and ingenuity.

54. Write an absolute value equation whose solutions are those numbers whose distance from 1 on the number line is 3. Solve the equation and check the results.

55. Write an absolute value inequality whose solutions are those numbers whose distances from -2 on the number line are greater than 6. Solve the inequality and graph the results on the number line.

2.2   Introduction to Applied Problems

Now that we are able to solve a variety of equations in one variable, we will apply that skill to solve applied problems.

Example 1   Standard Motors has developed three experimental automobiles that are designed to emphasize fuel efficiency and safety. The cars are destroyed during safety testing. The average weight of the cars is known to be 3000 pounds. It is known that Car 1 weighs 2500 pounds and Car 2 weighs 3200 pounds, but the information about the weight of Car 3 has been misplaced. What is the weight of Car 3?

Solution   Let x = the weight of Car 3. Then the average weight of the three cars can be expressed by

$$\frac{2500 + 3200 + x}{3}$$

which must be equal to 3000 from our given information. That is,

$$\frac{2500 + 3200 + x}{3} = 3000$$

$$5700 + x = 9000$$

$$x = 3300 \text{ pounds (weight of Car 3)} \quad \blacksquare$$

Example 1 indicates one direct use of algebra. An unknown quantity may be represented by a letter, say x. An equation involving the unknown (x) is set up from the information given in the statement of the problem. The equation is then solved to determine the value of x.

Example 2   Super Child Day-Care Center is planning to fence in a play area next to the main building. The shape is to be a rectangle that is twice as long as it is wide. One long side is to be the building itself. If there are 300 feet of fencing, what are the dimensions of the play area?

Solution   A sketch is generally useful for geometric problems (Figure 5). Let w = the width; then $2w$ = the length. The length of the fencing is $w + 2w + w$. This gives the equation

$$w + 2w + w = 300$$

$$4w = 300$$

$$w = 75 \text{ feet} \quad \text{(width)}$$

$$2w = 150 \text{ feet} \quad \text{(length)} \quad \blacksquare$$

Building

Width (w) Width (w)

Length $(2w)$

FIGURE 5

The next example involves motion. Such problems usually require the following relationship:

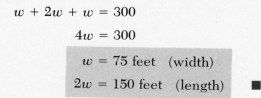

$$\text{distance} = \text{rate of speed} \cdot \text{time} \quad \text{or} \quad d = rt$$

For instance, a car traveling 50 miles per hour for 4 hours will cover a distance of $d = 50 \cdot 4 = 200$ miles. Note that the units of measurement must be consistent (*miles* and *miles* per hour; miles per *hour* and *hours*). A table of values is useful in solving motion problems.

Example 3 Joan and Phil drive toward each other from towns that are 270 miles apart. Joan left at noon. Phil left one hour later and drives 10 miles per hour faster. If they meet at 3 P.M., what was the rate of speed for each?

Solution If we let X = Joan's rate of speed, then Phil's rate is 10 miles per hour more, or $X + 10$. Joan travels from noon until 3 P.M., so her time is 3 hours, and Phil's time is 2 hours, since he left at 1 P.M. We can now fill in the following table:

	Joan	Phil
Rate (r)	X	$X + 10$
Time (t)	3	2
Distance (rt)	$3X$	$2(X + 10)$

Since we have expressions for the distances traveled by Joan and Phil, and we know the total distance, the equation is easy to determine:

Joan's distance + Phil's distance = 270

$$3X + 2(X + 10) = 270$$
$$3X + 2X + 20 = 270$$
$$5X = 250$$

$$X = 50 \text{ miles per hour} \quad \text{(Joan's rate)}$$
$$X + 10 = 60 \text{ miles per hour} \quad \text{(Phil's rate)}$$

■

Example 4 Two skilled craftsmen are considering pooling their talents to bid together on a construction job. The first one working alone can complete the job in 5 days. The second one working alone can complete the job in 3 days. If there is no loss of efficiency when they work together, should they bid on the job if a maximum of 2 days are permitted to complete the job?

Solution The first craftsman can complete $\frac{1}{5}$ of the job in 1 day and $X \cdot \frac{1}{5} = X/5$ of the job in X days. Likewise, the second craftsman can complete $\frac{1}{3}$ of the job in 1 day and $X \cdot \frac{1}{3} = X/3$ of the job in X days. Thus, the number of days (X) to complete *one* job is given by

$$\frac{X}{5} + \frac{X}{3} = 1$$

$$15 \cdot \frac{X}{5} + 15 \cdot \frac{X}{3} = 15 \cdot 1$$

$$3X + 5X = 15$$

$$8X = 15$$

$$X = \frac{15}{8} = \boxed{1\frac{7}{8} \text{ days}}$$

Yes, they should bid on the job. ∎

Example 5   A cook at Handi-Burger Restaurant wants to prepare a quart of orange drink. There is a can of concentrate that is 80% orange juice. How many ounces of concentrate and how many ounces of water should be mixed to obtain orange drink that is 30% orange juice?

Solution   Let X = the number of ounces of concentrate to be used. Then the number of ounces of water to be added is $32 - X$ (there are 32 ounces in 1 quart). Figure 6 can help organize our thinking.

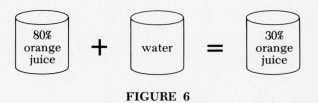

FIGURE 6

Pure orange juice: $0.80X + 0(32 - X) = 0.30(32)$

Multiplying the percent of orange juice by the volume gives us the amount of pure orange juice. (Note that water is 0% orange juice.) We now solve this equation for X.

$$0.80X + 0 = 0.30(32)$$

$$0.80X = 9.60$$

$$X = 12 \text{ ounces of concentrate}$$

$$32 - X = 20 \text{ ounces of water}$$ ∎

Conversion of Units

As the metric system becomes more and more a part of everyday life, it becomes more desirable to be able to convert units of measurement.

Example 6   If 1 inch = 2.54 centimeters, what is the height in centimeters of a person who is 5 feet 10 inches tall?

Solution   Since 1 foot = 12 inches, a person 5 feet 10 inches tall is $5 \cdot 12 + 10 = 60 + 10 = 70$ inches tall. If we let X = the person's height in centimeters, then 1 inch is to 2.54 centimeters as 70 inches is to X centimeters or

$$\frac{1}{2.54} = \frac{70}{X}$$

$$1X = 2.54(70) \qquad \left[\frac{a}{b} = \frac{c}{d} \text{ implies that } ad = bc \right]$$

$$X = 177.8 \text{ centimeters} \qquad \blacksquare$$

The equation we used in solving Example 6 $(1/2.54 = 70/X)$ is an example of a **proportion,** that is, a statement that two fractions are equivalent. Sometimes the fractions are called **ratios.** Thus the equation we used is based on the ratio of inches to centimeters remaining the same regardless of the length (or height) involved.

An alternate approach to converting units is based on **balancing units.** This is illustrated in the next example. Notice that the units cancel step-by-step, and the last value has the desired units of measurement.

Example 7   (a) Rework Example 6 by balancing units.
(b) Convert 3 miles to inches.

Solution   (a) 5 feet 10 inches = 70 inches, as before.

$$70 \ \cancel{\text{inches}} \cdot \frac{2.54 \text{ centimeters}}{1 \ \cancel{\text{inch}}} = \boxed{177.8 \text{ centimeters}}$$

(b) $3 \ \cancel{\text{miles}} \cdot \dfrac{5280 \ \cancel{\text{feet}}}{1 \ \cancel{\text{mile}}} \cdot \dfrac{12 \text{ inches}}{1 \ \cancel{\text{foot}}} = \boxed{190{,}080 \text{ inches}}$

What occurs is that a value is multiplied by a special form of 1 (12 inches = 1 foot is written as 12 inches/1 foot—this is sometimes called a **unit fraction**). $\blacksquare$

Given below are some key prefixes that are frequently used with the metric system.

kilo-   means   1000

centi-   means $\dfrac{1}{100}$

milli-   means $\dfrac{1}{1000}$

A centimeter is thus $\frac{1}{100}$ of a meter, or 100 centimeters = 1 meter. The basic units of measurement are

meter   for   length
liter   for   volume
gram   for   weight

Abbreviations are given by taking the first letter of the prefix (if present) followed by the unit; for instance, we use kg for *ki*logram and m for *m*eter. The actual equivalences, such as 1 in. = 2.54 cm, will be given in the exercises.

Variation

Variation is a way of expressing certain types of relationships between two or more variables. If an increase (decrease) in the magnitude of x results in an increase (decrease) in the magnitude of y, then y is said to vary directly as x or to be proportional to x. In the equation $y = 2x$, y varies directly as x; 2 is called the constant of variation in this equation. If an increase (decrease) in the magnitude of x results in a decrease (increase) in the magnitude of y, then y is said to vary inversely as x. In the equation $y = 5/x$, y varies inversely as x with 5 as the constant of variation.

If one variable varies directly as two or more variables, the relationship is described as joint variation. In the equation $y = 3xw$, y varies jointly as x and w with 3 as the constant of variation. The following are some of the usual expressions and the equivalent symbolic versions:

$$
\begin{array}{l}
y \textbf{ varies directly} \text{ as } x \longrightarrow y = kx \\[4pt]
y \text{ is } \textbf{proportional} \text{ to } x \longrightarrow y = kx \\[4pt]
y \textbf{ varies inversely} \text{ as } x \longrightarrow y = \dfrac{k}{x} \\[4pt]
y \textbf{ varies jointly} \text{ as } x \text{ and } w \longrightarrow y = kxw \\[4pt]
\left.\begin{array}{l} y \text{ varies directly as } m \text{ and inversely} \\ \text{as the } square \text{ of } d \end{array}\right\} \rightarrow y = \dfrac{km}{d^2}
\end{array}
$$

In each case k is called the **constant of variation.**

Problems involving variation usually give the type of variation in words and enough data to determine the constant of variation. New data are then applied to determine the value of one of the variables involved. Examples 8 and 9 will illustrate typical situations.

Example 8   If y varies directly as x, and if $y = 8$ when $x = 2$, what is the value of y when $x = 7$?

Solution $y = kx$ (set up the equation)

$8 = k \cdot 2$ (substitute for x and y)

$4 = k$ (determine the value of k)

Thus the variation is described by the equation

$$y = 4x$$

If $x = 7$, then $y = 4 \cdot 7 = \boxed{28}$.   ∎

Example 9   If y varies directly as the square of w and inversely as z, and if $y = 12$ when $w = 6$ and $z = 9$, what is the value of y when $w = 1$ and $z = 2$?

Solution $y = \dfrac{kw^2}{z}$ (set up the equation)

$12 = \dfrac{k \cdot 6^2}{9}$ (substitute for w, y and z)

$108 = 36k$

$3 = k$ (determine the value of k)

Thus the variation is described by the equation

$$y = \frac{3w^2}{z}$$

If $w = 1$ and $z = 2$, then $y = (3 \cdot 1^2)/2 = \boxed{\frac{3}{2}}$. ∎

In general, effective applications of algebra involve taking a clear statement of a problem, identifying the known and unknown quantities, setting up an equation based on our analysis of the problem, then using algebra to solve the equation.

SOLVING APPLIED PROBLEMS WITH ALGEBRA

1. Carefully read the statement of the problem.
2. Identify the known and unknown quantities. Assign variables as needed.
3. Set up an equation based on analysis of the problem.
4. Use algebra to solve the equation.
5. Check the solution to see whether it is a reasonable answer to the original problem.

Exercises 2.2

In Exercises 1–10 translate the phrase as an expression in x.
1. Twelve more than x.
2. Five less than twice x.
3. The square of the sum of x and 3.
4. The sum of the square of x and the square of 3.
5. The perimeter of a square with sides of length x.
6. The distance traveled by a car going 55 miles per hour for x hours.
7. The average of 69, 78, and x.
8. The fraction of a job completed in x days by a person who can complete the whole job in 10 days.
9. The number of cents in x quarters.
10. The number of ounces of pure acid in x ounces of a solution that is 36% acid.

In Exercises 11–30 identify the known and unknown quantities, set up an equation based on your analysis of the problem, use algebra to solve the equation, and check your answers.

Average and Total Problems (see Example 1)
11. If your first two test scores in algebra are 72 and 90, what score is required on your third test to have a test average of 85?
12. A salesperson's commissions increased by $960 in the second year, and the commissions during the third year were double that of the first year. If the average earnings for the three-year period were $15,000, what were the earnings in each year?

13. ABC Electronics doubled its order of microprocessors in the second third of the year as compared to the first third of the year. The order for the last third was double that of the second third. If the total order for the year was 1484 microprocessors, how many were ordered in each third?

14. Every six months the price of fishing equipment has increased by 10% over the price during the previous six months. At the beginning of this year the price of a deluxe fishing reel was $24.20. What was the price of the same reel at the beginning of last year?

Geometric Problems (see Example 2)

15. A novelty design is triangular in shape. The perimeter of the triangle is 24 inches, and the lengths of the sides in inches are three consecutive even integers. What are the lengths of the sides?

16. A safety rail is being considered for the sides of a rectangular skating rink. The length of the rink is 50 feet more than the width. If 340 feet of safety rail is needed, what are the dimensions of the rink?

17. Roadmonster cars are shipped to Japan in special protective crates. The length of the crate is two and one-half times the height of the crate, and the width is one and one-half times the height. What are the dimensions of the box-shaped crate if the perimeter of the rear end of the box is 20 feet?

18. A display window is twice as tall as it is wide. If 21 feet of caulk are required to seal the outside of the window, what are the dimensions of the window?

Motion Problems (see Example 3)

19. A radioactive medicine has been inadvertently placed in a shipment to a department store 300 miles away. If the truck travels 50 miles per hour, and if it left two hours ago, can a police car traveling 75 miles per hour catch the truck before the shipment arrives at its destination?

20. Cathy can row a boat at the rate of 3 miles per hour in still water. If it took her 2 hours to row upstream and 24 minutes to return, what is the rate of the river's current? How far did she row upstream?

21. A pilot for Tree Top Airlines flew north against a headwind of 35 miles per hour and made a trip in 4 hours. Returning at the same flying speed with a tailwind of 30 miles per hour, the pilot found that the trip took $3\frac{1}{2}$ hours. What was the flying speed of the plane? What was the distance between stops?

22. A mileage test consists of driving at a fixed rate of speed for 2 hours, increasing the speed by 10 miles per hour for 1 hour, then decreasing the latest speed by 5 miles per hour for 3 hours. If the distance traveled during the test is 295 miles, what are the rates of speed over each part of the test?

Mixture and Solution Problems (see Example 5)

23. An automatic coin counter at the bank indicates that $65 in dimes and quarters were processed. If there were 389 coins, how many were dimes and how many were quarters?

24. Two types of coffee are blended to make coffee with a wholesale price of $1.50 per pound. The first type wholesales for $1.20 per pound, and the

second wholesales for $1.65 per pound. If we need 45 pounds of the blend, how many pounds of each type should be used?

25. Flowers Unlimited wants to offer a bouquet of roses and greenery for $18.00. Roses cost $1.10 each, greenery costs $0.66 for each piece, and the arranger charges $1.50 per bouquet. If there are 17 pieces in the bouquet, how many are roses and how many are greenery?

26. How much water should be drained from a 12-quart radiator (filled with water) and replaced by a 75% antifreeze solution to meet engine protection requirements of 50% antifreeze?

27. A 76-ml solution of alcohol and water is 20% alcohol. How many milliliters of pure water should be added to make a solution that is 8% alcohol?

28. A quart of fruit punch is 10% fruit juice. How many ounces of pure fruit juice should be added to make fruit punch that is 25% fruit juice?

Work Problems (see Example 4)
29. Two machines can be used to complete one job in 3 hours. One of the machines is twice as fast as the other. How long would it take each machine working alone to complete the job?

30. If one executive can complete a report in 5 days and another executive can complete the same report in 4 days, how long will it take them if they work together?

Use the following information for Exercises 31–42:

1 foot = 12 inches	1 kilogram = 2.2 pounds
1 mile = 5280 feet	1 inch = 2.54 centimeters
1 quart = 32 ounces	1 quart = 0.946 liters

In Exercises 31–36 convert the units using the proportion method.
31. Convert 5 inches to centimeters.
32. Convert 5 centimeters to inches.
33. Convert 12 kilograms to pounds.
34. Convert 175 pounds to kilograms.
35. Convert $1\frac{2}{3}$ quarts to liters.
36. Convert 4 liters to quarts.

In Exercises 37–42 convert using the method of balancing units.
37. Convert 5 feet 5 inches to meters.
38. Convert 2000 meters to miles.
39. Convert 6.8 milliliters to ounces.
40. Convert 12 ounces to liters.
41. Convert 732 grams to pounds.
42. Convert 1 pound to kilograms.

In Exercises 43–52 set up an equation showing the relationship of the variables, solve for the constant of variation, and answer the question with the new data.
43. If y varies directly as x, and if $y = 12$ when $x = 4$, what is the value of y when $x = 5$?
44. If y is proportional to x, and if $y = \frac{1}{2}$ when $x = \frac{1}{6}$, what is the value of y when $x = 1$?

45. If y varies inversely as d, and if $y = 100$ when $d = 8$, what is the value of y when $d = 25$?
46. If y varies inversely as x, and if $y = 18$ when $x = \frac{1}{3}$, what is the value of y when $x = 2$?
47. If z varies directly as the square of x, and if $z = 64$ when $x = 2$, what is the value of z when $x = \frac{1}{8}$?
48. If w varies inversely as the square of d, and if $w = \frac{1}{1000}$ when $d = 100$, what is the value of w when $d = 500$?
49. If x varies directly as s and inversely as t, and if $x = 9$ when $s = 3$ and $t = 2$, what is the value of x when $s = 7$ and $t = 3$?
50. If z varies jointly as x and y, and if $z = 12$ when $x = 8$ and $y = 3$, what is the value of z when $x = 5$ and $y = 10$?
51. If y varies directly as m and inversely as the square of d, and if $y = 8$ when $m = 16$ and $d = 2$, what is the value of y when $m = 1$ and $d = 3$?
52. If d varies directly as b and as the square of c, and if $d = 20$ when $b = 64$ and $c = \frac{1}{4}$, what is the value of d when $b = 2$ and $c = 0.3$?

2.3 The Cartesian Coordinate System and Lines

Just as the number line is useful with statements in one variable, the **Cartesian coordinate system** is effective in working with a variety of statements in two variables.

The Cartesian coordinate system (see Figure 7) is constructed by placing a vertical number line and a horizontal number line on a plane so that their zero points coincide. The point where the two lines cross is called the **origin**. The horizontal line is called the **x-axis**, and the vertical line (with positive numbers upward) is called the **y-axis**. These lines divide the plane into four parts, called **quadrants**. The Roman numerals in Figure 7 indicate the number of each quadrant.

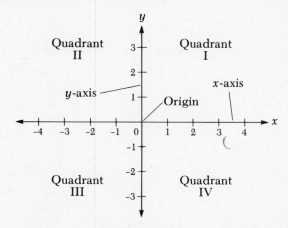

FIGURE 7

The points on the Cartesian plane correspond to ordered pairs of real numbers. The pairs are written in the form (x, y), such as $(2, 3)$ or $(-\frac{1}{2}, 0)$. The first number is called the **x-coordinate** or **abscissa**. The second number is called the **y-coordinate** or **ordinate**. For instance, $(2, 3)$ has an x-coordinate of 2 and a y-coordinate of 3.

The pair (a, b) is graphed or plotted on the Cartesian coordinate system by finding the position for a on the x-axis and then moving vertically $|b|$ units, upward if b is positive and downward if b is negative.

Example 1 Plot the points $(2, 3)$, $(0, 0)$, $(-1, 4)$, $(0, -2)$, $(3, 2)$, and $(3, -3)$ in the Cartesian coordinate system.

Solution See Figure 8. ■

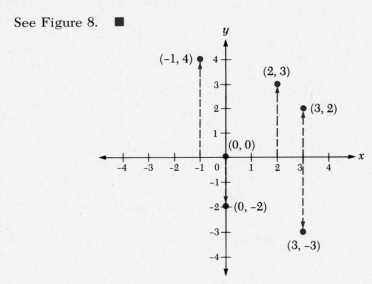

FIGURE 8

Example 2 Determine the ordered pairs (x, y) for the points labeled A, B, C, D, and E:

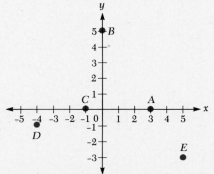

Solution   The coordinates are

$$A: (3, 0), \quad B: (0, 5), \quad C: (-1, 0),$$
$$D: (-4, -1), \quad E: (5, -3) \quad \blacksquare$$

A **first-degree equation in two variables** is an equation that can be written in the form $ax + by = c$, where a, b, and c are constants. Such equations are frequently called **linear equations in two variables.** The word "linear" refers to the graph of the set of solutions (x, y) of the equation, which turns out to be a straight line.

Example 3   Graph $\{(x, y) \mid x - 2y = 4\}$; that is, graph the solution set of $x - 2y = 4$.

Solution   First we determine at least three ordered pairs that satisfy the equation, say for $x = 0, 2,$ and 4:

x	y
0	$-2 \rightarrow (0, -2)$
2	$-1 \rightarrow (2, -1)$
4	$0 \rightarrow (4, 0)$

Usually, we pick a value of x (or y) and then determine the value of y (or x) from the equation; for instance, $x = 0$ yields $0 - 2y = 4$, and thus $y = -2$. Even though two distinct points will determine a line, three points will frequently uncover an error in our work. (If the three points are not in a line, check your calculations. Plotting a fourth point may also help.)

As shown in Figure 9, plot the points obtained on a Cartesian coor-

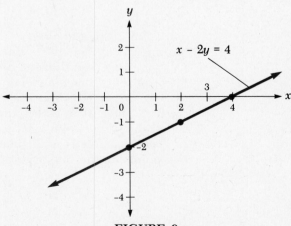

FIGURE 9

dinate system. Then connect the points with a line and extend it in both directions. ∎

Note that each point on a line corresponds to a solution of the appropriate linear equation, and each solution of a linear equation is a point on the appropriate line.

Some lines are particularly easy to graph. The equation $x = 2$ is a short form of $1x + 0y = 2$ and has the property that regardless of the value given to the variable y, the value of x is 2. Similarly, the equation $y = -3$ is a short form of $0x + 1y = -3$ and has the property that $y = -3$ regardless of the value given to x. Thus the graph of $x = 2$ is a vertical line two units to the right of the y-axis, and the graph of $y = -3$ is a horizontal line three units below the x-axis. Figure 10 illustrates these observations.

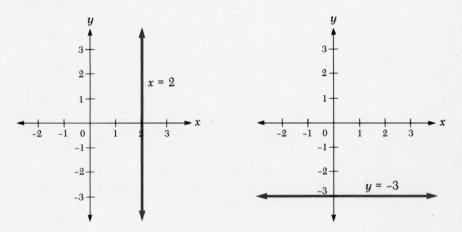

FIGURE 10

It is interesting to observe that the point $(2, -3)$ is the point where the line $x = 2$ crosses the line $y = -3$.

Lines other than those of the form $x =$ constant can be rewritten in the form $y = mx + b$, where m and b are constants. If $x = 0$, then $y = m \cdot 0 + b = b$, which indicates that the point $(0, b)$ is on the graph of the line. Since $(0, b)$ is the point where the line crosses the y-axis, it is called the **y-intercept**. The number m is called the **slope** of the line and is a measure of the inclination of the line. If (x_1, y_1) and (x_2, y_2) are two points on the line $y = mx + b$, then

$$y_1 = mx_1 + b$$

and

$$y_2 = mx_2 + b$$

Subtracting the top equation from the bottom equation gives

$$y_2 - y_1 = (mx_2 + b) - (mx_1 + b)$$
$$y_2 - y_1 = mx_2 + \not{b} - mx_1 - \not{b}$$
$$y_2 - y_1 = m(x_2 - x_1)$$
$$\frac{y_2 - y_1}{x_2 - x_1} = m$$

(see Figure 11). Thus the slope of a line is defined as

$$\text{slope} = \quad m = \frac{y_2 - y_1}{x_2 - x_1} = \frac{\text{change in } y}{\text{change in } x}$$

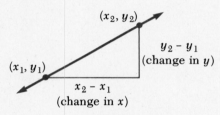

FIGURE 11

It is easy to see that if a line rises from left to right, its slope is positive, since

$$m = \frac{y_2 - y_1}{x_2 - x_1} = \frac{\text{positive change in } y}{\text{positive change in } x}$$

and, if a line falls from left to right, its slope is negative, since

$$m = \frac{y_2 - y_1}{x_2 - x_1} = \frac{\text{negative change in } y}{\text{positive change in } x}$$

(see Figure 12).

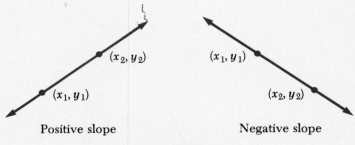

Positive slope   Negative slope

FIGURE 12

For a horizontal line,

$$m = \frac{\text{change in } y}{\text{change in } x} = \frac{0}{\text{change in } x} = 0$$

so the slope is always 0. For a vertical line the slope is undefined, since the slope would be

$$m = \frac{\text{change in } y}{\text{change in } x} = \frac{\text{change in } y}{0}$$

but division by 0 is undefined.

SLOPE–INTERCEPT FORM OF THE EQUATION OF A LINE

$$y = mx + b$$

where m = slope and b = y-intercept.

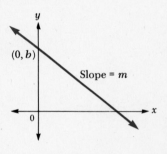

Example 4 (a) Determine the slope and y-intercept of the line $2x - 3y = 6$.
(b) What is the equation of a line whose slope is 3 and whose y-intercept is $\frac{6}{7}$?

Solution (a) Rewrite the equation in slope–intercept form:

$$2x - 3y = 6$$

$$-3y = -2x + 6$$

$$y = \frac{2}{3}x - 2$$

From this form it is easy to see that the slope is $\frac{2}{3}$ and the

y-intercept is -2 .

(b) If $m = 3$ and $b = \dfrac{6}{7}$, the form $y = mx + b$ gives

$$y = 3x + \frac{6}{7}$$ ∎

Since our information about a line may be given in other ways, we need to be familiar with other forms of the equation of a line.

If we are given the slope m of a line and one point (x_1, y_1) on the line, we can return to the definition of slope as a ratio of change in y-coordinates to change in x-coordinates. If (x, y) is a general point on a line and (x_1, y_1) is a particular (known) point on the line, then

$$\frac{y - y_1}{x - x_1} = m \quad \text{or} \quad y - y_1 = m(x - x_1)$$

This gives a useful form of the equation of a line.

POINT–SLOPE FORM OF THE EQUATION OF A LINE

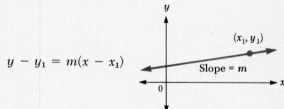

$$y - y_1 = m(x - x_1)$$

where (x_1, y_1) is a known point on the line and m is the slope of the line.

Example 5 What is the equation of a line with slope -7 and passing through the point $(6, 2)$?

Solution Using the point–slope form, we get

$$y - 2 = -7(x - 6)$$
$$y - 2 = -7x + 42$$
$$y = -7x + 44$$ ∎

When dealing with data from practical situations, we frequently have two pairs, (x_1, y_1) and (x_2, y_2). If it is suspected that the relationship between the variables x and y is linear, we can calculate the slope of the line (as the difference in y-coordinates divided by the corresponding difference in x-coordinates) and then use the point–slope form. This gives the two-point form:

TWO-POINT FORM OF THE EQUATION OF A LINE

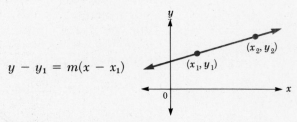

$$y - y_1 = m(x - x_1)$$

where $m = \dfrac{y_2 - y_1}{x_2 - x_1}$ and (x_1, y_1) and (x_2, y_2) are distinct points on the line with $x_1 \neq x_2$.

Example 6   Marvel Marketing Company arranges newspaper advertising for new consumer products. The relationship between dollars spent on newspaper advertising and the initial sales of a new product is linear. If $500 worth of advertising yields 100 sales and $1200 worth of advertising yields 240 sales, how many sales would result from spending $750?

Solution   Let x = the number of dollars spent on newspaper advertising and y = the initial sales of a product. Then (500, 100) and (1200, 240) are two points on the line. Using the two-point form, we obtain the equation of the line:

$$y - 100 = \frac{240 - 100}{1200 - 500}(x - 500)$$

$$y - 100 = \frac{140}{700}(x - 500)$$

$$y - 100 = \frac{1}{5}(x - 500)$$

$$y = \frac{1}{5}x - 100 + 100$$

$$y = \frac{1}{5}x$$

We now apply the equation and new data for dollars to determine the corresponding number of sales. If $x = 750$, then

$$y = \frac{1}{5} \cdot 750 = \boxed{150 \text{ sales}} \quad \blacksquare$$

The technique of Example 6 was to use known data to determine a mathematical model $(y = \frac{1}{5}x)$ of the relationship between dollars spent and sales. The projected amount to be spent (\$750) is fed into the model and generates a value for projected sales. How valid the result is depends on a variety of considerations such as the exact relationship between the variables (is it really linear?) and the accuracy of the known values.

The new value of x (\$750) is between the known values (\$500 and \$1200). In such a case, determining the corresponding value of y (150) is known as **interpolation**. Determining a value of y beyond the range of known values of x (e.g., for $x = \$100$ or $x = \$2000$) is known as **extrapolation**. Generally, interpolation is safer than extrapolation. (Why?) Is there an alternative to extrapolation if we want to estimate the population of the United States in the year 2000?

If two lines are parallel, then their slopes are the same, since their inclinations are the same. This fact can help us determine the equation of certain lines, as shown in the next example.

Example 7   Determine the equation of a line that is parallel to the line $y = 4x + 9$ and passes through the point $(5, -1)$.

Solution   The slope of the new line is 4, since the slope of $y = 4x + 9$ is 4. Since we know a point on the line, we use the point–slope form to obtain the desired equation:

$$y - (-1) = 4(x - 5)$$
$$y + 1 = 4x - 20$$
$$\boxed{y = 4x - 21} \quad \blacksquare$$

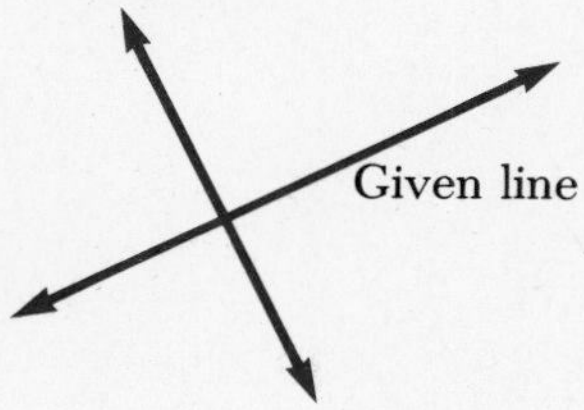

Normal line

Given line

FIGURE 13

A **normal line** to a given line is a line that is perpendicular to the given line (Figure 13). The slope, m_N, of the normal line is the negative reciprocal of the slope, m, of the given line; that is, $m_N = -\dfrac{1}{m}$.

Note: If $m = 0$, then m_N is undefined. This is not unexpected, since $m = 0$ indicates a horizontal line, so a line perpendicular to it would be a vertical line (which has undefined slope).

Example 8   Determine the equation of the line normal to $2x + 3y = 6$ at the point $(3, 0)$.

Solution   First the slope of $2x + 3y = 6$ must be found:

$$3y = -2x + 6$$

$$y = -\frac{2}{3}x + 2$$

The slope is $-\frac{2}{3}$, and so the slope of the normal line is $\frac{3}{2}$ (the negative reciprocal of $-\frac{2}{3}$). We also know that the normal line passes through the point $(3, 0)$, so we use the point–slope form to determine the equation of the normal line:

$$y - 0 = \frac{3}{2}(x - 3)$$

$$y = \frac{3}{2}x - \frac{9}{2} \qquad \blacksquare$$

Exercises 2.3

1. Plot the points $(1, 5)$, $(0, -3)$, $(-4, 3)$, and $(7, 0)$ on a Cartesian coordinate system.
2. Plot the points $(-2, 0)$, $(2, -5)$, $(0, 3\frac{1}{2})$, and $(-6, -2)$ on a Cartesian coordinate system.
3. Determine the ordered pairs corresponding to the points labeled A, B, C, and D on the following Cartesian coordinate system:

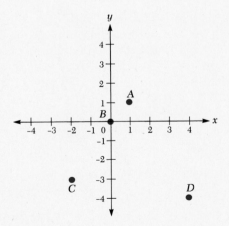

4. Determine the ordered pairs corresponding to the points labeled A, B, C, and D on the following Cartesian coordinate system:

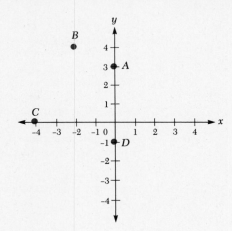

In Exercises 5–14 graph the line:

5. $\{(x, y) \mid 2x - y = 6\}$
6. $\{(x, y) \mid x + y = 4\}$
7. $\{(x, y) \mid 2x + 3y = 5\}$
8. $\{(x, y) \mid x = -2\}$
9. $\{(x, y) \mid 2y = 3\}$
10. $\{(x, y) \mid y = -6\}$
11. $\{(x, y) \mid x - 5 = 0\}$
12. $\{(x, y) \mid y = x - 3\}$
13. $\{(x, y) \mid -4x = 12 - 3y\}$
14. $\{(x, y) \mid x + 7y = -14\}$
15. Determine the slope and y-intercept of the lines in Exercises 5, 7, 9, 11, and 13.
16. Determine the slope and y-intercept of the lines in Exercises 6, 8, 10, 12, and 14.

In Exercises 17–22 determine the equation of the line having slope m and y-intercept b. Then graph the line.

17. $m = 4;\ b = 0$
18. $m = -2;\ b = 6$
19. $m = 0;\ b = -\frac{1}{2}$
20. $m =$ undefined
21. $m = \frac{2}{3};\ b = 3$
22. $m = -\frac{1}{4};\ b = \frac{6}{5}$

In Exercises 23–30 determine the equation of the line passing through the point P and having slope m. Then graph the line.

23. $P = (2, 5);\ m = 3$
24. $P = (-4, 0);\ m = 5$
25. $P = (-1, 5);\ m = -2$
26. $P = (0, 0);\ m = -1$
27. $P = (0, 2);\ m = \frac{5}{9}$
28. $P = (2, -3);\ m = \frac{1}{2}$
29. $P = (-7, -4);\ m = -\frac{1}{10}$
30. $P = (0, -1);\ m = -\frac{1}{42}$

In Exercises 31–38 determine the equation of the line through the two points. Then graph the line.

31. $(0, 0)$ and $(4, -2)$
32. $(-1, 0)$ and $(0, 6)$
33. $(2, 3)$ and $(1, 7)$
34. $(5, 1)$ and $(3, -3)$
35. $(-2, -2)$ and $(0, 5)$
36. $(10, -3)$ and $(0, 0)$
37. $(5, 8)$ and $(-3, 6)$
38. $(3, -\frac{1}{2})$ and $(1, 0)$

In Exercises 39–42 determine the equation of the line parallel to the given line and passing through the point P. Then graph both lines.

39. $x + y = 6;\ P = (3, -1)$
40. $x - 2y = -3;\ P = (0, 0)$
41. $5x + 2y = 1;\ P = (0, 3)$
42. $2x + 3y = -2;\ P = (-4, 5)$

In Exercises 43–46 determine the equation of the line normal to the given line and passing through the point P. Then graph both lines.

43. $x + y = 0; P = (2, 0)$ 44. $2x - 2y = 5; P = (1, 3)$

45. $3x - y = 6; P = (3, 3)$ 46. $5x + 10y = -4; P = (0, 0)$

47. If the relationship between number of employees and annual personnel paperwork is linear, compute the paperwork generated by 100 employees if 30 employees generate 2 tons of paperwork per year and 150 employees generate 10 tons of paperwork per year.

48. Estimate the population of Middle City in 1990 if population growth is linear, the population in 1970 was 17,000, and the population in 1982 was 21,200.

In Exercises 49–50 determine the equation of the line shown.

49. 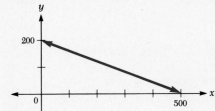   50.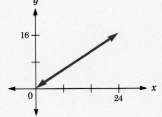

2.4 Functions and Relations

It is clear from the last section that the relationship between variables is important. Before we study these relationships further, some basic terminology is necessary.

DEFINITION A **relation** is a set of ordered pairs (x, y). The set of x-coordinates is called the **domain** of the relation, and the set of y-coordinates is called the **range** of the relation.

Example 1   (a) List the domain and range of

$$R = \{(-2, 7), (-1, 0), (0, 7), (5, 3), (5, 5), (10, 2)\}$$

(b) Graph R.

Solution   (a) Domain of $R = \{-2, -1, 0, 5, 10\}$
Range of $R = \{0, 2, 3, 5, 7\}$
(b) The graph appears in Figure 14. ∎

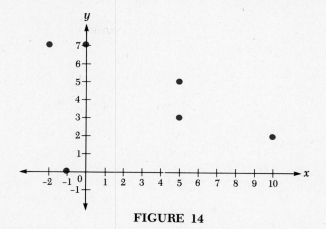

FIGURE 14

Other examples of relations are the set of ordered pairs constituting the line $5x - 2y = 1$, a circle of radius 2 about the origin in a Cartesian coordinate system, and the set of pairs of students' names from your class together with their ages.

Although relations are of interest in themselves, our primary interest is a special class of relations called functions. Recall from our study of lines, such as $y = 2x + 3$, that a value for y is obtained once a value of x has been assigned. In this sense, x is an independent variable, y is a dependent variable, and the stage is set for the following definition.

DEFINITION

A **function** is a relation with the property that for each value of the (independent) variable x, there is only one value of the (dependent) variable y.

An equivalent way of defining a function is as a relation in which no two ordered pairs have the same first coordinate but different second coordinates.

Example 2

Which of the following relations are functions?
(a) $f = \{(-1, 2), (2, 2), (3, 5), (6, 1)\}$
(b) $h = \{(0, 7), (1, 5), (1, 2), (3, -4)\}$

Solution

(a) f is a function, as the following correspondence shows:

$$
\begin{array}{cc}
x & y \\
-1 & \to 2 \\
2 & \to 2 \\
3 & \to 5 \\
6 & \to 1
\end{array}
$$

(b) h is not a function, since there are two values of y for $x = 1$:

$$
\begin{array}{rl}
x & y \\
0 \to & 7 \\
1 <& \begin{array}{l} 5 \\ 2 \end{array} \qquad \text{(violates the definition of function)} \\
3 \to & -4 \quad \blacksquare
\end{array}
$$

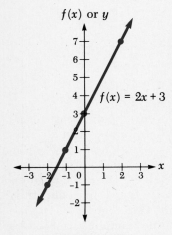

$f(x)$ or y

$f(x) = 2x + 3$

FIGURE 15

Notice that different values of x may have the same value of y in a function.

"Domain" and "range" have the same meaning with respect to functions as they have with relations.

An alternative to the set notation used in Example 2 is **functional notation**. We write $y = f(x)$ to indicate that y is the value associated with x by the function f. For instance, we would write $f(3) = 5$ for the pair $(3, 5)$ in Example 2(a). The definition of function merely says that the expression $f(x)$ is well defined; that is, for a fixed value of x, there is only one value of $f(x)$.

Lines of the form $y = mx + b$ can be easily rewritten in the functional form $f(x) = mx + b$. Here, as in the case of many functions, the domain of f is understood to be the set of all values of x for which the expression defining x (such as $mx + b$) makes sense.

Example 3 Let $f(x) = 2x + 3$.
(a) Compute $f(-2)$, $f(-1)$, $f(0)$, and $f(2)$.
(b) Graph f.

Solution (a) $f(-2) = 2 \cdot (-2) + 3 = -4 + 3 = \boxed{-1}$
$f(-1) = 2 \cdot (-1) + 3 = -2 + 3 = \boxed{1}$
$f(0) = 2 \cdot (0) + 3 = 0 + 3 = \boxed{3}$
$f(2) = 2 \cdot (2) + 3 = 4 + 3 = \boxed{7}$

(b) From (a) we have the points $(-2, -1)$, $(-1, 1)$, $(0, 3)$, and $(2, 7)$, which we plot and connect (Figure 15). $\blacksquare$

Example 4 (a) For $f(x) = 2x^2 - x - 5$, compute $f(-1)$ and $f(t)$.

(b) For $g(x) = \dfrac{x - 1}{2x + 3}$, compute $g(4)$ and $g(x - 1)$.

Solution (a) $f(-1) = 2(-1)^2 - (-1) - 5$

$$= 2 + 1 - 5 = \boxed{-2}$$

$$f(t) = \boxed{2t^2 - t - 5}$$

(b) $g(4) = \dfrac{4 - 1}{2 \cdot 4 + 3} = \boxed{\dfrac{3}{11}}$

$$g(x - 1) = \frac{(x - 1) - 1}{2(x - 1) + 3} = \boxed{\frac{x - 2}{2x + 1}} \quad \blacksquare$$

In Example 4 it is easy to see that f and g are functions, since each value of x leads to only one corresponding value of $f(x)$ or $g(x)$. Note that the domain of f is the set of all real numbers, but the domain of g is the set of all real numbers except $x = -\frac{3}{2}$, since $g(-\frac{3}{2})$ is undefined.

Although x and y are frequently used with functions, this choice of variables is arbitrary. For instance, $z = f(t)$ may define a function. The letter f is also an arbitrary choice to indicate a function. In most applications, letters that fit the problem may be used. For example, $d = f(t)$ may indicate that distance, d, is a function of time, t.

It is easy to identify which graphs have come from or describe functions. This involves the **vertical line test.** Each vertical line ($x = $ constant) can intersect the graph of a function in at most one point (see Figure 16).

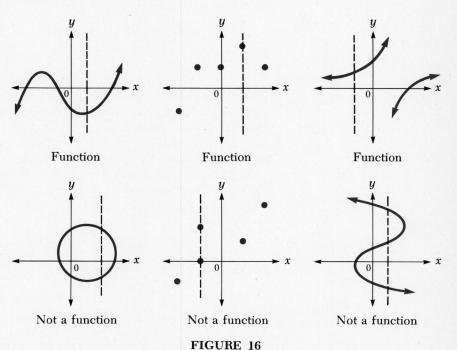

FIGURE 16

There are many occasions when we will want to combine two or more functions into a new function. This gives us an algebra of functions.

Addition of functions: $(f + g)(x) = f(x) + g(x)$
Subtraction of functions: $(f - g)(x) = f(x) - g(x)$
Multiplication of functions: $(fg)(x) = f(x) \cdot g(x)$

Division of functions: $\left(\dfrac{f}{g}\right)(x) = \dfrac{f(x)}{g(x)}$ if $g(x) \neq 0$

Composition of functions: $(f \circ g)(x) = f[g(x)]$

Example 5   For $f(x) = x^2$ and $g(x) = 2x - 3$, compute $f + g$, $f - g$, fg, f/g, $f \circ g$, and $g \circ f$.

Solution $(f + g)(x) = f(x) + g(x) = (x^2) + (2x - 3) = \boxed{x^2 + 2x - 3}$

$(f - g)(x) = f(x) - g(x) = (x^2) - (2x - 3) = \boxed{x^2 - 2x + 3}$

$(fg)(x) = f(x) \cdot g(x) = (x^2)(2x - 3) = \boxed{2x^3 - 3x^2}$

$\left(\dfrac{f}{g}\right)(x) = \dfrac{f(x)}{g(x)} = \boxed{\dfrac{x^2}{2x - 3}}$

$(f \circ g)(x) = f[g(x)] = f[2x - 3] = (2x - 3)^2 = \boxed{4x^2 - 12x + 9}$

$(g \circ f)(x) = g[f(x)] = g[x^2] = 2(x^2) - 3 = \boxed{2x^2 - 3}$ ∎

Note that the domain of $f + g$, $f - g$, and fg is the domain of f intersected with the domain of g. The domain of f/g is $\{x \mid x \in$ domain of f *and* $x \in$ domain of g *and* $g(x) \neq 0\}$, since $f(x)/g(x)$ is undefined if $g(x) = 0$. The domain of $f \circ g$ is $\{x \mid x \in$ domain of g and $g(x) \in$ domain of $f\}$. Also, $g \circ f$ is not the same as $f \circ g$.

Other applications of functions may include expressions involving functional notation (see Example 6).

Example 6   (a) For $f(x) = x^2$, compute $\dfrac{f(x + h) - f(x)}{h}$.

(b) For $g(x) = 2x^2 - 1$, compute $g(1) + g(2) + g(3)$.

Solution   (a) $\dfrac{f(x + h) - f(x)}{h} = \dfrac{(x + h)^2 - x^2}{h} = \dfrac{x^2 + 2xh + h^2 - x^2}{h}$

$= \dfrac{2xh + h^2}{h} = \boxed{2x + h}$

(b) $g(1) + g(2) + g(3) = (2 \cdot 1^2 - 1) + (2 \cdot 2^2 - 1) + (2 \cdot 3^2 - 1)$

$$= 1 + 7 + 17 = \boxed{25} \quad \blacksquare$$

This section has provided an introduction to functions and relations. Functions, relations, and associated equations and inequalities play a dominant role in the topics that follow in this book.

Exercises 2.4

In Exercises 1–8 list the domain and range of the relation. Then graph the relation.

1. $R = \{(-7, 2), (-3, 0), (5, -1), (-3, 6)\}$
2. $R = \{(-3, 1), (-1, 1), (0, 1), (4, 1)\}$
3. $R = \{(-4, 0), (-4, 4), (2, 3), (1, 9)\}$
4. $R = \{(5, 0), (0, 1), (0, 7)\}$
5. $R = \{(0, 0), (1, 1), (2, 4), (3, 9), (4, 16)\}$
6. $R = \{(1, -3), (2, -1), (3, 1), (4, 3), (5, 5)\}$
7. $R = \{(3, 0), (3, 1), (3, 2), (3, 3), (3, 4)\}$
8. $R = \{(1, 1), (2, 2), (4, 4), (9, 9)\}$
9. Which of the relations in Exercises 1–8 are functions?

In Exercises 10–15 compute the indicated functional values. Then graph the function.

10. $f(x) = 4x - 3$; $f(-1), f(0), f(1), f(2)$
11. $f(x) = x + 5$; $f(-2), f(-1), f(0), f(1)$
12. $g(x) = -2$; $g(1), g(2), g(3), g(4)$
13. $g(x) = \dfrac{x - 1}{3}$; $g(-2), g(1), g(4), g(7)$
14. $h(x) = (x + 2)^2$; $h(-4), h(-3), h(-2), h(-1), h(0)$
15. $h(x) = x^3$; $h(-2), h(-1), h(0), h(1), h(2)$

16. For $f(x) = x^2 - 2x + 3$, compute $f(-2), f(2)$, and $f(t)$.
17. For $g(x) = 2x^3 + 7$, compute $g(-3), g(0)$, and $g(z)$.
18. For $h(x) = \sqrt{x + 2}$, compute $h(-2), h(2)$, and $h(t - 2)$.
19. For $f(x) = \dfrac{5x - 1}{x + 3}$, compute $f(-1), f(3)$, and $f(x + 1)$.
20. What are the domains of the functions in Exercises 16–19?

In Exercises 21–28 use the vertical line test to identify which graphs describe functions.

21.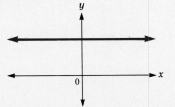

22.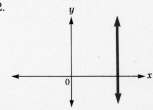

23.

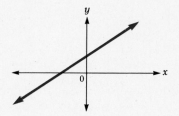

24.

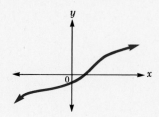

25.

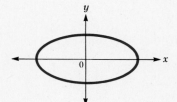

26.

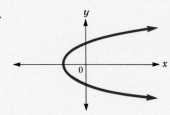

27.

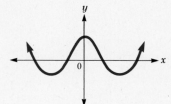

28.

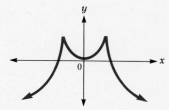

In Exercises 29–34 compute $f + g$, $f - g$, fg, $\dfrac{f}{g}$, and $f \circ g$.

29. $f(x) = x + 2$; $g(x) = x^2 + 1$
30. $f(x) = x^3$; $g(x) = 3x$

31. $f(x) = (x - 1)^2$; $g(x) = \dfrac{1}{x}$

32. $f(x) = \sqrt{x}$; $g(x) = x^2 - 2x + 3$
33. $f(x) = x^3 + 5x - 7$; $g(x) = 3x^2 - x - 1$

34. $f(x) = \dfrac{2x}{x - 5}$; $g(x) = \sqrt{x}$

35. Compute $g + f$, $g - f$, gf, g/f, and $g \circ f$ for the functions in Exercise 31, and compare the new results with the results of Exercise 31.
36. Compute $g + f$, $g - f$, gf, g/f, and $g \circ f$ for the functions in Exercise 32.

37. For $f(x) = 5x^2$, compute $\dfrac{f(x + h) - f(x)}{h}$.

38. For $f(x) = x^3$, compute $\dfrac{f(x + h) - f(x)}{h}$.

39. For $f(x) = 4x - 3$, compute $\dfrac{f(x + h) - f(x)}{h}$.

40.   For $g(x) = 2x^2 + x$; compute $\dfrac{g(x) - g(a)}{x - a}$.

41.   For $f(t) = (3t + 1)^2$, compute $\frac{1}{3}[f(0) + f(1) + f(2)]$.

42.   For $f(x) = 5x$, compute $f(x) \cdot f(-x)$.

43.   For $g(x) = x + 1$, compute $2[g(2)]^2 - g(2) + 5$.

44.   For $f(t) = \dfrac{2t - 1}{t + 3}$, compute $\dfrac{5}{f(4)}$.

45.   Using functional notation, express revenue (R) as a function of number of bicycles sold (x), where each bicycle sold generates \$110 of revenue.

46.   Using functional notation, express cost (C) as a function of number of bicycles produced (x), where each bicycle produced costs \$65 and fixed costs (independent of the number of bikes produced) are \$1200.

47.   What are appropriate domains for the functions in Exercises 45 and 46?

2.5   Special Functions

Certain functions appear in a variety of applications. These functions are usually given names that describe their nature.

One of these functions is the function $f(x) = |x|$, which is called the **absolute value function.**

Example 1   Graph $f(x) = |x|$.

Solution   An intuitive approach is to select a number of values of x, compute the corresponding $f(x)$, plot the resulting points, and then connect them as in Figure 17.

x	$f(x)$
-3	3
-2	2
-1	1
0	0
1	1
2	2
3	3

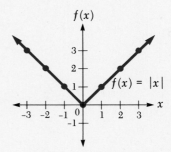

FIGURE 17

The graph shown is correct for this function. A more analytical approach would be to recall that

$$|x| = x \text{ for } x \geq 0$$

$$|x| = -x \text{ for } x < 0$$

Thus,

$$f(x) = x \text{ for } x \geq 0$$

$$f(x) = -x \text{ for } x < 0$$

Equivalently, the graph would coincide with the line $y = -x$ to the left of the y-axis and with the line $y = x$ to the right of the y-axis. This agrees with our intuitive results. ∎

The analysis following the graph in Example 1 provides the key to graphing functions that involve the absolute value of a linear expression, such as $|2x - 3|$. That is, the expression should be treated as $2x - 3$ where $2x - 3 \geq 0$ ($x \geq \frac{3}{2}$) and as $-(2x - 3) = -2x + 3$ where $2x - 3 < 0$ ($x < \frac{3}{2}$).

Another kind of function involves a variable exponent. An **exponential function** is a function of the form $y = a^x$, where a is a positive constant.

Example 2 Graph $y = 2^x$.

Solution With exponential functions it is generally best to choose enough values for x so that both positive and negative exponents are included. We compute the corresponding values of $f(x)$, plot the resulting points, and then connect them (Figure 18).

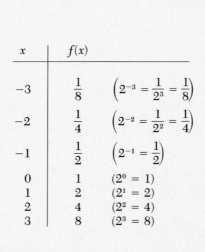

x	$f(x)$	
-3	$\frac{1}{8}$	$\left(2^{-3} = \frac{1}{2^3} = \frac{1}{8}\right)$
-2	$\frac{1}{4}$	$\left(2^{-2} = \frac{1}{2^2} = \frac{1}{4}\right)$
-1	$\frac{1}{2}$	$\left(2^{-1} = \frac{1}{2}\right)$
0	1	$(2^0 = 1)$
1	2	$(2^1 = 2)$
2	4	$(2^2 = 4)$
3	8	$(2^3 = 8)$

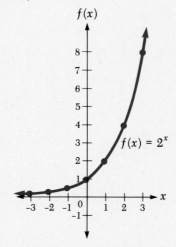

FIGURE 18

Note that the graph stays above the x-axis, since 2^x is positive for *all* values of x. ∎

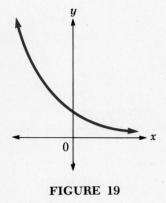

FIGURE 19

Exponential functions are of much practical use. For instance, if x represents time in hours (valid only for $x \geq 0$) and $f(x)$ represents the number of bacteria present in a culture, then $f(x) = 2^x$ models the bacterial growth for a strain that doubles in size every hour. Such growth is described as **exponential growth.**

If the function is $y = 2^{-x}$, the graph would look like that in Figure 19, which is described as **exponential decay.** The topic of exponential functions will be treated more fully in Chapter 6.

The **greatest integer function** (also known as the **staircase function**) is defined as follows:

$$f(x) = [[x]] = N, \text{ where } N \text{ is the unique integer such that}$$
$$N \leq x < N + 1$$

That is, N is the largest integer that is less than or equal to x. For instance,

$$[[1.2]] = 1 \quad \text{since} \quad 1 \leq 1.2 < 2$$
$$[[-2.4]] = -3 \quad \text{since} \quad -3 \leq -2.4 < -2$$
$$[[7]] = 7 \quad \text{since} \quad 7 \leq 7 < 8$$
$$\left[\left[\frac{15}{4}\right]\right] = 3 \quad \text{since} \quad 3 \leq \frac{15}{4} < 4$$

Example 3   Graph $f(x) = [[x]]$.

Solution   From the definition of x we can construct the following table:

x	$f(x)$
$-2 \leq x < -1$	-2
$-1 \leq x < 0$	-1
$0 \leq x < 1$	0
$1 \leq x < 2$	1
$2 \leq x < 3$	2

The graph is shown in Figure 20. A solid dot (●) indicates that a point is included, and an open dot (○) indicates that a point is omitted. ∎

The function that assigns the fee for mailing a letter or parcel with the U.S. Postal Service is a variation of the greatest integer function.

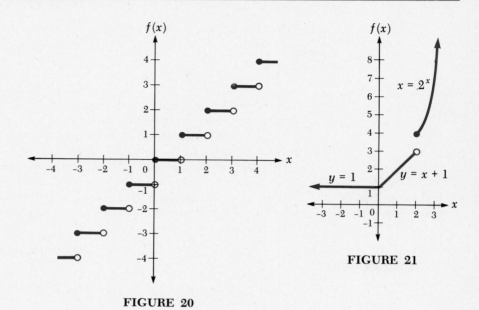

FIGURE 20

FIGURE 21

For instance, if x represents the weight (in ounces) of a letter or parcel, and if the fees for mailing are $0.25 for the first ounce and $0.20 for each additional ounce or part of an ounce, then the fee function is $f(x) = \$0.20[[x]] + \$0.25$.

Some functions are defined differently for different sets of x-values. Example 4 examines such a function.

Example 4 Graph

$$f(x) = \begin{cases} 1 & \text{for } x < 0 \\ x + 1 & \text{for } 0 \le x < 2 \\ 2^x & \text{for } x \ge 2 \end{cases}$$

Solution The function is graphed over the appropriate values of x in the ways that we have already learned. Figure 21 shows the graph. ∎

The final class of functions to be considered in this section is classified by special properties given in the next definition.

DEFINITION 1. A function $y = f(x)$ is said to be an **even function** if $f(-x) = f(x)$ for each x in the domain of f.
2. A function $y = f(x)$ is said to be an **odd function** if $f(-x) = -f(x)$ for each x in the domain of f.

The y-axis acts as a mirror for even functions, and the origin acts as a mirror for odd functions (see Figure 22). The functions $f(x) = |x|$ and $f(x) = x^2$ are even functions, and $f(x) = x^3$ is an odd function. Some functions are neither.

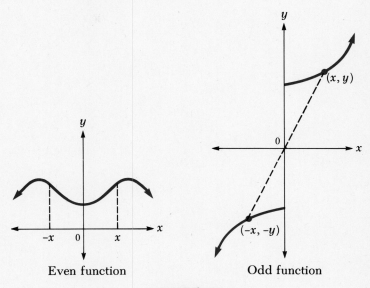

Even function   Odd function

FIGURE 22

Example 5 Show that $f(x) = 5x^3 - 2x$ is an odd function.

Solution $f(-x) = 5(-x)^3 - 2(-x)$
$= 5(-x^3) + 2x$
$= -5x^3 + 2x$
$= -(5x^3 - 2x) = -f(x)$

Since x represents an arbitrary number, the function is odd. ∎

Other classes of functions will be considered throughout the remainder of this book.

Exercises 2.5 In Exercises 1–22 graph the function.

1. $f(x) = |x| + 2$
2. $f(x) = |x + 2|$
3. $f(x) = 2|x|$
4. $f(x) = |x| + x$
5. $f(x) = |3x + 1|$
6. $f(x) = 3^x$
7. $f(x) = 10^x$
8. $f(x) = 2^{-x}$
9. $f(x) = 3^{-x}$
10. $f(x) = 2^{|x|}$

11. $f(x) = \dfrac{1}{5^x}$
   12. $f(x) = 2^x + 1$

13. $f(x) = 2^{x+1}$
   14. $f(x) = 2 \cdot (3^x)$

15. $f(x) = [[x]] - 2$
   16. $f(x) = 2[[x]]$

17. $f(x) = [[x]] + x$
   18. $f(x) = [[x - 1]]$

19. $f(x) = \begin{cases} 0 & \text{for } x < 0 \\ 2x + 3 & \text{for } x \ge 0 \end{cases}$

20. $f(x) = \begin{cases} -x & \text{for } x \le -1 \\ 2x + 3 & \text{for } -1 < x \le 2 \\ \frac{7}{2}x & \text{for } x > 2 \end{cases}$

21. $f(x) = \begin{cases} |x| & \text{for } x \le 3 \\ [[x]] & \text{for } 3 < x < 5 \\ 2 & \text{for } x \ge 5 \end{cases}$
   22. $f(x) = \begin{cases} 2^{-x} & \text{for } x < 0 \\ 2^x & \text{for } x \ge 0 \end{cases}$

In Exercises 23–30 classify the function as odd, even, or neither.

23. $g(x) = x^3$
   24. $f(x) = x^3 + x^2$

25. $h(x) = \dfrac{1}{x^2 - 1}$
   26. $g(x) = 5x$

27. $f(x) = 5x - 3$
   28. $h(x) = 2^x$

29. $g(x) = [[x]]$
   30. $f(x) = 3x^4 - x^2 + 17$

2.6 Inverse Functions

A function f makes an assignment such as $y = f(x)$. We may at times be interested in a possible reversal, that is, in finding a function g such that $x = g(y)$. This notion is made more formal by the following definition.

DEFINITION If $f = \{(x, y) \mid y = f(x)\}$ is a function, then the **inverse relation** associated with f is $g = \{(y, x) \mid y = f(x)\}$. If the relation g defined in this manner is a function, we say that g is the **inverse function** of f and frequently write f^{-1} to show that this function is related to the original function.

Note that the inverse relation is obtained by simply switching the components in each ordered pair; that is, (x, y) becomes (y, x).

If the function f has an inverse function f^{-1}, then $y = f(x)$ implies that

$$(f^{-1} \circ f)(x) = f^{-1}[f(x)] = f^{-1}[y] = x$$

and

$$(f \circ f^{-1})(y) = f[f^{-1}(y)] = f[x] = y$$

This shows that $f^{-1} \circ f$ is an "identity" function on the domain of f, since $(f^{-1} \circ f)(x) = x$, and $f \circ f^{-1}$ is an "identity" function on the range of f, since $(f \circ f^{-1})(y) = y$. It is also easy to see that the inverse function of f^{-1} is f.

Example 1   List the inverse relation for each of the following functions. Which functions have inverse functions?
(a) $f = \{(-1, 2), (0, 0), (1, 3), (2, 5)\}$
(b) $g = \{(-1, 2), (0, 2), (1, 3), (2, 5)\}$

Solution   (a) $\{(2, -1), (0, 0), (3, 1), (5, 2)\} = f^{-1}$ (inverse function of f).
(b) $\{(2, -1), (2, 0), (3, 1), (5, 2)\}$, not a function.   ∎

Since many functions are defined by a rule, such as $f(x) = 2x - 3$ or $y = 2x - 3$, we need a method other than reversing ordered pairs to obtain the inverse of a function if it exists. What we will do is simply replace y by x and x by y in the rule and then solve for y as an expression in x. If this process fails, there is no inverse function.

Example 2   Interchange the variables in each of the following functions. Which have inverse functions?
(a) $f(x) = 2x - 3$ (b) $f(x) = x^2$

Solution   (a) $f(x) = 2x - 3 \rightarrow y = 2x - 3$

$$x = 2y - 3 \qquad \text{(replace } y \text{ by } x \text{ and } x \text{ by } y\text{)}$$

$$-2y = -x - 3$$

$$y = \frac{1}{2}x + \frac{3}{2} \qquad \text{(a function)}$$

$$f^{-1}(x) = \frac{1}{2}x + \frac{3}{2} \qquad \text{(inverse of } f\text{)}$$

(b) $f(x) = x^2 \rightarrow y = x^2$

$$x = y^2 \qquad \text{(replace } y \text{ by } x \text{ and } x \text{ by } y\text{)}$$

$$y^2 = x \qquad \text{(interchange the sides of the equation)}$$

Since $y^2 = x$ yields $y = \sqrt{x}$ or $y = -\sqrt{x}$, this will generally give two values of y for each x. Thus $f(x) = x^2$ does not have an inverse function.   ■

Another way of seeing that a function has an inverse is to see if it is **one-to-one.** A function is one-to-one if different values of x have different values of y; that is, if $x_1 \neq x_2$, then $f(x_1) \neq f(x_2)$. This allows a value of y in the range of f to be "brought back" to a unique value of x in the domain of f, thus defining an inverse function.

If the graph of a function is given, the function is one-to-one (has an inverse function) if each horizontal line ($y = $ constant) crosses the graph of the function in at most one place (see Figure 23).

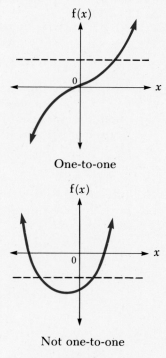

One-to-one

Not one-to-one

FIGURE 23

This **horizontal line test** gives a visual test for one-to-one functions. Now, by restricting the domain of some functions that are not one-to-one, we can obtain a new, restricted function that is one-to-one and thus has an inverse function. The two ideas together allow us to generate inverses to restricted functions by analyzing the graph of the function with the horizontal line test.

Example 3 Graph $y = x^2$ by computing y-values for $x = -2, -1, 0, 1, 2$ and connecting the graphs of the resulting pairs. Choose an appropriate restriction of the domain of $y = x^2$ so that the restricted function is one-to-one. Then determine the corresponding inverse function.

Solution

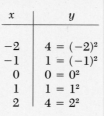

x	y
-2	$4 = (-2)^2$
-1	$1 = (-1)^2$
0	$0 = 0^2$
1	$1 = 1^2$
2	$4 = 2^2$

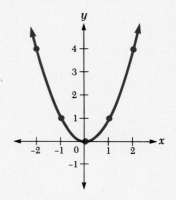

FIGURE 24

The graph appears in Figure 24. By looking at the graph and applying the horizontal line test, it is clear that the function is not one-to-one. If we restrict the domain to the nonnegative real numbers ($x \geq 0$), the restricted function is one-to-one. (The restriction to $x \leq 0$ would also have been valid.)

To compute the inverse of $y = x^2$ for $x \geq 0$, interchange variables and solve for y:

$$x = y^2 \quad \text{for } y \geq 0 \qquad \text{(replace } y \text{ by } x \text{ and } x \text{ by } y\text{)}$$

$$y^2 = x \qquad \text{(interchange sides)}$$

$$y = \sqrt{x} \qquad \text{(inverse function)}$$

We choose the positive square root, since $y \geq 0$ after interchanging the variables. ∎

One property of inverse functions is best illustrated by graphing a function and its inverse function on the same Cartesian coordinate system and then looking at the results relative to the line $y = x$ (see Example 4).

Example 4 Graph $y = x^2$ for $x \geq 0$ and $y = \sqrt{x}$ on the same Cartesian coordinate system.

Solution

$y = x^2$:

x	y
0	$0 = 0^2$
1	$1 = 1^2$
2	$4 = 2^2$

$y = \sqrt{x}$:

x	y
0	$0 = \sqrt{0}$
1	$1 = \sqrt{1}$
4	$2 = \sqrt{4}$

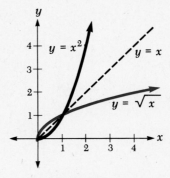

FIGURE 25

The graphs appear in Figure 25. ■

It is a reasonable guess from the graphs in Example 4 that a function and its inverse function are mirror images of each other about the line $y = x$. This is true, since the roles of x and y are merely interchanged for a function and its inverse.

The inverse function concept is important in the study of exponential and logarithmic functions in Chapter 6. It is also an important concept in trigonometry and the calculus.

Exercises 2.6

In Exercises 1–8 list the inverse relation for the function and determine whether it has an inverse function.

1. $f = \{(0, 1), (1, 2), (2, 3), (3, 4)\}$
2. $g = \{(0, 2), (1, 2), (2, 2), (3, 2)\}$
3. $h = \{(-1, 1), (0, 0), (1, 1), (2, 4)\}$
4. $f = \{(0, 0), (1, 1), (2, 4), (3, 9), (4, 16)\}$
5. $g = \{(-3, 3), (-1, 1), (1, 1), (3, 3), (5, 5)\}$
6. $h = \{(-3, 3), (-1, 1), (1, -1), (3, -3), (5, -5)\}$
7. $f = \{(4, 2), (9, 3), (16, 4), (25, 5)\}$
8. $g = \{(1, 1), (2, 2), (3, 0), (4, 1), (5, 2)\}$

In Exercises 9–18 interchange the variables for the function and determine whether it has an inverse function.

9. $f(x) = x + 2$
10. $f(x) = 5x + 9$
11. $f(x) = x^3$
12. $f(x) = 2x^2 + 1$
13. $f(x) = -3x$
14. $f(x) = \dfrac{1}{x}$
15. $f(x) = \sqrt{x - 2}$
16. $f(x) = \frac{1}{2}x - 4$
17. $f(x) = (x + 1)^2$
18. $f(x) = 2x^3 + 1$

In Exercises 19–26 use the horizontal line test to determine which of the graphs represent functions that are one-to-one.

19.

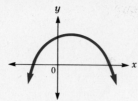

20.

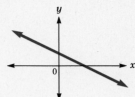

21.

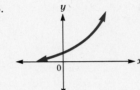

22.

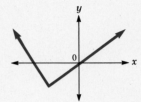

23.

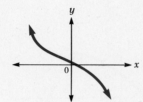

24.

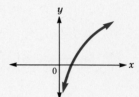

25.

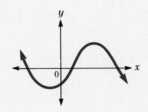

26.

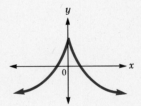

27. Form the inverse function of $f(x) = 2x$. Then verify that $f \circ f^{-1}$ and $f^{-1} \circ f$ are identity functions on the appropriate domains.

28. Form the inverse function of $f(x) = x - 6$. Then verify that $f \circ f^{-1}$ and $f^{-1} \circ f$ are identity functions on the appropriate domains.

29. Graph $f(x) = 2x$, its inverse function, and the line $y = x$ on the same Cartesian system.

30. Graph $f(x) = x - 6$, its inverse function, and the line $y = x$ on the same Cartesian system.

In Exercises 31–34 determine all appropriate restrictions of the domain of the function so that the restricted function will have an inverse function.

31.

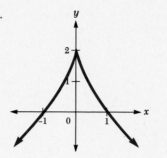

32.

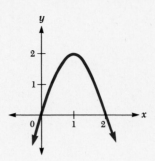

33.

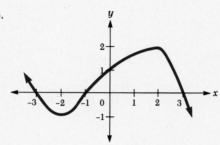

34.

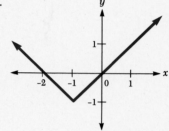

Key Terms and Formulas

Equation:
  identity
  conditional equation
  literal equation
  solution
  solve
Properties of equality
Inequality
Properties of inequality
Distance-rate-time formula:
  $d = rt$
Conversion of units
Proportion
Ratio
Balancing units (unit fractions)
Variation:
  varies directly
  proportional
  varies inversely
  varies jointly
  constant of variation

Cartesian coordinate system:
  origin
  x-axis
  y-axis
  quadrants
  x-coordinate (abscissa)
  y-coordinate (ordinate)
First-degree equation in two variables
Linear equation in two variables
y-intercept
Slope:

$$m = \frac{y_2 - y_1}{x_2 - x_1} = \frac{\text{change in } y}{\text{change in } x}$$

Slope–intercept form of the equation of a line: $y = mx + b$
Point–slope form of the equation of a line: $y - y_1 = m(x - x_1)$

Two-point form of the equation
of a line: $y - y_1 = m(x - x_1)$

with $m = \dfrac{y_2 - y_1}{x_2 - x_1}$

Interpolation
Extrapolation

Normal line: $m_N = -\dfrac{1}{m}$

Relation:
  domain
  range
Function:
  functional notation
  vertical line test
  algebra of functions
  composition of functions

Special functions:
  absolute value function
  exponential function
  exponential growth
  exponential decay
  greatest integer function
  (staircase function)
  even function
  odd function
Inverse relation
Inverse function
One-to-one function
Horizontal line test

Review Exercises

1. Show that $x = -1$ is a solution of $2x - 3 = 4(x - 2) + 7$.
2. Solve $5x + 2 = x - 6$.
3. Solve $3(2y - 5) = 4 - (y - 9)$.
4. Solve $\dfrac{2}{w} - \dfrac{5}{w} = \dfrac{3}{2}$.
5. Solve $\dfrac{x}{x - 1} + 2 = \dfrac{1}{x - 1}$.
6. Solve for x: $2x - 3y = 5$.
7. Solve for R: $\dfrac{1}{R} = \dfrac{1}{S} + \dfrac{1}{T}$.
8. Solve the inequality $1 - 2x < 5$. Then graph the solutions on the number line.
9. Solve the inequality $\dfrac{7x + 1}{4} \le x + 7$. Then graph the solutions on the number line.
10. Solve $|5x - 1| = 9$.
11. Solve the inequality $|2x + 3| < 9$. Then graph the solutions on the number line.
12. Three programmers earn an average of $22,000. If the first programmer earns $24,500 and the second programmer earns $19,000, how much does the third programmer earn?
13. Two parcels of land have one side in common. One parcel is in the shape of a square, and the other is in the shape of an equilateral triangle (three

equal sides). If 4200 feet of fencing are required to enclose the land and to separate the two parcels, what are the dimensions of each parcel of land?

14. Two air rescue planes take off in opposite directions from the same airport. The faster plane cruises at twice the speed of the slower plane. If the slower plane started at 12 noon and the faster plane at 12:30 P.M. and they are 600 miles apart at 2:00 P.M., what is the cruising speed of each plane?

15. If safe water has 5 parts per million of a germicide, how many gallons of germicide concentrate (105 parts per million) should be added to 1000 gallons of untreated water (0 parts per million) to make it safe?

16. Computer A can complete an insurance office's daily processing in 2 hours. Computer B can complete the same processing in 7 hours. If the two computers can effectively share their processing capabilities, how long will it take to complete the daily processing with both computers running at the same time?

17. Convert 19 kilograms to ounces.

18. Convert 1 mile to meters.

19. If A varies directly as s, and if $A = 28$ when $s = 7$, what is the value of A when $s = 1.5$?

20. If T varies jointly as P and V, and if $T = 300$ when $P = 30$ and $V = 5$, what is the value of T when $P = 10$ and $V = 95$?

21. Plot the points $(5, 0)$, $(-2, 2)$, $(-1, -3)$, and $(4, 3)$ on a Cartesian coordinate system.

22. Determine the ordered pairs corresponding to the points labeled A, B, C, and D on the following Cartesian system:

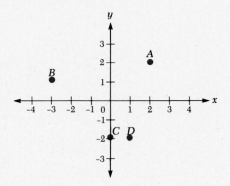

23. Graph the line $\{(x, y) \mid x = -1\}$.
24. Graph the line $\{(x, y) \mid x + 2y = 4\}$.
25. Identify the slope and y-intercept of the line $x - 3y = 6$.
26. Determine the equation of the line with slope $= -2$ and y-intercept $= 3$.
27. Determine the equation of the line passing through the point $(-1, 1)$ and having slope $= 3$.
28. Determine the equation of the line passing through the points $(4, 1)$ and $(2, 0)$.

29. Determine the equation of the line parallel to the line $y = 3x - 2$ and passing through the point $(5, -2)$.
30. Determine the equation of the line normal to the line $x - 2y = 4$ and passing through the origin.
31. The relationship between solid waste (garbage) and population is linear. If a population of 1 million generates 300 tons of solid waste per week, and if a population of $1\frac{1}{2}$ million generates 450 tons of solid waste, how much waste will be generated by a population of $2\frac{1}{4}$ million?
32. List the domain and range of the relation $R = \{(-3, 1), (0, 2), (0, -1), (5, 2)\}$.
33. Graph the relation $R = \{(2, -3), (2, -1), (2, 1), (2, 3)\}$.
34. For the function $f(x) = 2x + 1$, compute $f(-2)$, $f(0)$, $f(2)$, and $f(4)$. Then graph the function.
35. For the function $g(x) = x^3$, compute $g(-2)$, $g(-1)$, $g(0)$, $g(1)$, and $g(2)$. Then graph the function.
36. Use the vertical line test to identify which graph(s) describe functions:
    (a) 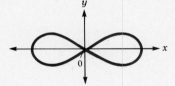 (b)

37. Use the vertical line test to identify which graph(s) describe functions:
    (a) 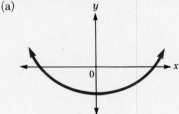 (b)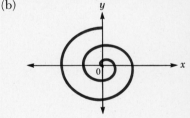

38. For $f(x) = 2x + 1$ and $g(x) = x^2 - 1$, compute $f + g$, $f - g$, fg, f/g, and $f \circ g$.
39. For $f(x) = 2x + 1$ and $g(x) = x^2 - 1$, compute $g + f$, $g - f$, gf, g/f, and $g \circ f$.
40. For $g(x) = 3x^2 + x$, compute $\dfrac{g(x + h) - g(x)}{h}$.
41. For $f(x) = (2x - 1)^2$, compute $\dfrac{f(0) + f(1) + f(2)}{3}$.
42. Using functional notation, express profit (P) as a function of number of lawnmowers sold (x), where each lawnmower costs \$120 and sells for \$165. The fixed costs are \$7500.
43. Graph $f(x) = 3|x|$.
44. Graph $f(x) = 5^x$.
45. Graph $f(x) = \frac{1}{2}[[x]]$.

46.   Graph $f(x) = \begin{cases} x^2 & \text{for } 0 \le x < 3 \\ 2x + 3 & \text{for } x \ge 3 \end{cases}$

47.   Classify the function $h(x) = (x^3 + x)^2$ as odd, even, or neither.

48.   List the inverse relation for $f = \{(2, 1), (3, 5), (4, 1)\}$. Does f have an inverse function?

49.   Interchange the variables for the function $f(x) = 2x - 7$. Does f have an inverse function?

50.   Use the horizontal line test to determine which graph(s) describe one-to-one functions:

(a)   (b)

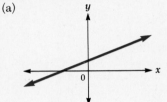   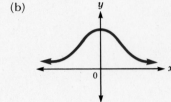

51.   Determine all appropriate restrictions of the domain of the function graphed below so that the restricted functions will have inverse functions.

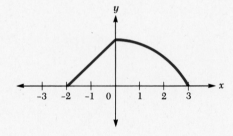

CHAPTER 3

First-
Degree
Systems

3.1 Systems of Linear Equations

Recall that the solution set S of a linear equation in two variables (x and y) is $S = \{(x, y) \mid ax + by = C\}$. We now turn our attention to solving *systems* of two linear equations in two variables.

DEFINITION — The **solution** of a linear system

$$A = \{(x, y) \mid a_1x + b_1y = C_1\}$$
$$B = \{(x, y) \mid a_2x + b_2y = C_2\}$$

is the set of all ordered pairs (x, y) that satisfy *both* equations.

If we let $A = \{(x, y) \mid a_1 x + b_1 y = C_1\}$ and $B = \{(x, y) \mid a_2 x + b_2 y = C_2\}$, then the intersection of their solution sets (denoted $A \cap B$) is the solution (set) of the system. There are three possible results:

1. $A \cap B = \varnothing$.
2. $A \cap B$ consists of a single ordered pair.
3. $A \cap B$ contains many ordered pairs.

Each possibility can be shown graphically.

CLASSIFICATION

1. There may be *no* ordered pairs that satisfy both equations. This occurs when the graphs of A and B are parallel lines [Figure 1(a)]. In this case the system of equations is said to be **inconsistent**.
2. There may be a *single* ordered pair (x_1, y_1) that satisfies both equations. This occurs when the lines intersect at a single point (x_1, y_1) [Figure 1(b)]. In this case the system is said to be consistent with a unique solution or simply **independent**.
3. The equations may have the same graph [Figure 1(c)]; hence all ordered pairs that satisfy A also satisfy B. In this case the system is consistent with infinitely many solutions and is commonly referred to as **dependent**.

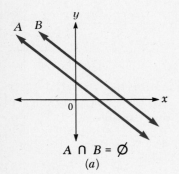

$A \cap B = \varnothing$
(a)

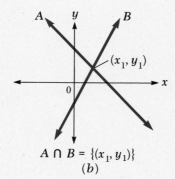

$A \cap B = \{(x_1, y_1)\}$
(b)

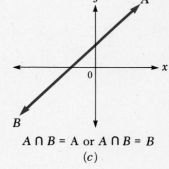

$A \cap B = A$ or $A \cap B = B$
(c)

FIGURE 1

We may classify any system by graphing both equations on the *same* coordinate system and observing the result. The next example shows this.

Example 1   Classify each of the following systems using graphs. (Notice that the formal set notation has been abbreviated but the letter names are retained for convenience.)

(a) A: $x + y = 2$ (b) A: $2x + 2y = 4$ (c) A: $x + y = 2$
     B: $x + y = 3$ B: $x + y = 2$ B: $x - y = 0$

Solution   (a) The graph is shown in Figure 2. Since the graphs are parallel lines, the system is   inconsistent   . Thus $A \cap B = \varnothing$.

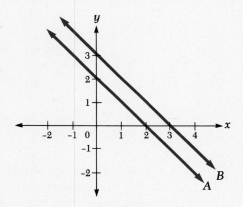

FIGURE 2

(b) The graph is shown in Figure 3. Since the graphs are the same line, the equations are   dependent   . Thus $A \cap B = A$ or $A \cap B = B$.

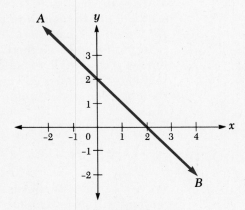

FIGURE 3

(c) The graph is shown in Figure 4. Since the graphs intersect at the single point (1, 1), the equations are independent . Thus $A \cap B = \{(1, 1)\}$. ∎

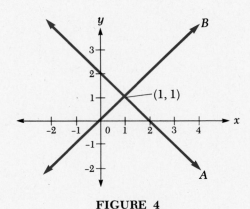

FIGURE 4

Although the graphical method shown in Example 1 is useful in classifying simple systems, it is not adequate for our purposes. For example, lines might appear to be parallel, but they must have the same slope or, ultimately, they will intersect.

Our goal in this section is to find the (unique) solution of an independent linear system. Often the solution cannot be determined precisely from a graph. You cannot distinguish the point (1.5, 2.5) from (1.5001, 2.5001) on a graph. Interpreting the graph only yields approximate values in most cases. Hence, a precise algebraic technique is required.

We will discuss two techniques in this section: the substitution method and the addition method. Three additional methods will be presented in Chapter 7.

In the first two methods, we will simplify the given system by reducing it to a single equation in one variable that can be solved by the methods of Chapter 2. The value found from the single equation can then be used to determine the corresponding value of the other variable.

The Substitution Method

We will begin by illustrating this method with an example. This method will be particularly useful in solving business problems (such as break-even analysis) and solving nonlinear systems (Chapter 5).

Example 2   Find the solution $(A \cap B)$ of the system

$$A: \quad 2x + 3y = 6$$
$$B: \quad x - y = 2$$

Solution   First we solve equation B for x:

$$A: \quad 2x + 3y = 6$$
$$B: \quad x \qquad = 2 + y$$

Now we use the expression for $x(2 + y)$ found in equation B to replace (or substitute for) x in equation A. This gives a single equation in y:

$$A: \quad 2(2 + y) + 3y = 6$$

Now solve for y:

$$A: \quad 4 + 2y + 3y = 6$$
$$4 + 5y = 6$$
$$5y = 2$$
$$y = \frac{2}{5}$$

The value of y is one coordinate of the solution. By replacing y by $\frac{2}{5}$ in equation B (or A), we have

$$B: \quad x = 2 + \frac{2}{5}$$

$$x = 2\frac{2}{5}$$

Therefore, $A \cap B = \boxed{\{(2\frac{2}{5}, \frac{2}{5})\}}$. ∎

You should check the accuracy of your solution by evaluating *both* equations at the point $(x_1, y_1) = (2\frac{2}{5}, \frac{2}{5})$. If your work is correct, both equations will be satisfied.

Check A: $\quad 2\left(2\frac{2}{5}\right) + 3\left(\frac{2}{5}\right) \stackrel{?}{=} 6 \qquad$ *Check* B: $\quad 2\frac{2}{5} - \frac{2}{5} \stackrel{?}{=} 2$

$$2\left(\frac{12}{5}\right) + \frac{6}{5} \stackrel{?}{=} 6 \qquad\qquad \frac{12}{5} - \frac{2}{5} \stackrel{?}{=} 2$$

$$\frac{24}{5} + \frac{6}{5} \stackrel{?}{=} 6 \qquad\qquad \frac{10}{5} \stackrel{?}{=} 2$$

$$\frac{30}{5} \stackrel{?}{=} 6 \qquad\qquad 2 \stackrel{\checkmark}{=} 2$$

$$6 \stackrel{\checkmark}{=} 6$$

Therefore, $x = 2\frac{2}{5}$, $y = \frac{2}{5}$ is a solution of the system.

The substitution method may be summarized as follows:

Step 1: Solve one of the equations for one of the variables.

Step 2: Substitute the expression found in Step 1 for that same variable in the *other* equation. This results in a single equation with one unknown.

Step 3: Solve the single equation found in Step 2. This gives one of the coordinates of the solution.

Step 4: Use the value found in Step 3 to determine the other coordinate.

Note: The value found by solving the single equation in one unknown must be substituted into one of the earlier equations containing two variables in order to determine the second coordinate.

Caution: If you erroneously substitute the value found in Step 1 into the equation in Step 1, you will get an identity (such as $2 = 2$) and will not be able to proceed to the solution of the system.

What happens if the system you are attempting to solve by the substitution method is inconsistent (and so has no solution) or is dependent (and so has infinitely many)?

We saw that the system

$$A: \quad x + y = 2$$
$$B: \quad x + y = 3$$

is inconsistent. Let us apply the substitution method.

Step 1: Solve B for y:

$$A: \quad x + y = 2$$
$$B: \quad \quad y = 3 - x$$

Step 2: Replace y in A:

$$A: \quad x + 3 - x = 2$$

Step 3: Solve:

$$A: \quad 3 = 2, \quad \text{which is false.}$$

If the system is inconsistent, you will always get a false statement.

We saw that the system

$$A: \quad 2x + 2y = 4$$
$$B: \quad x + y = 2$$

was dependent. Let us apply the substitution method.

Step 1: Solve *B* for *y*:

$$A: \quad 2x + 2y = 4$$

$$B: \qquad y = 2 - x$$

Step 2: Replace *y* in *A*:

$$A: \quad 2x + 2(2 - x) = 4$$

Step 3: Solve:

$$A: \quad 2x + 4 - 2x = 4$$

$$4 = 4, \quad \text{which is always true.}$$

If the system is dependent, you will always get an identity.

SYSTEM CLASSIFICATION

If the substitution method results in an *identity* (such as $4 = 4$), the equations are *dependent*.

If the substitution method results in a *false statement* (such as $2 = 3$), the equations are *inconsistent*.

Example 3 The marketing department of Data Instruments, Inc. has determined the following models of revenue and cost for the production and sale of *x* scientific calculators.

$$\text{Revenue:} \quad R = \$20x$$

$$\text{Cost:} \qquad C = \$12x + \$10,000$$

At what level of production and sales will cost be equal to revenue? What are the cost and the revenue?

Solution By substituting the value of *R* ($\$20x$) from the revenue equation for *C* in the cost equation, we have

$$\text{revenue} = \text{cost}$$

$$\$20x = \$12x + \$10,000 \qquad \text{Solve this equation for } x$$

$$\$8x = \$10,000$$

$$x = 1250 \text{ calculators.}$$

Thus $R = \$20(1250) = \$25,000 \ (=C)$. The cost and revenue will be $\$25,000$ for 1250 calculators produced and sold. This is called the

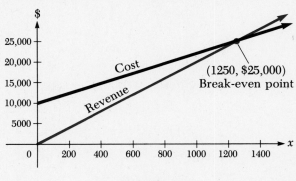

FIGURE 5

break-even point and can be seen graphically in Figure 5. If the company sells more than 1250 calculators, it will make a profit.   ■

The Addition Method

The addition method is based on the operation of addition of corresponding sides of two equations (expressed in standard form) to yield a *single equation* with *one unknown*. The goal of this method is the same as that for substitution, but the process is different.

Example 4   Solve the following system by the addition method:

$$A: \quad x + y = 3$$
$$B: \quad x - y = 1$$

Solution   First we add the corresponding members of the equations:

$$
\begin{array}{rl}
A: & x + y = 3 \\
B: & \underline{x - y = 1} \\
A + B: & 2x \quad\;\; = 4
\end{array}
$$

This gives a single equation in one unknown, which we solve:

$$x = 2$$

Now, use the value of $x \, (= 2)$, to determine the corresponding value of y. Replacing 2 for x in A (or B) gives

$$A: \quad 2 + y = 3$$

$$y = 1$$

Hence $A \cap B =$ {(2, 1)} .   ■

If we had replaced x by 2 in equation B, we would have found the same value, 1, for y. Remember, the solution solves both and therefore can be determined from either one.

If addition of equations in the system does not immediately eliminate one of the variables, we may use the multiplicative property of equality to obtain an equivalent system in which addition will eliminate one of the variables.

Note: The multipliers are chosen so that you will have *additive inverses* as coefficients of the variable to be eliminated.

Example 5   Solve the system

$$A: \quad 3x + 5y = 4$$
$$B: \quad 2x - 3y = -10$$

Solution   Suppose we choose to eliminate x. Then we can multiply A by -2 and B by $+3$:

$$-2A: \quad -6x - 10y = -8$$
$$\underline{3B: \quad +6x - 9y = -30}$$

Adding, we have

$$-19y = -38$$
$$y = 2$$

To determine x, replace 2 for y in A:

$$A: \quad 3x + 5(2) = 4$$
$$3x + 10 = 4$$
$$3x = -6$$
$$x = -2$$

Hence $x = -2$ and $y = 2$. ■

SYSTEM CLASSIFICATION

If the addition method leads to a *false statement* (such as $4 = 9$), the system is *inconsistent*.

If the addition method leads to an identity (such as $4 = 4$), the system is *dependent*.

Example 6   A testing specialist feels that part of a student's score is due to clerical errors because of the number of test items (x) and distractions due to the noise level (y). On a recent test, the total error was 6. The number of test items minus twice the noise level was 6. What number of items were on the test and what was the noise level?

Solution   Total error: $x + y = 6$
Interaction: $\underline{x - 2y = 6}$

To solve, multiply the error equation by 2 and add:

$$\begin{array}{ll} 2(\text{total error}): & 2x + 2y = 12 \\ \text{interaction:} & \underline{\ x - 2y = 6\ } \\ & 3x \quad\ = 18 \end{array}$$

$$x = 6 \text{ items}$$

The noise level y can be found from either equation:

$$\text{total error: } 6 + y = 6$$

$$y = 0 \text{ noise}$$

The test had 6 items and was taken in silence. ∎

Systems of Three Equations in Three Unknowns

The solution set S of a linear equation in three variables (x, y, and z) is $S = \{(x, y, z) \mid ax + by + cz = D\}$. For example, some solutions of $S = \{(x, y, z) \mid x + y - 2z = 0\}$ are $(1, 1, 1)$, $(1, 2, \frac{3}{2})$, $(2, -1, \frac{1}{2})$, and $(3, 1, 2)$.

DEFINITION   The *solution* of a linear system

$$A = \{(x, y, z) \mid a_1x + b_1y + c_1z = D_1\}$$

$$B = \{(x, y, z) \mid a_2x + b_2y + c_2z = D_2\}$$

$$C = \{(x, y, z) \mid a_3x + b_3y + c_3z = D_3\}$$

is the set of triples (x, y, z) that satisfy *all three* equations.

As with systems of two equations in two variables, there may be

1. no such triple (inconsistent),
2. exactly one triple (independent),
3. many triples (dependent).

In order to solve a system of three equations in three unknowns, we will use elimination by addition. Our goal is to reduce the system from three equations in three unknowns to two equations in two unknowns by elimination of a variable. Then we reduce the system of two equations in two unknowns to one equation in one unknown and determine the value of one of the variables. By replacing that value in a previous equation we can determine the value of another variable, and so on. This process can be illustrated in five steps. Study Example 7 carefully.

Example 7 Solve the following system by addition:

$$A: \quad x + \quad y - 2z = 0$$

$$B: \quad x - 2y + \quad z = 3$$

$$C: \quad 2x - \quad y + \quad z = 7$$

Solution *Step 1:* Decide which variable to eliminate first.

We choose to eliminate x.

Step 2: Select two different pairs of equations and perform the necessary operations to eliminate x from both pairs.

We select the pairs AB and AC.

Eliminate x from A and B: Eliminate x from A and C:

$$
\begin{array}{ll}
A: & x + y - 2z = 0 \\
-1B: & \underline{-x + 2y - z = -3} \\
& 3y - 3z = -3
\end{array}
\qquad
\begin{array}{ll}
-2A: & -2x - 2y + 4z = 0 \\
C: & \underline{2x - y + z = 7} \\
& -3y + 5z = 7
\end{array}
$$

Step 3: Step 2 reduced the system to two equations in two unknowns. Now eliminate one of the remaining variables (y or z).

We choose to eliminate y:

$$
\begin{array}{ll}
D: & 3y - 3z = -3 \\
E: & \underline{-3y + 5z = 7} \\
& 2z = 4
\end{array}
$$

$$\boxed{z = 2}$$

Step 4: Substitute the value of z in equation D or E and solve for y.

We choose to substitute $z = 2$ in equation D:

$$D: \quad 3y - 3(2) = -3$$

$$3y - 6 = -3$$

$$3y = -3 + 6$$

$$3y = 3$$

$$\boxed{y = 1}$$

Step 5: Substitute the values of y and z in equation A, B, or C and solve for x.

We choose to substitute the values in A.

$$A: \quad x + (1) - 2(2) = 0$$
$$x + 1 - 4 = 0$$
$$x - 3 = 0$$
$$x = 3$$

Hence the solution is $x = 3$, $y = 1$, and $z = 2$.
Check: Try all three values in *all three* equations.

$$A: \quad 3 + 1 - 2(2) \overset{?}{=} 0$$
$$4 - 4 \overset{?}{=} 0$$
$$B: \quad 3 - 2(1) + 2 \overset{?}{=} 3$$
$$3 - 2 + 2 \overset{?}{=} 3$$
$$3 \overset{?}{=} 3$$
$$C: \quad 2(3) - (1) + 2 \overset{?}{=} 7$$
$$6 - 1 + 2 \overset{?}{=} 7$$
$$7 \overset{?}{=} 7 \quad \blacksquare$$

SYSTEM CLASSIFICATION

If the equations D and E (in Step 3) are dependent, then the original system is dependent.

If the equations D and E (in Step 3) are inconsistent, then the original system is inconsistent.

Suppose you attempt to solve the system

$$A: \quad 2x + y + 2z = 2$$
$$B: \quad x + y + z = 1$$
$$C: \quad x + 2y + z = 4$$

by eliminating y using AB and AC. Step 3 yields the system

$$x + z = 1$$
$$x + z = 0 \quad \text{(actually } 3x + 3z = 0)$$

which is inconsistent. Therefore the original system is inconsistent.

If you attempt to solve the system

$$A: \quad x + \quad y + \quad z = 1$$

$$B: \quad -x + \quad y + 7z = 1$$

$$C: \quad 2x + 3y + 6z = 3$$

by eliminating x using AB and AC, then Step 3 yields the system

$$y + 4z = 1 \quad \text{(actually } 2y + 8z = 2\text{)}$$

$$y + 4z = 1$$

which is dependent. Therefore the original system is dependent.

When the system is dependent, you can find solutions of the original system by finding solutions for y and z (in the two-unknown system) and replacing these values in the original system to get x.

For example, in the y, z system above, $y = 1 - 4z$.

1. If $z = 1$, then $y = 1 - 4(1) = -3$. Replacing $z = 1$ and $y = -3$ in equation A gives $x = 3$. Hence, $(3, -3, 1)$ is a solution.
2. If $z = -1$, then $y = 1 - 4(-1) = 5$. Replacing $z = -1$ and $y = 5$ in equation A gives $x = -3$. Hence, $(-3, 5, -1)$ is a solution.
3. In general, for $z = c$ (any constant), $y = 1 - 4c$, and equation A yields $x + (1 - 4c) + c = 1$ or simply $x = 3c$. Hence, $(3c, 1 - 4c, c)$ is a solution for any c. For instance, if $c = 2$, then $x = 6$, $y = -7$, and $z = 2$.

Exercises 3.1

In Exercises 1–12 use graphs to classify each system as independent, dependent, or inconsistent. (See Example 1.)

1. $A: \quad x + y = 4$
   $B: \quad x - y = 2$
2. $A: \quad x - y = 3$
   $B: \quad x + y = 1$
3. $A: \quad x + y = 5$
   $B: \quad x + y = -3$
4. $A: \quad x - y = -2$
   $B: \quad x - y = 4$
5. $A: \quad 3x + 3y = 3$
   $B: \quad x + y = 1$
6. $A: \quad 3x - 3y = 6$
   $B: \quad 2x - 2y = 4$
7. $A: \quad 2x + y = 3$
   $B: \quad 4x - y = 0$
8. $A: \quad x + y = 1$
   $B: \quad x - 3y = 3$
9. $A: \quad 2x - y = 2$
   $B: \quad x - \frac{1}{2}y = 2$
10. $A: \quad x + 3y = 3$
    $B: \quad \frac{1}{3}x + y = 2$
11. $A: \quad 2x - y = 4$
    $B: \quad x - \frac{1}{2}y = 2$
12. $A: \quad x + 3y = 6$
    $B: \quad \frac{1}{3}x + y = 2$

In Exercises 13–20 use the substitution method to find the solution of the system. (See Example 2.)

13. $A: \quad x + y = 4$
    $B: \quad x - y = 2$
14. $A: \quad x - y = 3$
    $B: \quad x + y = 1$
15. $A: \quad 2x + y = 3$
    $B: \quad 4x - y = 0$
16. $A: \quad x + y = 1$
    $B: \quad x - 3y = 3$

17. A: $x + y = 3$
     B: $x - y = 3$
19. A: $3x + y = 1$
     B: $2x + 4y = 9$

18. A: $-x + y = 4$
     B: $x + y = -2$
20. A: $-2x + y = 2$
     B: $6x - y = -\frac{2}{3}$

In Exercises 21–24 use the substitution method to show that the system is either inconsistent or dependent.

21. A: $x - y = -2$
     B: $x - y = 4$
23. A: $3x + 3y = 3$
     B: $x + y = 1$

22. A: $2x - y = 2$
     B: $x - \frac{1}{2}y = 2$
24. A: $x + 3y = 6$
     B: $\frac{1}{3}x + y = 2$

In Exercises 25–30 use the addition method to find the solution of the system. (See Examples 4 and 5.)

25. A: $x + y = 4$
     B: $x - y = 2$
27. A: $2x + y = 1$
     B: $6x - y = 1$
29. A: $x + y = 5$
     B: $2x + 3y = 12$

26. A: $x - y = 3$
     B: $x + y = 1$
28. A: $x - 4y = 1$
     B: $-x + 10y = 2$
30. A: $3x - 2y = 22$
     B: $x - y = 9$

In Exercises 31–34 use the addition method to show that the system is either inconsistent or dependent.

31. A: $x + y = 5$
     B: $x + y = -3$
33. A: $3x - 3y = 6$
     B: $2x - 2y = 4$

32. A: $x + 3y = 3$
     B: $\frac{1}{3}x + y = 2$
34. A: $2x - y = 4$
     B: $x - \frac{1}{2}y = 2$

In Exercises 35–38 solve the system by any of the methods of this section.

35. A: $2x + y = -3$
     B: $x - 2y = 11$
37. A: $3x + 2y = 6$
     B: $5x + 6y = 12$

36. A: $x + y = 4$
     B: $2x + y = 1$
38. A: $3x - 7y = 1$
     B: $6x + 2y = -2$

39. The sum of two numbers is 18, and their difference is 12. Find the two numbers.

40. One number is twice as large as another number. Their sum is 21. Find the two numbers.

41. The perimeter of a rectangle is 22 units. If the width is doubled, the new perimeter will be 28 units. Find the length and width of the original rectangle.

42. A solution of 50% alcohol is combined with a solution of 20% alcohol to produce 100 cubic centimeters of a solution that is 26% alcohol. How many cubic centimeters of each solution were used?

43. A family has $5000 invested in two savings accounts. The regular savings account pays 6% a year, and the 90-day-notice account pays 8% a year. The interest at the end of the year is $340. How much was invested in each account?

44. The equation $y = mx + b$ is satisfied by the (x, y) values $(1, -2)$ and $(3, 4)$. Find the equation (solve for m and b).

45. In Example 6, suppose the sum of the number of test items and the noise level was 8, and the difference was 2. Find the number of test items and the noise level.

46. A company sells items for $30 each. It costs the company $12,000 to start production and each item costs $18 to complete. How many items must the company make and sell to break even?

47. A machine contained $30 in dimes and quarters. If there were 150 coins in the machine, find the number of each.

48. The perimeter of a rectangle is 32 meters. The difference in the length and width is 4 meters. Find the dimensions of the rectangle.

In Exercises 49–56 solve the system. If the system is independent, find the solution. If the system is dependent, find two numerical solutions and the general solution for any c.

49. A: $x - y + z = 7$
     B: $2x + y - z = 7$
     C: $x - 2y + 3z = 8$

50. A: $2x - y - z = 8$
     B: $x + 3y + z = 9$
     C: $x - y - z = 3$

51. A: $x + 4y - z = 10$
     B: $2x - y + 3z = 7$
     C: $x - y - z = 0$

52. A: $3x - y + z = 13$
     B: $x + 2y + z = 5$
     C: $x + 2y - z = 3$

53. A: $x + y + z = 6$
     B: $y + z = -1$
     C: $x - z = 2$

54. A: $x + y + z = 3$
     B: $x + z = 1$
     C: $2x + y + 2z = 2$

55. A: $x - y + z = 4$
     B: $y - 2z = 3$
     C: $x - z = 7$

56. A: $2x + y + z = 3$
     B: $x + y - z = 3$
     C: $x + z = 3$

3.2   Linear Inequalities and Linear Programming

We know that the solution set of a linear equation such as $x + y = 3$ graphs as a straight line. In this section we want to determine (and show graphically) the solution set of a **linear inequality** such as $x + y \leq 3$. The solution set will consist of the set $\{(x, y) | x + y = 3\} \cup \{(x, y) | x + y < 3\}$.

The graph of the equation $(x + y = 3)$ partitions the Cartesian plane into two **half-planes,** one on each side of the line. One of the half-planes is the solution set of the inequality $x + y < 3$. (The other half-plane is the solution set of $x + y > 3$.) The process for finding and graphing the solution of $x + y \leq 3$ is presented in the following discussion.

Step 1: Graph the equation $x + y = 3$. (See Figure 6.)

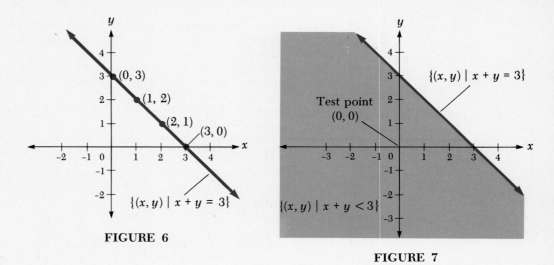

FIGURE 6

FIGURE 7

Step 2: Select any point in the plane (either half-plane) that is *not* on the line graphed in Step 1. The point selected is called a **test point.** If the test point satisfies the inequality ($x + y < 3$), then the entire half-plane containing that point is the solution set of the inequality. If the test point does not satisfy the inequality, then the other half-plane is the solution set.

Suppose you select $(0, 0)$ as your test point. Evaluate $x + y < 3$ at $(0, 0)$. Then $0 + 0 < 3$, or $0 < 3$, which is *true.* Since $(0, 0)$ satisfies the inequality, the half-plane containing $(0, 0)$ is the desired solution set. It is indicated graphically by shading, as in Figure 7.

The solution set of $x + y \leq 3$ consists of all points on and below the line, that is, $\{(x, y) | x + y = 3\} \cup \{(x, y) | x + y < 3\}$.

The test point method can always be used to determine the solution set of a linear inequality. However, if we put the statement in slope–intercept form, no test point will be needed. We will restrict the discussion to inequalities that are compounded statements containing both an equation (=) and an inequality (< or >).

Any linear inequality in two variables can be expressed in slope–intercept form $y \leq mx + b$ or $y \geq mx + b$ where m is the slope of the line and b is the y-intercept.

Case 1: The graph of the solution set of $y \geq mx + b$ consists of the half-plane above the line $y = mx + b$ and the line itself. (See Figure 8.)

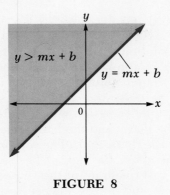

FIGURE 8

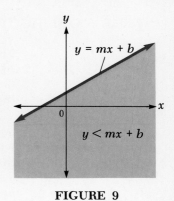

FIGURE 9

Case 2: The graph of the solution set of $y \leq mx + b$ consists of the half-plane below the line $y = mx + b$ and the line itself. (See Figure 9.)

Example 1 Graph the solution set of $2x - 3y \leq 6$.

Solution Put $2x - 3y \leq 6$ in slope–intercept form:

$$-3y \leq 6 - 2x$$

$$y \geq \frac{2}{3}x - 2$$

Now graph $y = \frac{2}{3}x - 2$ and shade the half-plane above the line (Figure 10). ■

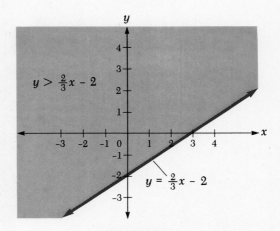

FIGURE 10

Systems of Linear Inequalities

To find the solution set of a system of two or more linear inequalities in two variables, graph *all* of them on the *same* coordinate system. The *intersection* of their solution sets (that is, the region common to all) is the solution of the system.

The following discussion will require that $x \geq 0$. (The solution set consists of the y-axis and the half-plane to the right of the y-axis.) We will also require that $y \geq 0$. (The solution set consists of the x-axis and the half-plane above the x-axis.) The solution set of the system $x \geq 0$, $y \geq 0$ is the first quadrant of the Cartesian plane, the origin, and the nonnegative x- and y-axes. (These restrictions are appropriate since our ultimate goal is to solve linear programming problems.)

Example 2   Graph the system

$$x \geq 0$$
$$y \geq 0$$
$$x + y \leq 4$$
$$2x - y \geq 2$$

Solution   We put $x + y \leq 4$ in slope–intercept form (obtaining $y \leq -x + 4$) and do the same to $2x - y \geq 2$ (obtaining $y \leq 2x - 2$). Now we graph both on the same coordinate system, as shown in Figure 11. (*Note:* If

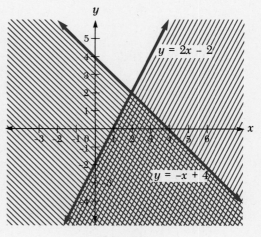

FIGURE 11

you use shading that is parallel to the lines, the intersection is easily seen.)

Since $x \geq 0$ and $y \geq 0$, the solution of the system of all four inequalities is a triangle located in the first quadrant, as shown in Figure 12. (*Note:* The vertex (2, 2) is found by solving the equations $y = -x + 4$ and $y = 2x - 2$ simultaneously by one of the methods presented earlier in this chapter.) ■

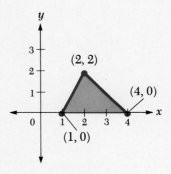

FIGURE 12

Linear Programming

In many business and scientific applications, the mathematician's objective is to optimize (that is, to maximize or minimize) some quantity of interest subject to several conditions (or **constraints**). If the constraints can be expressed as a *system of linear inequalities* and the objective as a *linear equation,* then a technique called **linear programming** can be employed to determine the optimum value of the objective.

Suppose you are the manager of a small company that produces two items, called Standard and Delux. Each Standard item costs $1 to produce, and each Delux item costs $2. Demand for these items is such that you do not want to produce more than 300 items (total). Because of financial considerations, the production cost must be $500 or less. Further, the number of Standard items must be less than or equal to the number of Delux items plus 100. You are subject to the following constraints:

Let x = the number of Delux items

y = the number of Standard items

Then

$$x \geq 0 \atop y \geq 0 \Big\}\quad \text{(The number of items produced} \atop \text{cannot be negative.)}$$

$$2x + 1y \leq 500 \qquad \text{(financial (cost) constraint)}$$

$$x + y \leq 300 \text{ (units)} \qquad \text{(demand constraint)}$$

$$y \leq x + 100 \text{ (units)} \qquad \text{(quantity constraint)}$$

The solution of the system of constraints is shown in Figure 13.

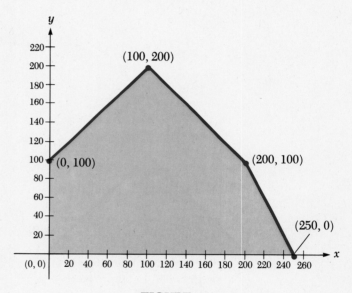

FIGURE 13

Suppose each Delux item sells for $3 and each Standard item sells for $2. You want to maximize your revenue subject to the constraints. That is, you want to

$$\text{maximize} \quad R = 3x + 2y$$

Problem: The solution of the system of constraints contains many combinations of numbers of items (x, y) that you might use. But is there one combination (x, y) that is best? In other words, is there one ordered pair (x, y) in the solution of the constraints that will maximize

R? If so, which one is it? To answer the questions, we need the Fundamental Principle of Linear Programming.

FUNDAMENTAL PRINCIPLE OF LINEAR PROGRAMMING

The optimum (maximum or minimum) value of the objective function (if it exists) occurs at one of the vertices of the solution of the constraints or along a boundary.

(Example 3 will illustrate the case where the optimum is at a single vertex, and Example 4 will illustrate the case where the optimum is along a boundary.)

Therefore, all you need to do is evaluate the objective equation $R = 3x + 2y$ at all the vertices of the solution set of the constraints and select the vertex that yields the largest value of R.

$$\text{At } (0, 0), R = 3(0) + 2(0) = 0$$

$$\text{At } (0, 100), R = 3(0) + 2(100) = 200$$

$$\text{At } (100, 200), R = 3(100) + 2(200) = 700$$

$$\text{At } (200, 100), R = 3(200) + 2(100) = 800$$

$$\text{At } (250, 0), R = 3(250) + 2(0) = 750.$$

Conclusion: You maximize revenue $R = 800$ (dollars) when you produce $x = 200$ Delux items and $y = 100$ Standard items.

Example 3 Small Computer Ltd. purchases microcomponents Al and Bg that cost $4 and $5 each, respectively. The production supervisor reports that a total of 30 or more units is needed for the next production run. Further, the number of Al units plus twice the number of Bg units is at least 40. How many Al and Bg components should be purchased to minimize cost subject to the constraints, and what is the minimum cost?

Solution Let x = the number of Al units

y = the number of Bg units

Objective: cost $C = 4x + 5y$ (dollars)

Constraints: $x \geq 0$

$y \geq 0$

$x + y \geq 30$

$x + 2y \geq 40$

Now find the solution of the system of constraints and determine all the vertices (Figure 14). Next, evaluate $C = 4x + 5y$ at each vertex.

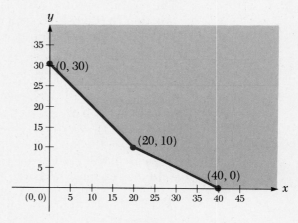

FIGURE 14

At $(0, 30)$, $C = 4(0) + 5(30) = 150$

At $(20, 10)$, $C = 4(20) + 5(10) = 130$

At $(40, 0)$, $C = 4(40) + 5(0) = 160$

Therefore the minimum cost is $C = 130$ (dollars) when the company purchases $x = 20$ Al components and $y = 10$ Bg components. ∎

The linear programming procedure can be summarized as follows:

Step 1: Write the stated problem in mathematical notation.
Step 2: Graph the system of constraints (inequalities).
Step 3: Determine all the vertices of the solution found in Step 2.
Step 4: Evaluate the objective at each vertex.
Step 5: Select the appropriate solution (if one exists).

Of course, a system of inequalities may not have a solution. An example is the system $x + y \geq 2$ and $x + y \leq 1$. Also, two vertices may produce equal optimum values of the objective. In this case all the points on the boundary between those vertices will also yield the same optimum value of the objective. This situation occurs when the slope of the objective equation is equal to the slope of one of the boundary lines of the solution of the constraints.

Example 4 Minimize $z = 2x + 4y$ subject to

$$x \geq 0$$

$$y \geq 0$$

$$3x + 2y \geq 8$$

$$x + 2y \geq 4$$

Solution First we graph the solution set of the constraints and identify all the vertices (Figure 15). Next we evaluate $z = 2x + 4y$ at each vertex:

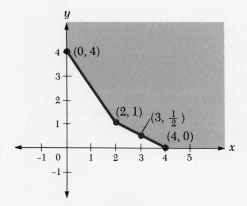

FIGURE 15

$$\text{At } (0, 4), z = 2(0) + 4(4) = 16$$

$$\text{At } (2, 1), z = 2(2) + 4(1) = 8$$

$$\text{At } (4, 0), z = 2(4) + 0 = 8$$

The minimum value of $z = 8$ occurs at $(0, 4)$ and $(2, 1)$. ■

If we choose any point on the boundary between $(4, 0)$ and $(2, 1)$ (such as $(3, \frac{1}{2})$), we will find again that $z = 8$.

Exercises 3.2

In Exercises 1–10 graph the solution set of the inequality. (See Example 1.)

1. $x + y \leq 4$
2. $x + y \geq -2$
3. $x - y \geq 2$
4. $x - y \leq -3$
5. $x + 2y \leq -4$
6. $x + 3y \geq 6$
7. $2x + 3y \geq -6$
8. $2x + 3y \leq -6$
9. $3x - y \leq 5$
10. $3x - y \geq 5$

In Exercises 11–18 graph the system of inequalities. (See Example 2.)

11. $x \geq 0$
    $y \geq 0$
    $x + y \leq 4$

12. $x \geq 0$
    $y \geq 0$
    $x + y \leq 6$

13. $x \geq 0$
    $y \geq 0$
    $x + y \leq 8$
    $x - y \geq 2$

14. $x \geq 0$
    $y \geq 0$
    $x + y \leq 8$
    $x - y \leq 2$

15. $x \geq 0$
    $y \geq 0$
    $x - y \geq 2$

16. $x \geq 0$
    $y \geq 0$
    $x - y \leq -3$

17. $x \geq 0$
    $y \geq 0$
    $x + y \leq 5$
    $2x + 3y \leq 12$

18. $x \geq 0$
    $y \geq 0$
    $x + y \geq 5$
    $2x + 3y \geq 12$

In Exercises 19–30 use graphs to solve the linear programming problem.

19. Maximize $z = x + y$
    subject to $x \geq 0$
    $y \geq 0$
    $3x + y \leq 10$
    $x + 2y \leq 10$

20. Minimize $z = x + y$
    subject to $x \geq 0$
    $y \geq 0$
    $2x + y \geq 10$
    $x + 3y \geq 10$

21. Maximize $z = 6x + y$
    subject to $x \geq 0$
    $y \geq 0$
    $3x + y \leq 10$
    $x + 2y \leq 10$

22. Minimize $z = x + 5y$
    subject to $x \geq 0$
    $y \geq 0$
    $2x + y \geq 10$
    $x + 3y \geq 10$

23. Maximize $z = x + 4y$
    subject to $x \geq 0$
    $y \geq 0$
    $3x + y \leq 10$
    $x + 2y \leq 10$

24. Minimize $z = 3x + y$
    subject to $x \geq 0$
    $y \geq 0$
    $2x + y \geq 10$
    $x + 3y \geq 10$

25. Maximize $z = x + 2y$
    subject to $x \geq 0$
    $y \geq 0$
    $x - 3y \geq -24$
    $2x + y \geq 8$
    $2x + 3y \geq 16$
    $5x + 3y \leq 60$

26. Maximize $z = 3x + y$ using the constraints of Exercise 25.
27. Maximize $z = 20x + 30y$ using the constraints of Exercise 25.
28. Minimize $z = x + 2y$ using the constraints of Exercise 25.
29. Minimize $z = 3x + y$ using the constraints of Exercise 25.

30. Minimize $z = 20x + 30y$ using the constraints of Exercise 25.

31. A company makes items A and B. Each A sells for \$2, and each B sells for \$1. The total production of A's and B's must not exceed 30 units. The number of A's plus twice the number of B's must be less than or equal to 40 units. Find the number of units of each product the company should produce to maximize its revenue. (Let x = the number of A's and y = the number of B's.)

32. In Exercise 31, if the company sells each A for \$2 and each B for \$3, how many units of each should it produce to maximize its revenue?

33. A company produces two similar items A and B. Both cost \$1 each to make. Market conditions indicate that the total number of items should not be less than 100 and not more than 500. Also, the number of A's minus the number of B's should not exceed 100 units. Find the number of A's and B's that would minimize cost.

34. In Exercise 33, if the company wanted to produce the same number of A's and B's, could it? Explain.

35. A furniture company makes two kinds of wooden table legs, one plain and the other containing carved scroll work. The plain legs take 2 hours on a lathe and 1 hour of sanding and carving to make. Each fancy leg requires 1 hour on the lathe and 4 hours of sanding and carving. The company has 4 lathes and 6 sanding and carving machines, all of which are available 12 hours per day. Each plain leg nets \$3 profit, and each fancy leg nets \$5 profit each. If the company can sell all it makes, how should it allocate production to maximize profit?

36. A person needs at least 10 units of vitamin x, 20 units of vitamin y, and 34 units of vitamin z. There are only two foods that contain these vitamins, A and B.

	Food A	Food B
Cost per ounce	6 cents	8 cents
Vitamin x per ounce	1	5
Vitamin y per ounce	5	2
Vitamin z per ounce	6	4

What combination of foods should a person eat to satisfy his or her vitamin requirements at the least cost?

37. A balanced diet should contain vitamins, protein, and minerals. Suppose the minimum daily requirements are 18 units of vitamins, 19 units of protein, and 7 units of minerals. Let the diet consist of vegetables and meat with the following content:

	Vegetable	Meat
Vitamins	3	1
Protein	4	3
Minerals	1	3

If the cost per unit of the vegetable is \$0.50 and the cost per unit of meat is \$1, find the number of units of each that should be bought to minimize cost and satisfy the requirements.

38. A person wants to invest all or part of \$10,000 in stocks that yield 5% and bonds that yield 3%. A financial advisor suggests that no more than \$6000

be put in bonds and $7000 or less in stock. Further, the amount invested in bonds should not exceed half the amount invested in stock. How should the person invest in order to maximize the return and satisfy the restrictions?

39. A company makes two items, A and B, that sell for $4 and $5, respectively. The number of A's plus twice the number of B's must not exceed 20. Twice the number of A's plus the number of B's must be less than or equal to 16. What number of A's and B's produces the maximum revenue?

40. A company makes two items, A and B, that cost $4 and $2, respectively. The total production of A's and B's must not exceed 20 units. The number of A's plus twice the number of B's must be less than or equal to 40 units. Demand is such that the company must produce a total of 5 or more units. Find the number of A's and B's that minimize cost. What is the minimum cost?

41. A chemical company wants to mix nitrogen and potash to make two kinds of fertilizer, N-Rich and P-Rich. Each bag of N-Rich contains 12 ounces of nitrogen and 4 ounces of potash. Each bag of P-Rich contains 2 ounces of nitrogen and 14 ounces of potash. The company has 3600 ounces of nitrogen and 2800 ounces of potash. If a bag of N-Rich sells for $3 and P-Rich for $2, how many bags of each type of fertilizer should be made to maximize the company's revenue? What is the maximum revenue?

42. In Exercise 41, if each bag of N-Rich sold for $10 and each bag of P-Rich for $1, how many bags of each type should be made? How many ounces of potash would be unused?

Key Terms and Formulas

Linear System

Independent

Dependent

Inconsistent

Consistent

Addition method

Substitution method

Inequality

Half-plane

Test point

Linear programming

Objective

Constraints

Vertex

Fundamental Principle of
    Linear Programming

Boundary

Review Exercises

In Exercises 1–6 classify the system by graphing.

1. $x + y = 1$
    $2x - y = 2$

2. $x + y = 1$
    $x + y = 3$

3. $\begin{aligned} 2x - y &= 3 \\ -4x + 2y &= 6 \end{aligned}$

4. $\begin{aligned} 2x - y &= 6 \\ -4x + 2y &= 6 \end{aligned}$

5. $\begin{aligned} x - 3y &= 2 \\ -x + y &= 4 \end{aligned}$

6. $\begin{aligned} x + 4y &= 8 \\ x - 2y &= -4 \end{aligned}$

In Exercises 7–14 solve the system by substitution if possible.

7. $\begin{aligned} x + y &= 1 \\ 2x - y &= 2 \end{aligned}$

8. $\begin{aligned} x + y &= 1 \\ x + y &= 3 \end{aligned}$

9. $\begin{aligned} 2x - y &= 3 \\ -4x + 2y &= 6 \end{aligned}$

10. $\begin{aligned} 2x - y &= 6 \\ -4x + 2y &= 6 \end{aligned}$

11. $\begin{aligned} x - 3y &= 2 \\ -x + y &= 4 \end{aligned}$

12. $\begin{aligned} x + 4y &= 8 \\ x - 2y &= -4 \end{aligned}$

13. $\begin{aligned} 3x + 4y &= 7 \\ 2x + 3y &= 5 \end{aligned}$

14. $\begin{aligned} 4x - 3y &= -1 \\ x + 5y &= 17 \end{aligned}$

In Exercises 15–23 solve the system by addition if possible.

15. $\begin{aligned} x + y &= 1 \\ 2x - y &= 2 \end{aligned}$

16. $\begin{aligned} x + y &= 1 \\ x + y &= 3 \end{aligned}$

17. $\begin{aligned} 2x - y &= 3 \\ -4x + 2y &= 6 \end{aligned}$

18. $\begin{aligned} 2x - y &= 6 \\ -4x + 2y &= 6 \end{aligned}$

19. $\begin{aligned} x - 3y &= 2 \\ -x + 4y &= 4 \end{aligned}$

20. $\begin{aligned} x + 4y &= 8 \\ x - 2y &= -4 \end{aligned}$

21. $\begin{aligned} 3x + 4y &= 7 \\ 2x + 3y &= 5 \end{aligned}$

22. $\begin{aligned} 4x - 3y &= -1 \\ x + 5y &= 17 \end{aligned}$

C 23. $\begin{aligned} 0.1x + 0.3y &= 4 \\ 0.2x - 0.4y &= -2 \end{aligned}$

24. In Exercises 1–23 you solved many problems by all methods. Comment on the relative merits of graphing, addition, and substitution.

In Exercises 25–28 solve the system by the addition method.

25. $\begin{aligned} x - 2y + 3z &= 11 \\ 3x + y + z &= 10 \\ x - y - 4z &= -4 \end{aligned}$

26. $\begin{aligned} x - 2y + 3z &= 4 \\ 3x + y + z &= 1 \\ x - y - z &= -9 \end{aligned}$

27. $\begin{aligned} x + y \quad\;\; &= 2 \\ y + z &= 3 \\ x \quad\;\; + z &= -5 \end{aligned}$

C 28. $\begin{aligned} 0.2x - 0.1y + 0.3z &= 2.1 \\ 0.1x + 0.1y - 0.2z &= -0.3 \\ 0.3x - 0.3y + 0.1z &= 0.3 \end{aligned}$

29. The equation $y = mx + b$ is satisfied by the (x, y) values $(1, 5)$ and $(-2, -1)$. Find the equation.

30. The equation $y = mx + b$ is satisfied by the (x, y) values $(1, -2)$ and $(4, 1)$. Find the equation.

31. A solution of 28% alcohol is combined with a solution of 8% alcohol to produce 80 cubic centimeters of solution that is 10% alcohol. How many cubic centimeters of each solution is used?

32. The perimeter of a rectangle is 54 meters. If the width is doubled, the new perimeter will be 68 meters. Find the length and width of the original rectangle.

33. A family has invested $20,000 in two securities. One pays 15% per year, and the other pays 12% per year. If the interest for one year is $2580, how much is invested in each?

34. In Exercise 33, if the interest for 1 year is $3000, how is the money invested?

In Exercises 35–42 solve the linear programming problem.

35. Maximize $z = 5x + 5y$
    subject to $x \geq 0$
    $y \geq 0$
    $x + y \leq 4$
    $2x - y \geq 2$

36. Maximize $z = 20x + y$
    subject to $x \geq 0$
    $y \geq 0$
    $x + y \leq 4$
    $2x - y \geq 2$

37. Maximize $z = x + 20y$
    subject to $x \geq 0$
    $y \geq 0$
    $x + y \leq 4$
    $2x - y \geq 2$

38. Comment on the effects of changing the objective function in Exercises 35, 36, and 37.

39. Minimize $z = 2x + 3y$
    subject to $x \geq 0$
    $y \geq 0$
    $3x + 2y \geq 12$
    $x + 2y \geq 8$

40. Minimize $z = x + 5y$
    subject to $x \geq 0$
    $y \geq 0$
    $3x + 2y \geq 12$
    $x + 2y \geq 8$

41. Minimize $z = 6x + y$
    subject to $x \geq 0$
    $y \geq 0$
    $3x + 2y \geq 12$
    $x + 2y \geq 8$

42. Minimize $z = 5x + 3y$
    subject to $x \geq 0$
    $y \geq 0$
    $x + y \leq 4$
    $2x - y \geq 2$

CHAPTER 4
Second-Degree Equations and Inequalities

4.1 Quadratic Functions and Equations

In this section we discuss three methods that are useful in solving a **quadratic equation** of the form $0 = Ax^2 + Bx + C$ where A, B, and C are constants and $A \neq 0$. The methods are (1) factoring, (2) completing the square, and (3) the quadratic formula. We then consider the general quadratic function $y = Ax^2 + Bx + C$ and discuss the characteristics of its graph.

Method 1: Factoring

The method of solving a quadratic equation by factoring depends on the well-known property that the product of two real numbers cannot be 0 unless at least one of the original numbers is zero. Thus if $A \cdot B = 0$, then $A = 0$ or $B = 0$ (or both). Of course, the factoring method is useful only when the quadratic equation can be factored.

Example 1   Solve for x: $x^2 + 3x + 2 = 0$

Solution   We factor the quadratic equation and replace the quadratic with its factors:

$$x^2 + 3x + 2 = (x + 1)(x + 2)$$

$$(x + 1)(x + 2) = 0$$

Next we equate each factor to 0:

$$x + 1 = 0 \quad \text{or} \quad x + 2 = 0$$

We solve each linear equation for x:

$$x = -1 \quad \text{or} \quad x = -2$$

Finally, we check these answers in the original equation:

$$
\begin{array}{ll}
x = -1: & x = -2: \\
(-1)^2 + 3(-1) + 2 \overset{?}{=} 0 & (-2)^2 + 3(-2) + 2 \overset{?}{=} 0 \\
+1 - 3 + 2 \overset{?}{=} 0 & 4 - 6 + 2 \overset{?}{=} 0 \\
0 \overset{?}{=} 0 & 0 \overset{?}{=} 0
\end{array}
$$

Thus the solutions to the equation are $\boxed{x = -1}$ and $\boxed{x = -2}$. ∎

Quadratic equations are useful in solving many types of applied problems. The next example demonstrates the process.

Example 2   The base of a triangle is 3 meters longer than its height, and the area of the triangle is 77 square meters. Find the base and height of the triangle.

Solution   Let x = the height, so that $x + 3$ = the base. The formula for the area of a triangle is

$$\text{area} = \frac{1}{2} \cdot \text{base} \cdot \text{height}$$

$$77 = \frac{1}{2}(x + 3)x$$

$$0 = x^2 + 3x - 154$$

To solve this equation, we factor and set the factors equal to 0:

$$0 = (x - 11)(x + 14)$$

$$x - 11 = 0 \quad \text{or} \quad x + 14 = 0$$

$$x = 11 \quad \text{or} \quad x = -14$$

The solution is $x = 11$, so the height is 11 and $x + 3 = 11 +$

$3 =$ 14 is the base . Note that $x = -14$ is not a solution because the

height of a triangle cannot be negative. ∎

The factoring method works for any quadratic equation that factors. The next example illustrates the method applied to a quadratic containing a common factor and to a quadratic in the form of the difference of two squares.

Example 3 Solve for y:
(a) $y^2 + 2y = 0$ (b) $y^2 - 9 = 0$

Solution (a) We take out the common factor, set the factors equal to 0, and solve:

$$y(y + 2) = 0$$
$$y = 0 \quad \text{or} \quad y + 2 = 0$$
$$y = 0 \quad \text{or} \quad y = -2$$

The solutions are $y = 0$ and $y = -2$.

(b) We show two ways to solve this equation.
Solution 1:

$$y^2 - 9 = 0$$
$$(y + 3)(y - 3) = 0 \qquad \text{(factor)}$$
$$y + 3 = 0 \quad \text{or} \quad y - 3 = 0 \qquad \text{(equate factors to 0)}$$
$$y = -3 \quad \text{or} \quad y = 3 \qquad \text{(solve each equation)}$$

Solution 2:

$$y^2 = 9 \qquad \text{(rewrite)}$$
$$\sqrt{y^2} = \pm\sqrt{9} \qquad \text{(take square root of each side)}$$
$$y = \pm 3 \qquad \text{(simplify)} \quad ∎$$

Example 3(b) illustrates an alternate method for solving a quadratic equation, which leads us to Method 2.

Method 2: Completing the Square

When a quadratic equation can be rewritten in the form $(ax + b)^2 = c$ (and it always can), the solutions can easily be determined, as the next example shows. Recall that $\sqrt{-1} = i$ and $-1 = i^2$.

Example 4 Solve for x: $(2x - 3)^2 = -18$.

Solution We take square root of each side and simplify:

$$2x - 3 = \pm\sqrt{-18}$$
$$2x - 3 = \pm\sqrt{-1 \cdot 9 \cdot 2}$$
$$2x - 3 = \pm 3\sqrt{2}i$$
$$2x = 3 \pm 3\sqrt{2}i$$
$$x = \frac{3}{2} \pm \frac{3\sqrt{2}}{2}i$$

Check:

$$\left[2\left(\frac{3}{2} \pm \frac{3\sqrt{2}}{2}i\right) - 3\right]^2 \stackrel{?}{=} -18$$
$$(3 \pm 3\sqrt{2}i - 3)^2 \stackrel{?}{=} -18$$
$$(\pm 3\sqrt{2}i)^2 \stackrel{?}{=} -18$$
$$-18 \stackrel{\checkmark}{=} -18$$

The solutions are $\dfrac{3}{2} \pm \dfrac{3\sqrt{2}}{2}i$. ■

Let us now extend the process used in Example 4 to a quadratic equation. As we will see in the next example, the quadratic must be properly "adjusted" or rewritten in the required form. The complete sequence of steps required for the method of completing the square will be given in Example 7.

Example 5 Solve for z: $z^2 + 6z + 5 = 0$.

Solution We begin by rewriting the left side so that it involves a perfect square trinomial: $z^2 + 6z + 5 = (z^2 + 6z + 9) - 4$. We can now factor the perfect square trinomial:

$$(z + 3)^2 - 4 = 0$$

$$(z + 3)^2 = 4$$

$$z + 3 = \pm 2$$

$$z = -3 \pm 2$$

$$z = -1 \quad \text{or} \quad z = -5$$

The solutions are $z = -1$ and $z = -5$. ∎

Note: The method used in Example 5 required a perfect square, $z^2 + 6z + 9 = (z + 3)^2$. Thus we needed to replace 5 with $9 - 4$. There is a very reasonable way to recognize how to accomplish this.

Consider a perfect square $(x + m)^2$ expanded to $x^2 + (2m)x + m^2$. When the coefficient of the x^2 term is 1, we see that the last term m^2 is $\left[\dfrac{1}{2}(2m) \right]^2$, where $2m$ is the coefficient of the middle term. This gives the formula for completing the square, as we see in the next example.

Example 6   Solve for x: $2x^2 + 6x - 7 = 0$.

Solution   *Step 1:* Divide by the coefficient of x^2:

$$x^2 + 3x - \frac{7}{2} = 0$$

Step 2: Rewrite by subtracting the constant term from both sides:

$$x^2 + 3x = \frac{7}{2}$$

Step 3: Compute the required constant, $\left(\dfrac{1}{2} \cdot 3 \right)^2$, and add to both sides:

$$x^2 + 3x + \left(\frac{3}{2} \right)^2 = \frac{7}{2} + \left(\frac{3}{2} \right)^2$$

Step 4: Factor and simplify:

$$\left(x + \frac{3}{2} \right)^2 = \frac{14 + 9}{4}$$

$$\left(x + \frac{3}{2} \right)^2 = \frac{23}{4}$$

Step 5: Solve:

$$x + \frac{3}{2} = \pm \frac{\sqrt{23}}{2}$$

$$x = -\frac{3}{2} \pm \frac{\sqrt{23}}{2}$$

Step 6: The solutions are $x = \dfrac{-3 \pm \sqrt{23}}{2}$. ∎

The technique used in Example 6 is called solving a quadratic equation by **completing the square.**
We now solve the general quadratic equation by this method.

Example 7 Solve the general quadratic equation $Ax^2 + Bx + C = 0$ by the method of completing the square.

Solution *Step 1:* Divide by the coefficient of x^2:

$$x^2 + \frac{B}{A} x + \frac{C}{A} = 0$$

Step 2: Rewrite by removing the constant term:

$$x^2 + \frac{B}{A} x = -\frac{C}{A}$$

Step 3: Compute the correct constant, $\left(\frac{1}{2} \cdot \frac{B}{A}\right)^2$, and add to both sides:

$$x^2 + \frac{B}{A} x + \left(\frac{B}{2A}\right)^2 = -\frac{C}{A} + \left(\frac{B}{2A}\right)^2$$

Step 4: Factor and simplify:

$$\left(x + \frac{B}{2A}\right)^2 = \frac{B^2 - 4AC}{4A^2}$$

Step 5: Solve:

$$x + \frac{B}{2A} = \pm \frac{\sqrt{B^2 - 4AC}}{2A}$$

$$x = -\frac{B}{2A} \pm \frac{\sqrt{B^2 - 4AC}}{2A}$$

Step 6: The solutions are

$$x = \frac{-B + \sqrt{B^2 - 4AC}}{2A} \quad \text{and} \quad x = \frac{-B - \sqrt{B^2 - 4AC}}{2A} \quad \blacksquare$$

Method 3: The Quadratic Formula

Using the results of Example 7, we have that the solutions to $Ax^2 + Bx + C = 0$ are

$$x = \frac{-B \pm \sqrt{B^2 - 4AC}}{2A}$$

This result is called the **quadratic formula.** We can use this formula directly, as the next example shows.

Example 8 Use the quadratic formula to solve for x in $3x^2 - 4x + 5 = 0$.

Solution We first determine the values of A, B, and C:

$$A = 3, \qquad B = -4, \qquad C = 5$$

Substitute into the formula

$$x = \frac{-B \pm \sqrt{B^2 - 4AC}}{2A}$$

$$x = \frac{-(-4) \pm \sqrt{(-4)^2 - 4(3)(5)}}{2(3)}$$

Simplify:

$$x = \frac{4 \pm \sqrt{16 - 60}}{6} = \frac{4 \pm \sqrt{-44}}{6} = \frac{4 \pm 2i\sqrt{11}}{6} = \frac{2 \pm i\sqrt{11}}{3}$$

Thus the solutions are $\boxed{x = \dfrac{2 \pm i\sqrt{11}}{3}}$, or $\boxed{x = \dfrac{2}{3} \pm \dfrac{\sqrt{11}}{3}i}$. $\blacksquare$

By examining the quadratic formula we see that the expression $B^2 - 4AC$ actually determines the *nature of the roots,* that is, whether the roots will be real and equal ($B^2 - 4AC = 0$), real and unequal ($B^2 - 4AC > 0$), or complex ($B^2 - 4AC < 0$). The value $B^2 - 4AC$ is called the **discriminant** of the quadratic equation.

We now consider the general quadratic function $y = Ax^2 + Bx + C$

and the properties of its graph. The three methods previously discussed allow us to solve for the values of x which produce a y-value of 0. These x-values are called the **zeros** or **roots** of the equation. When graphing the equation these x-values are also called **x-intercepts**. Conversely, the **y-intercept** is produced by substituting an x-value of 0 in the equation. The next two examples illustrate the fact that the graph of a quadratic always has a U-shaped curve which goes upward when $A > 0$ and downward when $A < 0$. The vertex of the graph occurs when $x = -B/(2A)$.

The example which follows shows that $A < 0$ gives a downward graph. It also shows that the formula $x = -B/(2A)$ gives the x-coordinate of the vertex.

Example 9   Graph $y = 1 - 2x - 3x^2$.

Solution   *Step 1:* Solve for the x-intercepts. (Set $y = 0$ and solve.)

$$1 - 2x - 3x^2 = 0$$

$$(1 - 3x)(1 + x) = 0$$

$$1 - 3x = 0 \quad \text{or} \quad 1 + x = 0$$

$$x = \frac{1}{3} \quad \text{or} \quad x = -1$$

Step 2: Solve for the y-intercept. (Substitute $x = 0$ in the equation.)

$$y = 1 - 2 \cdot 0 - 3 \cdot 0^2$$

$$y = 1$$

Step 3: Solve for the vertex by completing the square [or by using $x = -B/(2A)$].

$$y = 1 - 2x - 3x^2$$

$$\frac{y}{-3} = x^2 + \frac{2x}{3} - \frac{1}{3}$$

$$\frac{y}{-3} + \frac{1}{3} + \left(\frac{1}{3}\right)^2 = x^2 + \frac{2}{3}x + \left(\frac{1}{3}\right)^2$$

$$\frac{y}{-3} = \left(x + \frac{1}{3}\right)^2 - \frac{4}{9}$$

$$y = -3\left(x + \frac{1}{3}\right)^2 + \frac{4}{3}$$

The largest value for y will occur when $x + \frac{1}{3} = 0$, so $x = -\frac{1}{3}$,

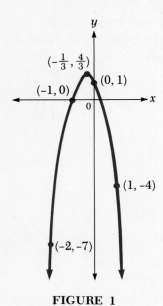

$(-\frac{1}{3}, \frac{4}{3})$

$(0, 1)$

$(-1, 0)$

$(1, -4)$

$(-2, -7)$

FIGURE 1

$y = \frac{4}{3}$ is the vertex. (Note that $x = -B/(2A)$ directly gives $x = -\frac{1}{3}$, so that completing the square is not necessary.)

Step 4: Build a table of values by substituting a range of x-values into the equation.

x	y
-2	-7
-1	0
0	1
1	-4
2	-15

Step 5: Plot the intercepts, vertex, and additional table values. Then join the points plotted, as in Figure 1. ∎

Next we consider an example where $A > 0$. Further, we use the formula $x = -B/(2A)$ (without completing the square) to get the vertex.

Example 10   Graph $y = x^2 + 3x - 4$.

Solution   *Step 1:* Solve for the x-intercepts by setting $y = 0$ and using one of the three previous methods to solve for the x-values.

$$0 = x^2 + 3x - 4$$

$$0 = (x - 1)(x + 4)$$

$$x - 1 = 0 \quad \text{or} \quad x + 4 = 0$$

$$x = 1 \quad \text{or} \quad x = -4$$

Step 2: Solve for the y-intercept by setting $x = 0$.

$$y = 0^2 + 3 \cdot 0 - 4$$

$$y = -4$$

Step 3: Solve for the vertex using $x = -B/(2A)$ and substituting to get the y-value.

$$x = \frac{-3}{2 \cdot 1} = -\frac{3}{2}$$

$$y = \left(-\frac{3}{2}\right)^2 + 3\left(-\frac{3}{2}\right) - 4$$

$$y = \frac{-25}{4}$$

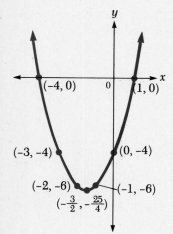

FIGURE 2

Step 4: Substitute additional values into the equation to build a table of values. Use the vertex as the middle value in the chart.

x	y
-4	0
-3	-4
-2	-6
$-\dfrac{3}{2}$	$-\dfrac{25}{4}$
-1	-6
0	-4
1	0

Step 5: Plot the x-intercepts, y-intercept, and additional points from Step 4. Then join the points plotted, as in Figure 2. ∎

Recalling our discussion of the discriminant of a quadratic equation, we have the following:

1. $B^2 - 4AC > 0$ means two real roots.
2. $B^2 - 4AC = 0$ means one real root.
3. $B^2 - 4AC < 0$ means no real roots.

Since the roots of the equation are the x-intercepts, we can have the six general types of graphs, as shown in Figure 3.

We have shown how to find the roots of a quadratic equation and how to graph a quadratic. Our final example in this section shows how to find the quadratic if the roots are known.

Example 11   Find a quadratic equation which has roots $x = 2$ and $x = -3$.

Solution   The process we use is the reverse of the steps used in solving a quadratic by factoring. First, write the roots:

$$x = 2 \quad \text{and} \quad x = -3$$

Then rewrite each as a linear equation and multiply:

$$x - 2 = 0 \quad \text{and} \quad x + 3 = 0$$

$$(x - 2)(x + 3) = 0$$

Expand to obtain the quadratic equation:

$$x^2 + x - 6 = 0 \quad ∎$$

$A > 0, B^2 - 4AC < 0$

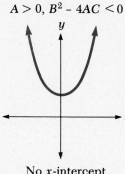

No x-intercept

$A > 0, B^2 - 4AC = 0$

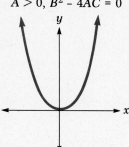

One x-intercept

$A > 0, B^2 - 4AC > 0$

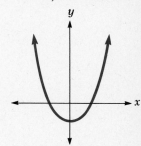

Two x-intercepts

$A < 0, B^2 - 4AC < 0$

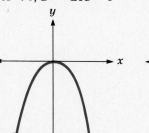

No x-intercept

$A < 0, B^2 - 4AC = 0$

One x-intercept

$A < 0, B^2 - 4AC > 0$

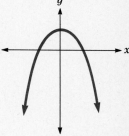

Two x-intercepts

FIGURE 3

In general, a quadratic equation having roots r_1 and r_2 is $x^2 - (r_1 + r_2)x + r_1 r_2 = 0$. For example, if $r_1 = 5$ and $r_2 = 6$, then a quadratic equation with these roots is $x^2 - 11x + 30 = 0$ ($11 = 5 + 6$, $30 = 5 \cdot 6$).

Exercises 4.1

In Exercises 1–30 solve the quadratic equation by factoring.

1. $x^2 + 7x + 12 = 0$
2. $x^2 + 7x + 10 = 0$
3. $y^2 + 4y = -3$
4. $x^2 - 7x + 10 = 0$
5. $x^2 - 15x + 56 = 0$
6. $z^2 - 7z = -12$
7. $x^2 - 3x - 18 = 0$
8. $x^2 + 3x - 28 = 0$
9. $y^2 + 3y = 18$
10. $x^2 - 6x + 9 = 0$
11. $x^2 + 8x + 16 = 0$
12. $x^2 + 1 = -2x$
13. $z^2 - 4z = 0$
14. $2x^2 - 3x = 0$
15. $2x^2 = -7x$
16. $2y^2 + 3y + 1 = 0$
17. $3x^2 + 11x = -6$
18. $2z^2 - 17z + 35 = 0$

19. $3x^2 = 11x - 6$
20. $4x^2 - 7x - 15 = 0$
21. $5x^2 - 2 = -9x$
22. $x^2 - 16 = 0$
23. $y^2 = 25$
24. $25x^2 - 9 = 0$
25. $4x^2 = 1$
26. $z^2 = 5$
27. $2x^2 - 3x = 0$
28. $x^2 = -1$
29. $(x - 2)^2 = 9$
30. $(2x - 1)^2 = -4$

In Exercises 31–40 solve by completing the square.
31. $x^2 + 4x - 5 = 0$
32. $y^2 + 2y - 8 = 0$
33. $2z^2 + z - 3 = 0$
34. $3x^2 + 5x = -2$
35. $2z^2 = 3z + 2$
36. $x^2 + 2x - 1 = 0$
37. $y^2 = y - 1$
38. $3z^2 + 2z = -1$
39. $x^2 - 7 = 0$
40. $2x^2 - 3x = 0$

In Exercises 41–52
(a) determine the nature of the roots using the discriminant;
(b) solve the equation using the quadratic formula.
41. $x^2 = 8x - 15$
42. $2y^2 + 5y + 2 = 0$
43. $2z^2 + 5z = 12$
44. $3x^2 = 13x + 10$
45. $3y^2 + 6 = -11y$
46. $6z^2 - 7z + 2 = 0$
47. $x^2 + 2x - 1 = 0$
48. $x^2 - 2x - 1 = 0$
49. $2y^2 = -3y + 1$
50. $x^2 = 6x - 10$
51. $x^2 + x + 1 = 0$
52. $2z^2 = -3$

In Exercises 53–60
(a) find the x-intercepts (if any);
(b) find the y-intercept;
(c) find the vertex;
(d) draw the graph.
53. $y = x^2 - 4x - 5$
54. $y = x^2 - 6x - 7$
55. $y = x^2 - x - 6$
56. $y = x^2 + 3x - 4$
57. $y = 5 + 4x - x^2$
58. $y = 2 - x - x^2$
59. $y = x^2 + x + 4$
60. $y = -10 + 2x - x^2$

In Exercises 61–75 write the equation and solve.
61. The base of a triangle is 4 meters longer than its height, and the area of the triangle is 6 square meters. Find the base and height of the triangle.
62. The base of a triangle is 5 meters shorter than its height, and the area of the triangle is 25 square meters. Find the base and height of the triangle.
63. The length of a rectangle exceeds its width by 4 units, and the area of the rectangle is 21 square units. Find the length and width of the rectangle.
64. The width of a rectangle is 5 units less than its length, and the area is 36 square units. Find the length and width of the rectangle.
65. Find two consecutive positive integers the sum of whose squares is 25.
66. Find two consecutive negative integers the sum of whose squares is 41.
67. Find two numbers whose sum is 10 and whose product is 24.
68. Find two numbers whose sum is 4 and whose product is $3\frac{3}{4}$.
69. Find the dimensions of a square whose diagonal is one unit longer than a side. (*Hint:* Use the Pythagorean theorem.)

70. Find the dimensions of a square whose diagonal is 2 units longer than a side.
71. A polygon having x sides has $\frac{1}{2}x(x-3)$ diagonals. How many sides has a polygon with 9 diagonals?
72. If a polygon has 54 diagonals, how many sides does it have?
73. Is there a polygon having 12 diagonals? Explain your answer.
74. The cost c per unit of producing x items is $c = x^2 - 6x + 12$.
    (a) Find the minimum cost.
    (b) Graph the equation for $x \geq 0$.
75. The cost c per unit of producing x items is $c = x^2 - 8x + 20$.
    (a) Find the minimum cost.
    (b) Graph the equation for $x \geq 0$.

4.2   Radical Equations and Quadratic Forms

Frequently we encounter equations that can be either reduced or rewritten in the form of a quadratic equation and hence can be solved by the methods of the previous section.

The first of these is a **radical equation,** that is, an equation containing a variable which is under a radical sign. The solution depends on the property that if two numbers A and B are equal, then $A^2 = B^2$, and further, the solutions of $A^2 = B^2$ are $A = B$ and $A = -B$. We show how this property is used in solving the next example.

Example 1 Solve for x: $\sqrt{x + 2} = x$.

Solution We treat the equation as $A = B$ and square both sides:
$$(\sqrt{x + 2})^2 = x^2$$

Next, we simplify, rewrite in quadratic form, and use a method of Section 4.1 to solve. (In this case the equation factors.)

$$x + 2 = x^2$$
$$0 = x^2 - x - 2$$
$$0 = (x - 2)(x + 1)$$
$$x = 2 \quad \text{or} \quad x = -1$$

It is important to check the results. (As explained earlier, the solutions of $A^2 = B^2$ are $A = B$ and $A = -B$, so that it is possible that not all of the answers obtained are true solutions.)

Check $x = 2$:   *Check* $x = -1$:

$\sqrt{2 + 2} \overset{?}{=} 2$ $\sqrt{-1 + 2} \overset{?}{=} -1$

$\sqrt{4} \overset{?}{=} 2$ $\sqrt{1} \overset{?}{=} -1$

$2 \overset{\checkmark}{=} 2$ $1 \neq -1$

$x = 2$ is a solution $x = -1$ is *not* a solution

The solution is $x = 2$.   ■

An x-value obtained in solving an equation which is not a true solution to the problem is sometimes called an **extraneous solution.**

The method of solving a radical equation by squaring both sides works most efficiently when a radical term is first isolated. The process of squaring both sides of the equation should be continued until all radicals are eliminated.

Example 2   Solve for x: $\sqrt{3x + 4} - \sqrt{x} = 2$

Solution   We rewrite the equation, isolating a radical term:

$$\sqrt{3x + 4} = \sqrt{x} + 2$$

Now we square both sides and simplify:

$$(\sqrt{3x + 4})^2 = (\sqrt{x} + 2)^2$$
$$3x + 4 = x + 4\sqrt{x} + 4$$

Since the resulting equation contains a radical, we isolate it, square again, and simplify:

$$2x = 4\sqrt{x}$$
$$(2x)^2 = (4\sqrt{x})^2$$
$$4x^2 = 16x$$

We can rewrite this as a quadratic equation and solve:

$$4x^2 - 16x = 0$$
$$4x(x - 4) = 0$$
$$x = 0 \quad \text{or} \quad x = 4$$

Check x = 0:   *Check x = 4:*

$$\sqrt{3 \cdot 0 + 4} - \sqrt{0} \overset{?}{=} 2 \qquad \sqrt{3 \cdot 4 + 4} - \sqrt{4} \overset{?}{=} 2$$

$$\sqrt{4} - 0 \overset{?}{=} 2 \qquad\qquad \sqrt{16} - 2 \overset{?}{=} 2$$

$$2 \overset{\checkmark}{=} 2 \qquad\qquad\qquad 4 - 2 \overset{?}{=} 2$$

$x = 0$ is a solution $\qquad\qquad\qquad 2 \overset{\checkmark}{=} 2$

$$x = 4 \text{ is a solution}$$

The solutions are $\boxed{x = 0}$ and $\boxed{x = 4}$ ■

    The second type of equation we consider is one which requires substitution. The equation $y - 3\sqrt{y} + 2 = 0$ is very similar to $x^2 - 3x + 2 = 0$, as we see in the next example.

Example 3   Solve for y: $y - 3\sqrt{y} + 2 = 0$.

Solution   Since $y = (\sqrt{y})^2$, we can rewrite the equation as

$$(\sqrt{y})^2 - 3(\sqrt{y}) + 2 = 0$$

Next we substitute x for $\sqrt{y}$ and solve for x using a quadratic method:

$$x^2 - 3x + 2 = 0$$

$$(x - 1)(x - 2) = 0$$

$$x = 1 \quad \text{or} \quad x = 2$$

Now we resubstitute $\sqrt{y}$ for x and solve for y:

$$\sqrt{y} = 1 \quad \text{or} \quad \sqrt{y} = 2$$

$$y = 1 \quad \text{or} \quad y = 4$$

Check y = 1:   *Check y = 4:*

$$1 - 3\sqrt{1} + 2 \overset{?}{=} 0 \qquad 4 - 3\sqrt{4} + 2 \overset{?}{=} 0$$

$$1 - 3 + 2 \overset{?}{=} 0 \qquad 4 - 3 \cdot 2 + 2 \overset{?}{=} 0$$

$$0 \overset{\checkmark}{=} 0 \qquad\qquad 4 - 6 + 2 \overset{?}{=} 0$$

$y = 1$ is a solution $\qquad\qquad 0 \overset{\checkmark}{=} 0$

$$y = 4 \text{ is a solution}$$

The solutions are $\boxed{y = 1}$ and $\boxed{y = 4}$.   ■

Note: The key to recognizing when this method is appropriate is the following: The power of the variable of one term in the expression must be twice the power of the variable in another term.

Example 4   Rewrite each of the following as a quadratic equation:
(a) $y^3 - 4y^{3/2} + 5 = 0$ (b) $(y + 1)^2 + 6(y + 1) - 7 = 0$
(c) $y^{2/3} - 7y^{1/3} + 8 = 0$

Solution   (a) Since $(y^{3/2})^2 = y^3$, we have $(y^{3/2})^2 - 4(y^{3/2}) + 5 = 0$, or let $x = y^{3/2}$

to obtain $x^2 - 4x + 5 = 0$.

(b) Let $x = y + 1$ to obtain $x^2 + 6x - 7 = 0$.

(c) Since $(y^{1/3})^2 = y^{2/3}$, we have $(y^{1/3})^2 - 7(y^{1/3}) + 8 = 0$ or let $x = y^{1/3}$

to obtain $x^2 - 7x + 8 = 0$.   ■

Quadratic forms are useful in many areas. The following example illustrates a result about higher-degree equations, which we will consider in a later chapter.

Example 5   Find the four fourth roots of 625.

Solution   We will solve the equation $x^4 = 625$. We can rewrite this as a quadratic equation by letting $y = x^2$:

$$x^4 - 625 = 0 \quad \text{gives} \quad y^2 - 625 = 0$$

Now we solve for y:

$$(y - 25)(y + 25) = 0$$
$$y = 25 \quad \text{or} \quad y = -25$$

Finally, we replace y by x^2 and solve:

$$x^2 = 25 \qquad x^2 = -25$$
$$x = \pm 5 \qquad x = \pm 5i$$

Thus the four roots are $5, -5, 5i$ and $-5i$.   ■

Note that the number of roots (four) in Example 5 is the same as the degree of the equation.

A quadratic equation may also occur in solving a fractional equation.

Example 6 Solve for x: $\dfrac{x}{x-1} + \dfrac{1}{x+2} = \dfrac{3}{x^2+x-2}$.

Solution The L.C.D. is $(x-1)(x+2)$. We multiply both sides of the equation by the L.C.D., simplify, and solve the resulting quadratic equation:

$$(x-1)(x+2)\frac{x}{x-1} + (x-1)(x+2)\frac{1}{x+2}$$

$$= (x-1)(x+2)\frac{3}{x^2+x-2}$$

$$(x+2)x + (x-1)\cdot 1 = 3$$

$$x^2 + 2x + x - 1 = 3$$

$$x^2 + 3x - 4 = 0$$

$$(x-1)(x+4) = 0$$

$$x = 1 \quad \text{or} \quad x = -4$$

Check $x = 1$:

$$\frac{1}{1-1} + \frac{1}{1+2} \overset{?}{=} \frac{3}{1^2+1-2}$$

Since the undefined quantity $1/0$ results, $x = 1$ is *not* a solution.

Check $x = -4$:

$$\frac{-4}{-4-1} + \frac{1}{-4+2} \overset{?}{=} \frac{3}{(-4)^2-4-2}$$

$$\frac{-4}{-5} + \frac{1}{-2} \overset{?}{=} \frac{3}{10}$$

$$\frac{8}{10} + \frac{-5}{10} \overset{?}{=} \frac{3}{10}$$

$$\frac{3}{10} \overset{\checkmark}{=} \frac{3}{10}$$

$$x = -4 \text{ is a solution}$$

The solution is $\boxed{x = -4}$. ∎

Exercises 4.2 In Exercises 1–43 solve the equation and check your solutions.

1. $\sqrt{x} - 3 = 0$ 2. $\sqrt{y} + 5 = 0$

3. $\sqrt{z} - 4 = 0$ 4. $\sqrt{x} + 2 = 8$

5. $\sqrt{y} - 6 = 7$ 6. $\sqrt{z} + 7 = 13$

7. $\sqrt{x + 3} = x + 1$

8. $\sqrt{x - 4} = x - 4$

9. $\sqrt{3y + 4} = y$

10. $\sqrt{x - 2} + x = 2$

11. $\sqrt{2x + 8} - x = 0$

12. $x + \sqrt{x + 5} = 1$

13. $\sqrt{x^2 - 7x - 4} = 2$

14. $\sqrt{x - 1} + \sqrt{x + 2} = 1$

15. $\sqrt{3x + 1} - 2 = \sqrt{2x - 6}$

16. $\sqrt{4x + 17} + \sqrt{x + 1} = 4$

17. $\sqrt{2x - 2} - \sqrt{x} = 1$

18. $\sqrt{3x - 3} = \sqrt{x} + 1$

19. $x - 3\sqrt{x} + 2 = 0$

20. $y - 2\sqrt{y} - 15 = 0$

21. $z + \sqrt{z} - 12 = 0$

22. $x^4 - 13x^2 + 36 = 0$

23. $y^4 - 5y^2 + 4 = 0$

24. $z^4 - 10z^2 + 9 = 0$

25. $x^4 + 3x^2 = 4$

26. $y^4 + 10y^2 = -9$

27. $z^4 = -5z^2 - 4$

28. $x^3 - 5x^{3/2} = -4$

29. $x^{1/2} - 3x^{1/4} + 2 = 0$

30. $x^3 - 3x^{3/2} + 2 = 0$

31. $(y + 1)^2 - 5(y + 1) + 6 = 0$

32. $(x - 2)^2 - 2(x - 2) - 15 = 0$

33. $2(z + 1)^2 - 7(z + 1) + 6 = 0$

34. $\dfrac{x}{x + 4} + \dfrac{2}{x - 3} = \dfrac{20}{(x + 4)(x - 3)}$

35. $\dfrac{2x}{x + 3} - \dfrac{1}{x - 1} = \dfrac{-4}{(x + 3)(x - 1)}$

36. $\dfrac{x}{x - 4} + \dfrac{1}{x + 1} = \dfrac{-5}{(x - 4)(x + 1)}$

37. $\dfrac{x}{x + 1} - \dfrac{2}{x + 2} = \dfrac{2}{x^2 + 3x + 2}$

38. $\dfrac{3x}{x + 5} + \dfrac{2}{x - 1} = \dfrac{20}{x^2 + 4x - 5}$

39. $(2y - 3)^2 - 6(2y - 3) + 7 = 0$

40. $(x - 5) + 2\sqrt{x - 5} = 8$

41. $3z^{1/2} - 5z^{1/4} - 2 = 0$

42. $(z - 2)^2 - 3 = -2(z - 2)$

43. $\dfrac{2}{3y - 1} + \dfrac{3y}{2y - 5} = 0$

44. Find the four fourth roots of 16.
45. Find the four fourth roots of 81.
46. Find the four fourth roots of 1.
47. Find two numbers whose sum is 20 and whose product is 96.
48. Find two numbers whose sum is 12 and whose product is 35.

4.3 Quadratic Inequalities

In the previous two sections we saw how to find the solution for x in a quadratic equation such as $0 = x^2 + 3x - 4$. In fact, the x-intercepts of the graph of $y = x^2 + 3x - 4$ were the solutions to the given quadratic equation, since y was replaced by 0.

We now extend our discussion by allowing y to assume a range of values. We define a **quadratic inequality** as any one of the following:

$$Ax^2 + Bx + C \le 0$$

$$Ax^2 + Bx + C < 0$$

$$Ax^2 + Bx + C \ge 0$$

$$Ax^2 + Bx + C > 0$$

In solving $Ax^2 + Bx + C \le 0$, we want all values of x for which the inequality is true. We show two methods for solving the inequality.

Example 1 Solve for x: $x^2 + 3x - 4 \le 0$

Solution We graph the quadratic equation $y = x^2 + 3x - 4$ (Figure 4).

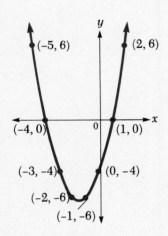

x	y
-5	6
-4	0
-3	-4
-2	-6
-1	-6
0	-4
1	0
2	6

FIGURE 4

From the table of values and the graph, we read the x-values which produce a y-value of 0 or less than 0. These are all x-values less than or equal to 1 and, at the same time, greater than or equal to -4. Thus the solution is $\{x \mid -4 \le x \le 1\}$. ■

Notation: It is sometimes convenient to express certain sets of real numbers in a form called **interval notation.** For instance, the solution set from Example 1 could be written in the form $[-4, 1]$. The notation involves two numbers separated by a comma, with the left number less than or equal to the right number. Square brackets, "[" on the left or "]" on the right, indicate that the number next to the bracket is included in the set; a parenthesis indicates that the number next to the parenthesis is *not* included in the set. The following are the types of intervals we will encounter in Example 2: (a, b) means the interval of real numbers x with $a < x < b$; (a, ∞) means the interval of real numbers x with $x > a$; and, $(-\infty, b)$ means the interval of real numbers x with $x < b$.

Example 2 Solve the problem from Example 1 a second way.

Solution First we solve $x^2 + 3x - 4 = 0$:

$$(x - 1)(x + 4) = 0$$

$$x = 1 \quad \text{or} \quad x = -4$$

Next we check a point in each of the intervals created by the solutions of the quadratic equation. These are $(-\infty, -4)$, $(-4, 1)$, and $(1, +\infty)$.

Check -5 from $(-\infty, -4)$:

$$(-5)^2 + 3(-5) - 4 \overset{?}{\le} 0$$

$$25 - 15 - 4 \overset{?}{\le} 0$$

$$6 \not\le 0$$

$(-\infty, -4)$ is *not* part of the solution.

Check 0 from $(-4, 1)$:

$$0^2 + 3(0) - 4 \overset{?}{\le} 0$$

$$-4 \overset{\checkmark}{\le} 0$$

$(-4, 1)$ *is* part of the solution.

Check 2 from $(1, \infty)$:

$$2^2 + 3(2) - 4 \overset{?}{\le} 0$$

$$4 + 6 - 4 \overset{?}{\le} 0$$

$$6 \not\le 0$$

$(1, \infty)$ is *not* part of the solution

Thus the solution is $\boxed{\{x \mid -4 \le x \le 1\}}$. ∎

Computational Note: For problems that factor as in Step 1 of Example 2 above, we can plot each factor and indicate where each is positive and where each is negative (Figure 5). Now for quadratic expressions that are less than or equal to 0, the factors must have opposite signs. From the diagram we see that this is the interval $-4 \le x \le 1$.

Once a solution is known for one of the inequalities, as in Example 2, the other three types are easily found, as we see in the next example.

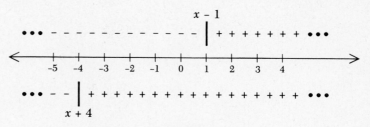

FIGURE 5

Example 3 Solve each of the following, using the results of Example 2:

(a) $x^2 + 3x - 4 < 0$ (b) $x^2 + 3x - 4 \geq 0$

(c) $x^2 + 3x - 4 > 0$

Solution (a) The solution is $\{x \mid -4 < x < 1\}$. This is obtained from Example 2, deleting the values $x = -4$ and $x = 1$, since they produce 0.

(b) The solution is $\{x \mid x \geq 1 \text{ or } x \leq -4\}$. All points that do not work in part (a) must work in part (b).

(c) The solution is $\{x \mid x > 1 \text{ or } x < -4\}$, the same values as in part (b) except for the end values. As in part (a), they are the values that produce 0. ■

Exercises 4.3 In Exercises 1–39 solve the inequality.

1. $x^2 - 1 \geq 0$ 2. $x^2 - 1 \leq 0$

3. $x^2 - 1 > 0$ 4. $x^2 - 1 < 0$

5. $x^2 \geq 0$ 6. $x^2 \leq 0$

7. $x^2 > 0$ 8. $x^2 < 0$

9. $x^2 + 1 \geq 0$ 10. $x^2 + 1 \leq 0$

11. $x^2 + 1 > 0$ 12. $x^2 + 1 < 0$

13. $x^2 - 4 \leq 0$ 14. $x^2 - 4 > 0$

15. $x^2 - 9 \geq 0$ 16. $x^2 - 9 < 0$

17. $x^2 - 16 > 0$ 18. $x^2 - 16 \leq 0$

19. $x^2 - 25 < 0$ 20. $x^2 - 25 \geq 0$

21. $x^2 - x - 6 \leq 0$ 22. $x^2 - x - 6 > 0$

23. $x^2 + x \leq 20$ 24. $x^2 + x \geq 20$

25. $x^2 + x < 20$ 26. $x^2 + x > 20$

27. $x^2 \leq -3x + 10$ 28. $x^2 < -3x + 10$

29. $x^2 \geq -3x + 10$ 30. $x^2 > -3x + 10$

31. $x^2 + x + 1 \geq 0$ 32. $x^2 + x + 1 > 0$

33. $x^2 + x + 1 < 0$ 34. $(2x - 3)(x + 1) \leq 0$

35. $(2x - 3)(x + 1) > 0$
36. $-x^2 - x + 6 \geq 0$
37. $-x^2 - x + 6 < 0$
38. $-x^2 + 6x - 6 > 0$
39. $-x^2 + 6x - 6 \leq 0$
40. The profit P derived from the sale of x items is $P = x^2 - 1000x + 160{,}000$. Find the values of x for which $P > 0$.
41. In Exercise 40, what numbers of items sold produce a loss $(P < 0)$?
42. If profit $P = x^2 - 600x + 50{,}000$, find the values of x for which the profit is nonnegative $(P \geq 0)$.
43. In Exercise 42, find the values of x for which P is negative.
44. A projectile is shot upward with an initial velocity of 64 feet per second. The height y of a projectile after t seconds is given by $y = 64t - 16t^2$.
    (a) What is the domain?
    (b) Find the values of t for which $y \geq 0$.
    (c) At what time does the projectile reach its maximum height? (*Hint:* Consider the role of the vertex.)
45. The initial velocity of a projectile is 96 feet per second.
    (a) Write the equation that represents the height y after t seconds. (*Hint:* Consider the role of velocity in the equation in Exercise 44.)
    (b) Find the values of t for which $y \geq 0$.
    (c) At what time will the projectile reach its maximum height?

4.4 Conic Sections

In this section we will briefly consider the general forms and graphs of a (1) circle, (2) parabola, (3) ellipse, and (4) hyperbola. These are usually called **conic sections** or **conics,** since the graph of each is the intersection of one or more right circular cones and a plane (see Figure 6). Each satisfies a particular form of the general quadratic equa-

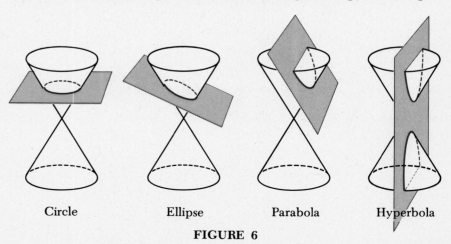

Circle   Ellipse   Parabola   Hyperbola

FIGURE 6

tion in two variables x and y, which is $Ax^2 + Bxy + Cy^2 + Dx + Ey + F = 0$. The variables $A, B, C, D, E,$ and F are real numbers with $A \neq 0$, $B \neq 0,$ or $C \neq 0$. For our discussion we will assume we are working in an xy-plane. We will also assume that $B = 0$ in the above general form, except in the last example of this section.

The Circle

We begin by considering the circle and its properties.

DEFINITION
A **circle** is the set of all points (x, y) which are a fixed distance R from a fixed point (h, k). The value R is called the **radius** of the circle, and the point (h, k) is called the **center** of the circle. (See Figure 7.)

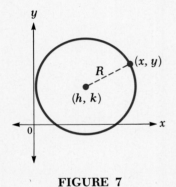

FIGURE 7

The general equation satisfied by the points on a circle is

$$(x - h)^2 + (y - k)^2 = R^2$$

where (h, k) is the center of the circle and R is the radius.

This equation follows from the Pythagorean theorem and the distance formula for two points in the plane. We state both without supplying proofs. The Pythagorean theorem states that $a^2 + b^2 = c^2$ for a right triangle labeled as in Figure 8. From this, the distance d from (x_1, y_1) to (x_2, y_2) satisfies (see Figure 9) $d^2 = (x_2 - x_1)^2 + (y_2 - y_1)^2$, or

$$d = \sqrt{(x_2 - x_1)^2 + (y_2 - y_1)^2}$$

If the center of a circle and the radius are known, the equation is easy to find.

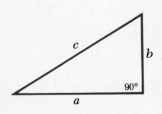

FIGURE 8

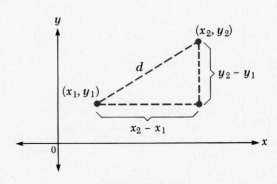

FIGURE 9

Example 1 Find the equation of the circle:
(a) with center $(2, -3)$ and radius 5.
(b) with center at the origin and radius 4.

Solution
(a) We use the general equation with $(h, k) = (2, -3)$ and $R = 5$. This gives $(x - 2)^2 + (y - (-3))^2 = 5^2$, or

$$(x - 2)^2 + (y + 3)^2 = 25$$

(b) Again, we use the general equation with $(h, k) = (0, 0)$ and $R = 4$. This gives $(x - 0)^2 + (y - 0)^2 = 4^2$, or

$$x^2 + y^2 = 16 \quad \blacksquare$$

Note that in part (a) we could expand $(x - 2)^2 + (y + 3)^2 = 25$ to get $(x^2 - 4x + 4) + (y^2 + 6y + 9) = 25$ or $x^2 - 4x + y^2 + 6y = 12$. Our next example shows how to verify that the graph of such an equation is a circle by reversing these steps.

Example 2 Show that the graph of $x^2 + y^2 - 2x + 6y - 6 = 0$ is a circle. Find its center and radius.

Solution We group the like variables and complete the square for each variable:

$$(x^2 - 2x) + (y^2 + 6y) = 6$$

$$(x^2 - 2x + \underline{1}) + (y^2 + 6y + \underline{9}) = 6 + \underline{1} + \underline{9}$$

Now we rewrite, as in the general form:

$$(x - 1)^2 + (y + 3)^2 = 16$$

We can now see that the graph of $(x - 1)^2 + (y + 3)^2 = 16$ is a *circle* with center $(1, -3)$ and radius 4 . $\blacksquare$

In Example 2 we found the value of the radius squared ($R^2 = 16$). This resulting sum must be positive to give a circle. A value of 0 for R^2 would mean that the graph is a single point (h, k). A negative value of R^2 would imply that there are no real points satisfying the equation.

The Parabola

The next conic to be considered is the parabola.

DEFINITION

A **parabola** is the set of all points (x, y) which are of equal distance from a fixed point (a, b) and a fixed line $Ax + By + C = 0$. The point (a, b) is called the **focus,** and the fixed line $Ax + By + C = 0$ is called the **directrix.** The point (h, k) which is midway between the focus and directrix is on the parabola and is called the **vertex.**

The general equation satisfied by a parabola (with directrix parallel to the x-axis or y-axis) is one of the following:

$$(y - k)^2 = 4P(x - h)$$

or

$$(x - h)^2 = 4P(y - k)$$

where P is the distance from the focus to the vertex. If we expand the last equation and solve for y, we will recognize that it is a quadratic equation, as discussed in Section 4.1. Thus the graph will be a U-shaped curve which opens upward if $P > 0$ and downward if $P < 0$. Similarly the other equation is quadratic in the y-variable and will be a U-shaped curve (on its side) which will open to the right if $P > 0$ and to the left if $P < 0$. (See Figure 10.)

In the next example we use the definition of parabola to generate its equation.

Example 3

Write the equation of a parabola with focus $(3, 2)$ and directrix $x = -1$ by using the definition.

Solution

We start by drawing a sketch, plotting the focus $(3, 2)$ and the directrix line $x = -1$ (Figure 11). We then locate the point midway between the focus and directrix, which is $(1, 2)$. This is the vertex. The definition requires a point (x, y) on the parabola to satisfy

$$\underbrace{(x - 3)^2 + (y - 2)^2}_{\substack{\text{distance from} \\ (x, y) \text{ to } (3, 2)}} = \underbrace{(x - (-1))^2 + (y - y)^2}_{\substack{\text{distance from } (x, y) \text{ to} \\ \text{line, using } (-1, y) \text{ as} \\ \text{the closest point}}}$$

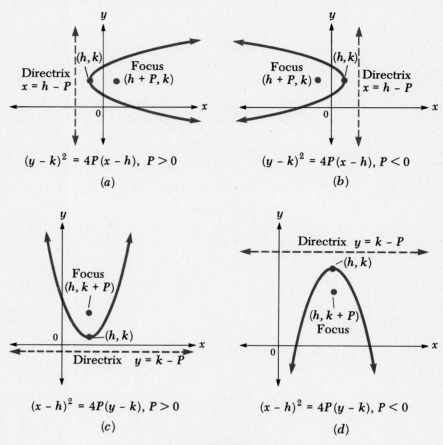

$$(y - k)^2 = 4P(x - h), \ P > 0$$

(a)

$$(y - k)^2 = 4P(x - h), P < 0$$

(b)

$$(x - h)^2 = 4P(y - k), P > 0$$

(c)

$$(x - h)^2 = 4P(y - k), P < 0$$

(d)

FIGURE 10

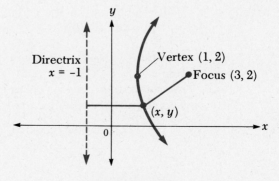

FIGURE 11

Simplifying gives

$$x^2 - 6x + 9 + y^2 - 4y + 4 = x^2 + 2x + 1$$
$$y^2 - 4y = 8x - 12$$

We complete the square on the quadratic expression and rewrite:

$$y^2 - 4y + \underline{4} = 8x - 12 + \underline{4}$$
$$(y - 2)^2 = 8x - 8$$
$$(y - 2)^2 = 8(x - 1)$$

Note that the distance from the vertex (1, 2) to the focus (3, 2) is $P = 2$. Thus our final result is

$$(y - 2)^2 = 4 \cdot (2)(x - 1)$$

$$\underset{k}{\nearrow} \qquad \underset{P}{\nearrow} \quad \underset{h}{\nearrow}$$

which is of the form $(y - k)^2 = 4P(x - h)$ for $P = 2$ and vertex $(h, k) = (1, 2)$. ■

The result in Example 3 required a great deal of effort to achieve. Let us work the example a second time with less effort by using the general form directly.

Example 4   Solve Example 3 using the general equation of a parabola.

Solution   To use the general equation, we need the values of P and the vertex (h, k). We first draw a sketch, plotting the focus (3, 2) and directrix line $x = -1$ (Figure 12). Then we find the midway point, which is the vertex. The value of (h, k) is (1, 2).

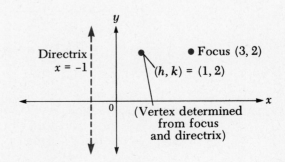

FIGURE 12

The value of P is the distance from the vertex $(1, 2)$ to the focus $(3, 2)$, so $P = 2$. We write the equation and substitute (refer to Figure 10 to find which general form to use). Using $(y - k)^2 = 4P(x - h)$, we have $(y - 2)^2 = 4 \cdot 2(x - 1)$, or

$$(y - 2)^2 = 8(x - 1) \qquad \blacksquare$$

We now introduce some additional terms which are used for parabolas (see Figure 13).

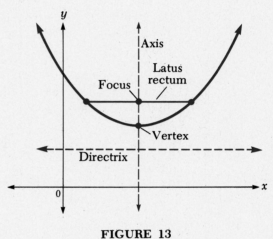

FIGURE 13

DEFINITION   The **axis** of a parabola is the line through the focus and the vertex. The **latus rectum** is the chord drawn through the focus which is perpendicular to the axis of the parabola.

The length of the latus rectum is always $|4P|$, as the next example demonstrates.

Example 5   Write the equation of the parabola whose axis is the y-axis, whose vertex is at the origin, and which passes through $(-4, 2)$. Calculate the length of the latus rectum.

Solution   A sketch appears in Figure 14. The equation must have the form (refer

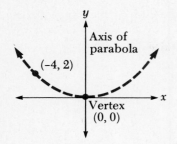

FIGURE 14

to Figure 10)

$$(x - h)^2 = 4P(y - k) \quad \text{with } (h, k) = (0, 0)$$

The point $(-4, 2)$ satisfies this equation, so that $(-4)^2 = 4P(2)$. Solving this equation gives

$$P = \frac{16}{8} = 2$$

Thus the equation is

$$x^2 = 8y$$

The focus must lie on the y-axis and be at distance 2 from $(0, 0)$. Thus the focus is $(0, 2)$. Observe that the latus rectum must join the points $(-4, 2)$ and $(4, 2)$. This chord has length 8, which we note is the value of $|4P|$. Thus the equation is $x^2 = 8y$ and the length of the latus rectum is 8 . ■

An equation can be shown to be that of a parabola by rewriting it in one of the general forms of a parabola.

Example 6 Show that $y^2 + 4x = -8$ is a parabola.

Solution We rewrite as

$$y^2 = -4x - 8$$

$$y^2 = -4(x + 2)$$

$$y^2 = 4(-1)(x + 2)$$

Thus the equation is a *parabola* with vertex $(-2, 0)$ and $P = -1$. ■

The Ellipse

The third conic to be discussed is the ellipse. The definition requires two fixed points F_1 and F_2 and a fixed constant K.

DEFINITION An **ellipse** is the set of all points $P = (x, y)$ such that the distance d_1 from P to F_1 added to the distance d_2 from P to F_2 is the constant K. (See Figure 15.)

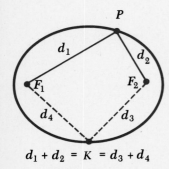

$d_1 + d_2 = K = d_3 + d_4$

FIGURE 15

The two points F_1 and F_2 are the **foci**. The line which goes through F_1 and F_2 is called the **major axis**.

If we let $A = K/2$, let $2C$ be the distance between the foci, and let $B^2 = A^2 - C^2$, then the general equation satisfied by an ellipse is

$$\frac{(x - h)^2}{A^2} + \frac{(y - k)^2}{B^2} = 1$$ (if major axis is parallel to the x-axis)

or

$$\frac{(y - k)^2}{A^2} + \frac{(x - h)^2}{B^2} = 1$$ (if major axis is parallel to the y-axis)

Let us derive the equation for an ellipse where the major axis is the x-axis and the origin (h, k) is midway between the foci.

Example 7 Using the definition, write the equation of an ellipse with foci $(3, 0)$ and $(-3, 0)$ with constant (required by definition) $K = 10$.

Solution A sketch appears in Figure 16. The point $(5, 0)$ belongs to the ellipse because the distance from $(5, 0)$ to $(3, 0)$ is 2 and the distance from $(5, 0)$ to $(-3, 0)$ is 8, and this totals $10 = K$. Similarly, $(-5, 0)$ is on the ellipse. (Thus $A = K/2 = \frac{10}{2} = 5$.) From the definition of an ellipse,

$$\underbrace{\sqrt{(x - 3)^2 + (y - 0)^2}}_{\substack{\text{distance from} \\ (x, y) \text{ to } (3, 0)}} + \underbrace{\sqrt{(x + 3)^2 + (y - 0)^2}}_{\substack{\text{distance from} \\ (x, y) \text{ to } (-3, 0)}} = \underbrace{10}_{\substack{\text{fixed constant} \\ K}}$$

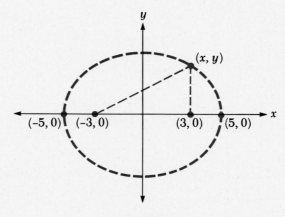

FIGURE 16

We expand and simplify this equation to obtain

$$x^2 - 6x + 9 + y^2 = 100 - 20\sqrt{x^2 + 6x + 9 + y^2} + x^2 + 6x + 9 + y^2$$

$$-12x - 100 = -20\sqrt{x^2 + 6x + 9 + y^2}$$

$$144x^2 + 2400x + 10{,}000 = 400(x^2 + 6x + 9 + y^2)$$

$$10{,}000 - 3600 = 400x^2 - 144x^2 + 400y^2$$

$$6400 = 256x^2 + 400y^2$$

$$1 = \frac{256x^2}{6400} + \frac{400y^2}{6400}$$

$$1 = \frac{x^2}{25} + \frac{y^2}{16}$$

$$1 = \frac{x^2}{5^2} + \frac{y^2}{4^2}$$

We have seen that $A = 5$. The value of $2C$ is the distance from $(-3, 0)$ to $(3, 0)$, so $C = 3$. Thus $B^2 = A^2 - C^2 = 25 - 9 = 16$. We conclude

that $\boxed{1 = \dfrac{x^2}{5^2} + \dfrac{y^2}{4^2}}$ is in the general form for the equation of an *el-lipse*. ■

The graph of $x^2/A^2 + y^2/B^2 = 1$ fits inside the rectangle created by the lines $x = \pm A$ and $y = \pm B$. The graph of $x^2/5^2 + y^2/4^2 = 1$ is shown in Figure 17. The line joining $(0, 4)$ and $(0, -4)$, which is per-

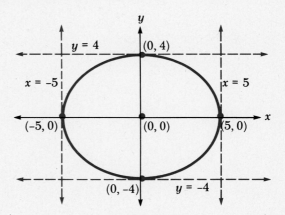

FIGURE 17

pendicular to the major axis, is called the **minor axis.** The point (h, k), here $(0, 0)$, is called the **center** of the ellipse. The points $(\pm 5, 0)$ and $(0, \pm 4)$ are called **vertices.**

Let us rework Example 7 in a shortened form.

Example 8   Work Example 7 using the general equation.

Solution   The value of $2C$ is the distance from $(-3, 0)$ to $(3, 0)$, so $C = 3$. The value of A is $K/2$, so $A = \frac{10}{2} = 5$. Then we can compute

$$B^2 = A^2 - C^2$$

$$B^2 = 25 - 9 = 16$$

Now we write the equation and substitute:

$$\frac{x^2}{A^2} + \frac{y^2}{B^2} = 1$$

$$\frac{x^2}{25} + \frac{y^2}{16} = 1$$

Thus the equation is $\dfrac{x^2}{25} + \dfrac{y^2}{16} = 1$.   ∎

As with the circle and parabola, we can verify that the graph of an equation is an ellipse by rewriting the equation in the general form. Thus $x^2 + 4y^2 + 2x - 3 = 0$ is the equation of an ellipse, since it can be rewritten as

$$\frac{(x + 1)^2}{4} + \frac{y^2}{1} = 1$$

The Hyperbola

The last conic to be discussed is the hyperbola. As in the definition of an ellipse, the hyperbola requires two fixed points F_1 and F_2 and a fixed constant K.

DEFINITION   A **hyperbola** is the set of all points $P = (x, y)$ such that the difference of the distances d_1 (from P to F_1) and d_2 (from P to F_2) is the constant K. (See Figure 18.)

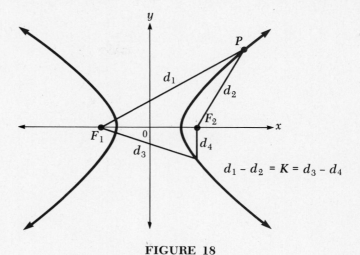

FIGURE 18

The graph of a hyperbola will consist of two separate parts or **branches,** as shown in Figure 18. The line through the foci points F_1 and F_2 is called the **transverse axis.**

The general equation of a hyperbola is

$$\frac{(x - h)^2}{A^2} - \frac{(y - k)^2}{B^2} = 1$$

[if the transverse axis is parallel to the x-axis; see Figure 19(a)]

or

$$\frac{(y - k)^2}{A^2} - \frac{(x - h)^2}{B^2} = 1$$

[if the transverse axis is parallel to the y-axis; see Figure 19(b)]

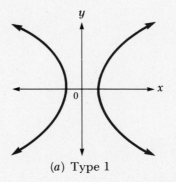

(a) Type 1

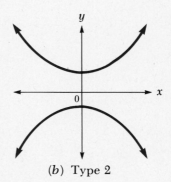

(b) Type 2

FIGURE 19

As for the ellipse, $A = K/2$ and $2C$ is the distance between foci. However, we have $B^2 = C^2 - A^2$ (not $A^2 - C^2$ as for the ellipse). For a Type 1 hyperbola having center $(h, k) = (0, 0)$, the lines $y = \pm B/A$ are asymptotes. (Again, with center $(h, k) = (0, 0)$, the lines $y = \pm A/B$ are asymptotes for a Type 2 hyperbola.)

Example 9   Sketch the graph of the hyperbola $\dfrac{y^2}{25} - \dfrac{x^2}{16} = 1$.

Solution   We know that $A^2 = 25$ and $B^2 = 16$, so $A = \pm 5$ and $B = \pm 4$. The equations of the asymptotes are

$$y = \pm \frac{5}{4} x$$

The asymptotes are easy to construct. From the term $y^2/25$ plot the points $(0, \pm 5)$, and from the term $x^2/16$ plot $(\pm 4, 0)$. Then, just as for an ellipse, construct the rectangle from the lines $y = \pm 5$ and $x = \pm 4$. The asymptotes always pass through the corners. The points $(0, \pm 5)$, which satisfy the equation, are called **vertices.** Then we complete the graph, as in Figure 20. ∎

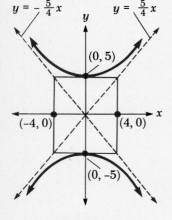

FIGURE 20

For our last example we consider a case where the coefficient (B) of the xy term is not 0. In order to recognize the shape of the graph, we state (without proof) the following theorem:

THEOREM 1   The graph of $Ax^2 + Bxy + Cy^2 + Dx + Ey + F = 0$, if it exists, can be identified as:
(a) An ellipse if $B^2 - 4AC < 0$. (The graph could be an isolated point.)
(b) A parabola if $B^2 - 4AC = 0$. (The graph could be a line or two parallel lines.)
(c) A hyperbola if $B^2 - 4AC > 0$. (The graph could be two intersecting lines.)

Example 10   Graph $xy = 4$.

Solution   From Theorem 1, the graph will be a hyperbola, since $B^2 - 4AC = 1^2 - 4(0)(0) = 1 > 0$, which fits case (c). We build a table of values that satisfy the equation.

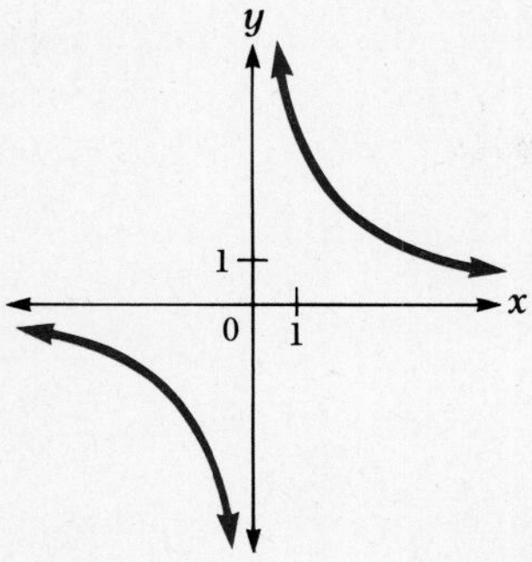

x	y
4	1
2	2
1	4
-4	-1
-2	-2
-1	-4
0	undefined

After plotting these points, we can sketch the hyperbola (Figure 21). ∎

FIGURE 21

Exercises 4.4

In Exercises 1–8 write the equation of the circle described and sketch the graph.

1. Center at origin, radius = 5.
2. Center at origin, radius = 3.
3. Center = (0, 2), radius = 4.
4. Center = $(-1, 3)$, radius = 2.
5. Center at origin, (2, 3) belongs to graph.
6. Center at origin, $(-3, 4)$ belongs to graph.
7. Center = $(-1, -2)$, (2, 4) belongs to graph.
8. Center = $(3, -4)$, $(-1, -1)$ belongs to graph.

In Exercises 9–16 show that the graph of the equation is a circle. Find the center and radius of the circle. Sketch the graph.

9. $x^2 + y^2 - 9 = 0$
10. $x^2 - 25 = -y^2$
11. $x^2 + 2x + y^2 = 35$
12. $x^2 + y^2 - 4y = 21$
13. $x^2 + y^2 + 6x - 14y + 54 = 0$
14. $x^2 + y^2 - 10x + 18y + 90 = 0$
15. $2x^2 + 2y^2 = 4x + 5y - 2$
16. $4x^2 + 4y^2 = 6x - 4y - 1$

In Exercises 17–24 write the equation of the parabola described and sketch the graph. Label the focus, vertex, and directrix.

17. Focus = (0, 2), directrix $x = -2$.
18. Focus = (3, 0), directrix $y = -3$.
19. Focus = (2, 1), directrix $y = 4$.
20. Focus = (2, 1), directrix $x = 4$.
21. Focus = $(-2, 1)$, vertex = $(-2, 3)$.
22. Focus = $(3, -2)$, vertex = $(0, -2)$.
23. Latus rectum = 8, vertex = origin, opens downward.
24. Contains $(-2, -8)$, vertex = origin, opens downward.

In Exercises 25–32 show that the graph of the equation is a parabola. Find the vertex and focus. Sketch the graph.

25. $x^2 - 8y = 0$
26. $16x + y^2 = 0$
27. $y^2 = 8x + 4$
28. $x^2 + 8 = 16y$
29. $y^2 - 4y - 2x = 0$
30. $x^2 + 6x = 3 + 6y$
31. $4x^2 - 4x - 8y + 3 = 0$
32. $9y^2 - 6y - 12x = 8$

In Exercises 33–42 write the equation of the ellipse described. Sketch the graph. Label the center, foci, and vertices.

33. Foci (0, 3) and (0, −3), one vertex at (0, −5).
34. Foci (0, 4) and (0, −4), one vertex at (0, 5).
35. Foci (4, 0) and (−4, 0), constant (required by definition) $K = 12$ (see Example 7).
36. Foci (5, 0) and (−5, 0), constant (required by definition) $K = 12$ (see Example 8).
37. Center at origin, length of major axis = 8, length of minor axis = 4, minor axis = x-axis.
38. Center at origin, major axis = x-axis, length of major axis = 14, length of minor axis = 6.
39. Center at origin, vertices at (0, 7) and (9, 0).
40. Center at origin, vertices at (0, −6) and (−4, 0).
41. Center = (2, 3), focus (6, 3), vertex (7, 3).
42. Center = (2, 3), focus (2, 7), vertex (2, 8).

In Exercises 43–52 show that the graph of the equation is an ellipse. Find the center, foci, and vertices. Sketch the graph.

43. $4x^2 + 9y^2 = 36$
44. $9x^2 + 4y^2 = 36$
45. $x^2 + 4y^2 = 1$
46. $4x^2 + y^2 = 1$
47. $3x^2 + 4y^2 = 12$
48. $5x^2 + 4y^2 = 20$
49. $x^2 + 2x + 9y^2 = 35$
50. $4x^2 + y^2 + 6y = 27$
51. $9x^2 - 18x + 4y^2 + 16y = 11$
52. $4x^2 + 16x + 9y^2 - 54y + 61 = 0$

In Exercises 53–60 write the equation of the hyperbola described. Sketch the graph. Label the foci and vertices.

53. Center = origin, vertex = (2, 0), focus = (3, 0).
54. Center = origin, vertex = (3, 0), focus = (4, 0).
55. Vertices at (0, 3) and (0, −3), foci at (0, 5) and (0, −5).
56. Vertices at (0, 4) and (0, −4), foci at (0, 6) and (0, −6).
57. Center = (1, 2), vertex = (5, 2), length of transverse axis = 10.
58. Center = (3, −2), vertex = (3, −6), length of transverse axis = 12.
59. Vertices at (6, 2) and (−2, 2), foci at (7, 2) and (−3, 2).
60. Vertices at (1, −1) and (−7, −1), foci at (2, −1) and (−8, −1).

In Exercises 61–70 show that the graph of the equation is a hyperbola. Find the center, foci, and vertices. Sketch the graph.

61. $x^2 - y^2 = 9$
62. $y^2 - x^2 = 16$
63. $x^2 - y^2 = -25$
64. $y^2 - x^2 = -4$
65. $9x^2 - 4y^2 = 36$
66. $4x^2 - 9y^2 = 36$
67. $4(x - 1)^2 - 9(y + 2)^2 = -36$
68. $9(x + 3)^2 - 4(y - 5)^2 = -36$
69. $xy = 9$ (graph only)
70. $xy = -4$ (graph only)

In Exercises 71–74 classify the type of conic section. (Use Theorem 4.1.) Do not graph.

71. $4x^2 + 24xy + 11y^2 = 100$
72. $5x^2 - 2xy + 5y^2 - 12 = 0$
73. $x^2 - 4xy + 4y^2 - 4x + 8 = 0$
74. $xy - x + y = 0$

In Exercises 75–84 classify (by recognition of the form or by Theorem 4.1) and graph.

75. $x + y^2 = 4$
76. $x^2 = 4 - y^2$
77. $x^2 - y^2 = 4$
78. $x^2 = 4y^2 + 4$
79. $4y^2 + x^2 = 4$
80. $x^2 - 4y = 4$
81. $9x^2 - 4y = 18x + 23 - y^2$
82. $9x^2 - y^2 - 18x + 4y = 31$
83. $x^2 - 4y - 2x = 31 - y^2$
84. $y^2 + 6 = 2x + 4y$

4.5 Solving Systems Involving Quadratics

In Chapter 3 we studied systems of linear equations. In Section 3.1 the solutions were found by one of three methods: graphing, substitution, or addition. In this section we want to extend the use of these methods to systems of two equations of which one (or both) is a conic section. We will use graphing to determine the number of roots and their approximate values. However, we will use the other methods (substitution or addition) to find the exact solutions.

Let us begin by considering systems consisting of a line and a conic. These systems will have two real solutions, one real solution, or no real solution. (See Figure 22.) We will solve these systems by applying the substitution method—using the linear equation to substitute into the conic—as the next example shows.

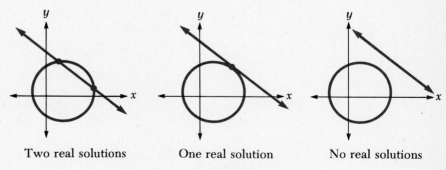

Two real solutions   One real solution   No real solutions

FIGURE 22

Example 1   Solve the system

$$x + y = 1 \quad \text{(line)}$$

$$x^2 + 4y^2 = 4 \quad \text{(ellipse)}$$

Solution   By sketching the graphs we see that there are two real solutions (Fig-

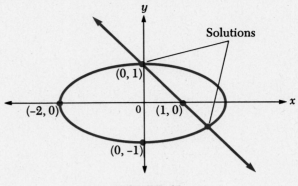

FIGURE 23

ure 23). Using the linear equation, we solve for one variable in terms of the other:

$$x + y = 1$$

$$x = 1 - y$$

Now we substitute $x = 1 - y$ into the conic $x^2 + 4y^2 = 4$, simplify, and solve for y:

$$\begin{cases} x = 1 - y \\ x^2 + 4y^2 = 4 \end{cases}$$

$$(1 - y)^2 + 4y^2 = 4$$

$$1 - 2y + y^2 + 4y^2 = 4$$

$$5y^2 - 2y - 3 = 0$$

$$(5y + 3)(y - 1) = 0$$

$$y = -\frac{3}{5} \quad \text{or} \quad y = 1$$

Finally, we solve for x in the equation we used earlier. (If the other equation is used, extraneous roots may be produced.)

$$x = 1 - y$$

$$y = -\frac{3}{5} \qquad \qquad y = 1$$

$$x = 1 - \left(-\frac{3}{5}\right) \qquad x = 1 - 1$$

$$x = \frac{8}{5} \qquad \qquad x = 0$$

The solutions are $\left(\dfrac{8}{5}, -\dfrac{3}{5}\right)$ and $(0, 1)$. (You should check these roots in both equations.) ∎

The substitution method is very useful in application problems involving the area of a rectangle.

Example 2 Find the dimensions of a rectangle which has area 56 square centimeters and perimeter 30 centimeters.

Solution Let l = the length and w = the width of the rectangle. Then area = $l \cdot w$ and perimeter = $2l + 2w$. We set up the two equations, using these relationships:

$$l \cdot w = 56 \quad \text{(hyperbola)}$$

$$2l + 2w = 30 \quad \text{(line)}$$

We sketch a graph of the system (Figure 24) and solve the linear equation for one of the variables.

$$2l + 2w = 30$$

$$2l = 30 - 2w$$

$$l = 15 - w$$

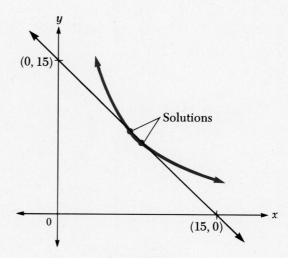

FIGURE 24

Substituting $l = 15 - w$ into the conic $l \cdot w = 56$ and simplifying gives

$$(15 - w) \cdot w = 56$$

$$15w - w^2 = 56$$

$$0 = w^2 - 15w + 56$$

$$0 = (w - 7)(w - 8)$$

$$w = 7 \quad \text{or} \quad w = 8$$

From the equation $l = 15 - w$ we have

$$
\begin{array}{c|c}
w = 7 & w = 8 \\
l = 15 - 7 = 8 & l = 15 - 8 = 7
\end{array}
$$

Thus the substitution method gives two solutions: $w = 7$ and $l = 8$, or $w = 8$ and $l = 7$. Since width is always less than or equal to length, the only solution which fits our application is

$$\boxed{\text{width} = 7} \quad \text{and} \quad \boxed{\text{length} = 8} \quad \blacksquare$$

The substitution method can also be used when both equations are conics. A hyperbola such as $xy = 4$ can easily be rewritten as $x = 4/y$. Similarly, a parabola such as $x - 2y^2 = -2$ can easily be rewritten as $x = 2y^2 - 2$. We demonstrate this by using one of these conics in the next example.

Example 3 Solve the system

$$x - 2y^2 = -2 \quad \text{(parabola)}$$

$$x^2 + y^2 + 4x = 1 \quad \text{(circle)}$$

Solution Graphing the system, we see that there are two real solutions (Figure 25). We solve the parabola for the variable x:

$$x = 2y^2 - 2$$

Substituting into the other equation and simplifying yields

$$x^2 + y^2 + 4x = 1$$

$$(2y^2 - 2)^2 + y^2 + 4(2y^2 - 2) = 1$$

$$4y^4 - 8y^2 + 4 + y^2 + 8y^2 - 8 = 1$$

$$4y^4 + y^2 - 5 = 0$$

$$(4y^2 + 5)(y^2 - 1) = 0$$

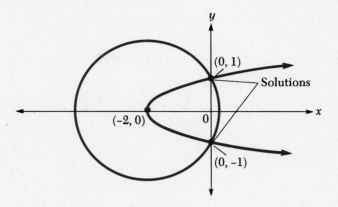

FIGURE 25

$$y^2 = -\frac{5}{4} \quad \bigg| \quad y^2 = 1$$
$$\text{(no real solution)} \quad \bigg| \quad y = \pm 1$$
$$y = \pm \frac{\sqrt{5}}{2}\, i \quad \bigg|$$

Solving for x gives

$$x = 2y^2 - 2$$

$$y = \pm \frac{\sqrt{5}}{2}\, i \quad \bigg| \quad y = \pm 1$$

$$x = -\frac{5}{2} - 2 \quad \bigg| \quad x = 2 - 2$$

$$x = -\frac{9}{2} \quad \bigg| \quad x = 0$$

The real solutions are $(0, 1)$ and $(0, -1)$. The complex solutions are $\left(-\frac{9}{2}, \frac{\sqrt{5}}{2}\, i\right)$ and $\left(-\frac{9}{2}, -\frac{\sqrt{5}}{2}\, i\right)$. ■

The previous example could have been solved in an easier way. The algebraic method is used to eliminate a variable which is common to both equations in a system. The resulting equation in one variable can then be solved.

Example 4   Use the algebraic method to eliminate the y^2 term from the system

$$x - 2y^2 = -2$$
$$x^2 + y^2 + 4x = 1$$

Solution   We multiply the second equation by 2 so that the y^2 terms will add to 0:

$$x - 2y^2 = -2 \qquad \text{(no change)}$$
$$2x^2 + 2y^2 + 8x = 2 \qquad \text{(times 2)}$$

Adding the two equations gives

$$2x^2 + 9x = 0 \quad \blacksquare$$

In Example 4 the system could be solved using the quadratic obtained from the algebra method. We show the full process in the next example.

Example 5   Solve the system

$$x - 2y^2 = -2$$
$$x^2 + y^2 + 4x = 1$$

Solution   By graphing the system (see Example 3 and Figure 26) we see that there are two real solutions. We solve the quadratic equation obtained in Example 4.

$$2x^2 + 9x = 0$$
$$x(2x + 9) = 0$$
$$x = 0 \quad \text{or} \quad x = -\frac{9}{2}$$

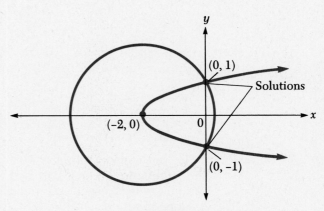

FIGURE 26

Next we substitute into *either* of the original conics. From the graph it is obvious that $x = -\frac{9}{2}$ will not yield a real solution.

$$x - 2y^2 = -2$$

$$x = 0 \qquad\qquad x = -\frac{9}{2}$$

$$-2y^2 = -2 \qquad -\frac{9}{2} - 2y^2 = -2$$

$$y^2 = 1 \qquad\qquad -2y^2 = -2 + \frac{9}{2}$$

$$y = \pm 1 \qquad\qquad y^2 = -\frac{5}{4}$$

$$\text{(no real solution)}$$

$$y = \pm \frac{\sqrt{5}}{2} i$$

The real solutions are $(0, 1)$ and $(0, -1)$. The complex solutions are $\left(-\frac{9}{2}, \frac{\sqrt{5}}{2} i\right)$ and $\left(-\frac{9}{2}, -\frac{\sqrt{5}}{2} i\right)$. ■

For our final example we use the algebraic method without including the graph.

Example 6 Solve the system

$$x^2 - y^2 = 3$$
$$2x^2 + 3y^2 = 11$$

Solution We must eliminate one of the variables; we choose the y^2 term.

$$
\begin{array}{ll}
3x^2 - 3y^2 = 9 & \text{(times 3)} \\
2x^2 + 3y^2 = 11 & \text{(no change)} \\
\hline
5x^2 \qquad\quad = 20 & \text{(add)} \\
\\
x^2 = 4 & \text{(solve for } x) \\
x = \pm 2 &
\end{array}
$$

Next we solve for y, using *either* of the original equations.

$$x^2 - y^2 = 3$$

$$
\begin{array}{c|c}
x = 2 & x = -2 \\
4 - y^2 = 3 & 4 - y^2 = 3 \\
y^2 = 1 & y^2 = 1 \\
y = \pm 1 & y = \pm 1
\end{array}
$$

The real solutions are

$$(2, 1), (2, -1), (-2, 1) \text{ and } (-2, -1) \quad \blacksquare$$

As noted earlier, the solutions should be checked to rule out extraneous roots (sketches of the conics are recommended). A system consisting of two conics may have 0, 1, 2, 3, or 4 real roots, as shown in Figure 27.

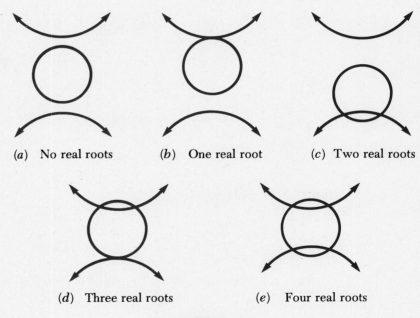

(a) No real roots   (b) One real root   (c) Two real roots

(d) Three real roots   (e) Four real roots

FIGURE 27

Exercises 4.5 In Exercises 1–18 use the substitution method to solve for the (real) solutions. Check each of your answers in both equations.

1. $x + y = 1$
    $x^2 - y^2 = 4$

2. $x + y = 4$
    $x^2 + y^2 = 16$

3. $x^2 + 4y^2 = 4$
    $x + y = 2$

4. $x - y = 2$
   $4x^2 - y^2 = 16$

5. $x + 2y = 3$
    $y = x^2 + 1$

6. $x = y^2 - 1$
   $x - 2y = -2$

7. $x + y = 5$
   $x^2 + y^2 = 4$

8. $x - y = 6$
   $4x^2 + y^2 = 4$

9. $x^2 + y^2 - 4x - 6y + 9 = 0$
        $y - x = 3$

10. $x^2 - 4y^2 + 4x = 0$
         $x - 2y = 1$

11. $x + y = 4$
       $xy = 4$

12. $xy = 6$
    $3x - y = 3$

13. $xy = 9$
    $x^2 + y^2 = 18$

14. $xy = 4$
    $x^2 + y^2 = 8$

15. $xy = 9$
    $x^2 + y^2 = 4$

16. $xy = 4$
    $x^2 + y^2 = 1$

17. $x^2 = y + 4$
    $x^2 + y^2 = 4$

18. $y = x^2 + 4$
    $x^2 + y^2 = 26$

In Exercises 19–28 use the addition method to find the (real) solutions. Check each of your answers in both equations.

19. $x^2 + y^2 = 4$
    $x^2 - y^2 = 4$

20. $x^2 + y^2 = 9$
    $x^2 - y^2 = -9$

21. $x^2 + 2y^2 = 8$
    $3x^2 + y^2 = 9$

22. $x^2 + y^2 = 25$
    $2x^2 + y^2 = 34$

23. $x^2 + 2y^2 = 8$
    $x^2 - y^2 = -4$

24. $3x^2 + y^2 = 4$
    $x^2 + 2y^2 = 3$

C 25. $x^2 = y - 4$
       $x^2 + y^2 = 4$

26. $y = x^2 + 4$
    $x^2 + y^2 = 26$

27. $36x^2 + 36y^2 = 13$
    $12x^2 - 24y = -5$

28. $16x - 16y^2 = 7$
     $x^2 - 4y^2 = 0$

In Exercises 29–34 solve by either method.

29. The sum of two numbers is 14. The difference of their squares is 28. Find the two numbers.

30. The difference of two positive numbers is 4. The difference of their squares is 64. Find the two numbers.

31. The area of a rectangle is 99 square inches, and its perimeter is 40 inches. Find the length and width.

32. The area of a rectangle is 132 square centimeters, and its perimeter is 46 centimeters. Find the length and width.

33. A rectangle has length twice as long as its width. The length of the diagonal is $4\sqrt{5}$ feet. Find the dimensions of the rectangle.

34. The length of a rectangle is 1 centimeter longer than its width. The length of the diagonal is 5 centimeters. Find the dimensions of the rectangle.

Key Terms and Formulas

Quadratic equation:
$Ax^2 + Bx + C = 0$
Factoring
Completing the square
Quadratic formula:
$$x = \frac{-B \pm \sqrt{B^2 - 4AC}}{2A}$$
Discriminant: $B^2 - 4AC$
Conic section:
circle: $(x - h)^2 + (y - k)^2 = R^2$
parabola: $(y - k)^2 = 4P(x - h)$ or $(x - h)^2 = 4P(y - k)$
ellipse: $\dfrac{(x - h)^2}{A^2} + \dfrac{(y - k)^2}{B^2} = 1$
hyperbola: $\dfrac{(x - h)^2}{A^2} - \dfrac{(y - k)^2}{B^2} = 1$ or $\dfrac{(y - k)^2}{A^2} - \dfrac{(x - h)^2}{B^2} = 1$

Quadratic graph:
x-intercept(s)
vertex
y-intercept
Extraneous solution
Quadratic inequality
Interval notation
Quadratic system

Review Exercises

In Exercises 1–6 solve by factoring.
1. $x^2 + 7x - 30 = 0$
2. $2x^2 - 50 = 0$
3. $3x^2 = 27x$
4. $4x^2 - 49 = 0$
5. $2x^2 - 12 = -5x$
6. $(x - 3)^2 = 25$

In Exercises 7–12 solve by completing the square.
7. $x^2 - 5 = 0$
8. $x^2 + 3x - 4 = 0$
9. $x^2 + x = 1$
10. $3x^2 + 6x + 8 = 0$
11. $3x^2 = 4x$
12. $9x^2 + 5 = 12x$

In Exercises 13–18 solve by the quadratic formula.
13. $3x^2 = -5$
14. $x^2 + 5 = 6x$
15. $x^2 + 8x + 18 = 0$
16. $9x^2 - 12x = -9$
17. $4x^2 - 7x = 0$
18. $2x^2 = 2x + 1$

In Exercises 19–22 graph the quadratic function. Label the x-intercepts, y-intercept, and vertex.
19. $y = x^2 - 2x - 15$
20. $y = 2 - x - 3x^2$
21. $y = 8 - 4x^2$
22. $y = 3x^2 - 7x$

In Exercises 23–26 write the equation for the problem and solve it.
23. The base of a triangle is 5 inches longer than its height, and the area of the triangle is 18 square inches. Find the base and the height of the triangle.

24. The length of a rectangle exceeds its width by 8 centimeters and the area of the rectangle is 84 square centimeters. Find the length and width of the rectangle.
25. Find two consecutive natural numbers the sum of whose squares is 145.
C 26. Find two numbers whose sum is 1 and whose product is -241.5.

In Exercises 27–36 solve the equation and check your solutions.

27. $2\sqrt{x} - 5 = 0$
28. $x - \sqrt{6 - x} = 0$
29. $\sqrt{x^2 - 3x} = -2$
30. $2\sqrt{2x} - \sqrt{x + 2} = 2$
31. $\sqrt{7x + 4} = \sqrt{3x + 7} + 1$
32. $2x - 5\sqrt{x} = 3$
33. $2(x + 1)^2 + (x + 1) - 3 = 0$
34. $x^{1/2} + 4x^{1/4} - 5 = 0$
35. $\dfrac{x}{x - 3} - \dfrac{2}{x + 2} = \dfrac{3x + 16}{x^2 - x - 6}$
36. $\dfrac{3x}{x - 4} + \dfrac{x + 1}{x + 2} = \dfrac{20x - 8}{x^2 - 2x - 8}$
37. Find the four fourth roots of $\frac{1}{16}$.
38. Find the four fourth roots of 9.

In Exercises 39–45 solve the inequality.

39. $x^2 - 4 \geq 0$
40. $x^2 + 2 < 0$
41. $x^2 - 7x - 8 > 0$
42. $x^2 - 7x - 8 \geq 0$
43. $(2x - 1)(x + 5) \leq 0$
C 44. $x^2 + 0.9x - 7.36 \geq 0$
C 45. $x^2 - 8x + 8.16 < 0$

46. If profits are given by $P = x^2 - 1000x - 750,000$, find the values of x for which the profit is nonnegative.

In Exercises 47–50 write the equation of the circle and sketch the graph.
47. Center at the origin and radius = 10.
48. Center at (3, 0) and radius = 2.
49. Center at (1, 1), (2, 4) belongs to the graph.
50. Center at $(-2, -3)$, $(4, -3)$ belongs to the graph.

In Exercises 51–54 write the equation of the parabola and sketch the graph.
51. Focus at (0, 1), directrix $y = -3$.
52. Focus at (1, 3), directrix $x = -1$.
53. Focus at (1, 3), vertex at $(1, -3)$.
54. Focus at (0, 0), vertex at $(-4, 0)$.

In Exercises 55–58 write the equation of the ellipse and sketch the graph.
55. Foci at (2, 0) and $(-2, 0)$, one vertex at (3, 0).
56. Center at origin, vertices at $(0, -4)$ and (6, 0).
57. Center at (1, 3), focus at (1, 7), vertex at (1, 8).
58. Center at (1, 3), focus at (5, 3), vertex at (6, 3).

In Exercises 59–62 write the equation of the hyperbola and sketch the graph.
59. Center at origin, vertex at $(0, -2)$, focus at $(0, -3)$.
60. Vertices at (0, 5) and $(0, -5)$, foci at (0, 6) and $(0, -6)$.
61. Vertices at (1, 2) and $(-3, 2)$, foci at (2, 2) and $(-4, 2)$.
62. Center at $(1, -4)$, vertex at (1, 0), length of transverse axis = 10.

In Exercises 63–70 identify the conic section by rewriting the given equation in standard form and then sketching the graph.
63. $x^2 - 16y = 0$
64. $x^2 - 16y^2 = 25$

65. $x^2 + 16y^2 = 25$

66. $x^2 - 35 = 2y - y^2$

67. $8x = 4y^2 + 4y - 3$

68. $4x^2 + 8y^2 = 64$

69. $2x^2 + 3y^2 + 4x - 12y + 8 = 0$

70. $2x^2 - 3y^2 - 4x + 12y - 16 = 0$

In Exercises 71–74 solve the system of equations and check your solutions.

71. $x^2 + y^2 = 25$
    $x + 3y = 15$

72. $x^2 + y^2 = 16$
    $x^2 - y^2 = 4$

73. $xy = 9$
    $x^2 + y^2 = 18$

74. $3x^2 + 4y^2 = 12$
    $2x^2 - y^2 = 8$

75. The sum of the squares of two positive numbers is 289. The difference of the squares is 161. Find the two numbers.

CHAPTER 5

Higher-Degree Equations and Rational Functions

5.1 The Factor Theorem and Remainder Theorem

In this chapter we will study equations of the form $p(x) = 0$ where $p(x)$ is a polynomial. The Factor Theorem and the Remainder Theorem, which we discuss in this section, are basic to our study. We begin with the following definition.

DEFINITION An expression of the form

$$p(x) = a_n x^n + a_{n-1} x^{n-1} + \cdots + a_1 x + a_0$$

where $a_0, a_1, \ldots, a_n$ are numbers, $a_n \neq 0$, and n is a positive integer, is called a **polynomial of degree n.** The numbers $a_0, a_1, \ldots, a_n$ are called the **coefficients** of $p(x)$, and a_n is called the **leading coefficient.** An equation of the form $p(x) = 0$, where $p(x)$ is a polynomial, is called a **polynomial equation.** If x_0 is a number such that $a_n x_0^n + a_{n-1} x_0^{n-1} + \cdots + a_1 x_0 + a_0 = 0$, then x_0 is called a **root** of the equation or a **zero** of the polynomial $p(x) = a_n x^n + a_{n-1} x^{n-1} + \cdots + a_1 x + a_0$.

Example 1 If $p(x) = 4x^3 - 5x^2 - 7x + 2$, then $p(x)$ is a polynomial of degree 3 with leading coefficient 4. Furthermore, $p(2) = 4(2)^3 - 5(2)^2 - 7(2) + 2 = 32 - 20 - 14 + 2 = 0$. Since $p(2) = 0$, 2 is a zero of $p(x)$ or a root of the equation $4x^3 - 5x^2 - 7x + 2 = 0$. ∎

Recall that in division of numbers, the following relationship holds:

$$\text{dividend} = (\text{quotient})(\text{divisor}) + \text{remainder}$$

A similar relationship holds for polynomials. We state the following theorem without proof.

THEOREM 1 If $p(x)$ is a polynomial and c is a number, then there is a unique polynomial $q(x)$ and a number R such that for all x,

$$p(x) = q(x)(x - c) + R$$

The polynomial $q(x)$ is called the **quotient,** and R is called the **remainder** in the division of $p(x)$ by $(x - c)$.

Example 2 If $p(x) = 4x^3 - 10x^2 + 7x + 4$ is divided by $(x - 2)$, we obtain

$$
\begin{array}{r}
4x^2 - 2x + 3 \quad \leftarrow \text{quotient} \\
x - 2\overline{)4x^3 - 10x^2 + 7x + 4} \\
4x^3 - 8x^2 \\
\hline
-2x^2 + 7x \\
-2x^2 + 4x \\
\hline
+3x + 4 \\
+3x - 6 \\
\hline
+10 \leftarrow \text{remainder}
\end{array}
$$

Hence the quotient is $q(x) = 4x^2 - 2x + 3$, and the remainder is $R = 10$. Moreover,

$$\underbrace{4x^3 - 10x^2 + 7x + 4}_{p(x)} = \underbrace{(4x^2 - 2x + 3)}_{q(x)}\underbrace{(x - 2)}_{\text{divisor}} + \underbrace{10}_{R} \quad ∎$$

Example 3   If $p(x) = 3x^2 + 2x - 2$ is divided by $(x + \frac{1}{2})$, we get

$$
\begin{array}{r}
3x \ + \ \frac{1}{2} \quad \leftarrow \text{quotient} \\
x + \tfrac{1}{2}\overline{)3x^2 + 2x - 2} \\
\underline{3x^2 + \tfrac{3}{2}x} \\
\tfrac{1}{2}x - 2 \\
\underline{\tfrac{1}{2}x + \tfrac{1}{4}} \\
-\tfrac{9}{4} \leftarrow \text{remainder}
\end{array}
$$

Hence

$$
3x^2 + 2x - 2 = \underbrace{\left(3x + \frac{1}{2}\right)}_{q(x)}\underbrace{\left(x + \frac{1}{2}\right)}_{\text{divisor}} \underbrace{- \frac{9}{4}}_{R} \quad \blacksquare
$$

If $p(x)$ is divided by $(x - c)$, then by Theorem 1,

$$
p(x) = q(x)(x - c) + R
$$

Hence $p(c) = q(c)(c - c) + R = q(c)(0) + R = 0 + R = R$,   the remainder. We have proved the following theorem.

THEOREM 2   *The Remainder Theorem.*
If a polynomial $p(x)$ is divided by $(x - c)$, then the remainder is equal to $p(c)$.

Example 4   (a) If $p(x) = 4x^3 - 10x^2 + 7x + 4$ is divided by $(x - 2)$, the remainder is

$$
p(2) = 4(2)^3 - 10(2)^2 + 7(2) + 4 = \boxed{10}
$$

(b) If $p(x) = 3x^2 + 2x - 2$ is divided by $(x + \frac{1}{2}) = (x - (-\frac{1}{2}))$, the remainder is

$$
p\left(-\frac{1}{2}\right) = 3\left(-\frac{1}{2}\right)^2 + 2\left(-\frac{1}{2}\right) - 2 = \boxed{-\frac{9}{4}} \quad \blacksquare
$$

Notice that these results agree with the results obtained in Examples 2 and 3.

The next result follows from the Remainder Theorem.

THEOREM 3   *The Factor Theorem.*
Let $p(x)$ be a polynomial. Then $(x - c)$ is a factor of $p(x)$ if and only if $p(c) = 0$. That is, $(x - c)$ is a factor of $p(x)$ if and only if c is a zero of $p(x)$.

Example 5   Is $(x - 2)$ a factor of $p(x) = 3x^3 - 2x^2 - 4x - 8$?

Solution   We compute $p(2) = 3(2)^3 - 2(2)^2 - 4(2) - 8 = 24 - 8 - 8 - 8 = 0$. Since $p(2) = 0$, it follows from the Factor Theorem that $(x - 2)$ is a factor of $p(x)$.   ∎

Example 6   Is $(x + 1)$ a factor of $p(x) = 4x^5 + 3x^3 + x^2 + 2x + 4$?

Solution   Since $(x + 1) = (x - (-1))$, we must find $p(-1)$.
$$p(-1) = 4(-1)^5 + 3(-1)^3 + (-1)^2 + 2(-1) + 4$$
$$= -4 - 3 + 1 - 2 + 4 = -4 \neq 0$$
Hence $(x + 1)$ is not a factor of $p(x)$.   ∎

Example 7   Find a polynomial $p(x)$ of degree 3 that has zeros 1, -2, and 3.

Solution   By the Factor Theorem, $p(x)$ has factors $(x - 1)$, $(x + 2)$, and $(x - 3)$. Hence
$$p(x) = (x - 1)(x + 2)(x - 3)$$
$$= x^3 - 2x^2 - 5x + 6$$
Note that for any constant c, $c(x - 1)(x + 2)(x - 3)$ also has zeros 1, -2, and 3.   ∎

Exercises 5.1   In Exercises 1–10 use the Remainder Theorem to find the remainder when the first expression is divided by the second.
1. $x^4 + 3x^3 - 5x^2 + 2x - 5$, $x - 1$
2. $x^4 + 2x^3 - 7x^2 - 8x + 10$, $x - 2$
3. $x^4 + 2x^3 - 7x^2 - 8x + 16$, $x + 2$
4. $x^4 + 2x^3 - 7x^2 - 8x + 12$, $x + 3$
5. $x^4 + 2x^3 - 7x^2 - 8x + 20$, $x - 1$
6. $x^7 + 3x^5 + 2x + 7$, $x + 1$

7. $2x^5 + 3x^4 - 8x^2 - 10x + 28$, $x + 2$
8. $2x^3 + 3x^2 - 4x - 5$, $x - \frac{1}{2}$
9. $x^n - c^n$, $x - c$
10. $x^n + c^n$ (where n is odd), $x + c$

In Exercises 11–20 use the Factor Theorem to show that the second expression is a factor of the first.

11. $x^4 + 2x^3 - 7x^2 - 8x + 12$, $x - 2$
12. $x^4 + 2x^3 - 7x^2 - 8x + 12$, $x + 2$
13. $3x^7 + 4x^6 + 5x^4 - 3x^2 - 2x - 7$, $x - 1$
14. $2x^3 - 2x + 48$, $x + 3$
15. $5x^7 - 3x^2 - 2x$, $x - 1$
16. $5x^2 - \frac{5}{4}$, $x + \frac{1}{2}$
17. $3x^3 + 2x^2 + 2x + \frac{5}{9}$, $x + \frac{1}{3}$
18. $x^n - c^n$, $x - c$
19. $x^n + c^n$ (where n is odd), $x + c$
20. $3x^{100} - 7x^{50} + 2x + 2$, $x - 1$

In Exercises 21–25 find the quotient $q(x)$ and the remainder R when the first expression is divided by the second.

21. $x^4 + 3x^3 - 5x^2 + 2x - 5$, $x - 1$
22. $2x^3 - 2x + 48$, $x + 3$
23. $3x^3 - 2x^2 + 7x - 3$, $x - 2$
24. $4x^2 - 5x + 1$, $x + \frac{1}{2}$
25. $x^3 - 30$, $x - 3$
26. Find a polynomial of degree 3 having zeros -3, 1, and -1.
27. Find a polynomial of degree 4 having zeros -1, 1, 2, and -2.
28. Find a polynomial of degree 3 having zeros 4, -2, and $\frac{1}{2}$.
29. Find a polynomial of degree 5 having zeros $\sqrt{2}$, $-\sqrt{2}$, 0, 1, -4.

5.2 Synthetic Division

Synthetic division is a method that simplifies the long division process when a polynomial is divided by a divisor of the form $x - c$. To illustrate the process, let us divide $3x^3 - 8x^2 + 9x + 2$ by $x - 2$.

$$
\begin{array}{r}
3x^2 - 2x\ \ + 5 \\
x - 2\overline{)\,3x^3 - 8x^2 + 9x +\ \ 2} \\
\underline{3x^3 - 6x^2} \\
-2x^2 + 9x \\
\underline{-2x^2 + 4x} \\
5x +\ \ 2 \\
\underline{5x - 10} \\
12
\end{array}
$$

First we may omit all x's and allow the position of the numbers to indicate the powers of x. We obtain

$$
\begin{array}{r}
3 - 2 + 5 \\
1 - 2 \overline{)\, 3 - 8 + 9 + 2 } \\
\underline{\textcircled{3} - 6 } \\
-2 + \textcircled{9} \\
\underline{\boxed{-2} + 4 } \\
5 + \textcircled{2} \\
\underline{\textcircled{5} - 10} \\
12
\end{array}
$$

The numbers circled may be omitted, since they are repeated. Further, the coefficients of the divisor will always be of the form $1 - c$, and so the 1 may be omitted. We obtain

$$
\begin{array}{r}
3 - 2 + 5 \\
-2 \overline{)\, 3 - 8 + 9 + 2 } \\
\underline{-6 } \\
-2 \\
\underline{ 4 } \\
5 \\
\underline{ -10} \\
12
\end{array}
$$

In order to make the notation compact, we move the numbers up as follows:

$$
\begin{array}{r}
3 - 2 + 5 \\
-2 \overline{)\, 3 - 8 + 9 + 2 } \\
\underline{-6 + 4 - 10} \\
-2 + 5 + 12
\end{array}
$$

Notice that if the first term of the dividend were written on the last row as follows

$$
\begin{array}{r}
3 - 2 + 5 \\
-2 \overline{)\, 3 - 8 + 9 + 2 } \\
\downarrow -6 + 4 - 10
\end{array}
$$

last row $\rightarrow\ 3 - 2 + 5 + \boxed{12}\ \leftarrow$ remainder

then the last row would contain the quotient along with the remainder. Since the first row is repeated in the last row, it is no longer needed. Our final simplification yields

$$
\begin{array}{r|rrrr}
\underline{-2|} & 3 & -8 & +9 & +2 \\
& \downarrow & -6 & +4 & -10 \\
\hline
& 3 & -2 & +5 & \boxed{+12}\ \leftarrow \text{remainder}
\end{array}
$$

$$
\underbrace{}_{\substack{\text{coefficients of} \\ \text{the quotient}}}
$$

Notice that in obtaining the last row we subtracted. We may simplify this by changing the -2 to $+2$ and then adding. Our division now becomes

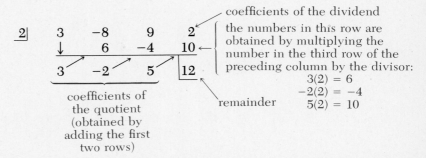

The procedure consisting of all other steps combined with this last simplification is called **synthetic division.**

Note: The coefficients of the dividend are always written in order of decreasing powers of x. The divisor is always of the form $x - c$. Furthermore, we use 0 to denote any missing powers of x. For example, the coefficients of $5x^6 + 3x^4 - 2x^2 + 7x + 1$ are written

$$5 \quad 0 \quad 3 \quad 0 \quad -2 \quad 7 \quad 1$$

missing x^3 term

missing x^5 term

Example 1 Divide $2x^3 - x^2 - 8x - 7$ by $x - 3$.

Solution

$$
\begin{array}{r|rrrr}
3 & 2 & -1 & -8 & -7 \\
  &   & 6 & 15 & 21 \\
\hline
  & 2 & 5 & 7 & \boxed{14} \leftarrow \text{remainder}
\end{array}
$$

coefficients of
the quotient

Hence the quotient is $2x^2 + 5x + 7$ with a remainder of 14. Notice that

$$2x^3 - x^2 - 8x - 7 = \underbrace{(2x^2 + 5x + 7)}_{\text{quotient}}\underbrace{(x - 3)}_{\text{divisor}} + \underbrace{14}_{\text{remainder}} \quad \blacksquare$$

Example 2 Divide $4x^4 - 16x^2 + 5x + 3$ by $x + 2$.

Solution Notice that $x + 2 = x - (-2)$, so $c = -2$. Our division takes the form

$$
\begin{array}{r|rrrrr}
-2 & 4 & 0 & -16 & 5 & 3 \\
 & & -8 & 16 & 0 & -10 \\
\hline
 & 4 & -8 & 0 & 5 & \boxed{-7} \leftarrow \text{remainder}
\end{array}
$$

Hence our quotient is $4x^3 - 8x^2 + 0x + 5$, or simply $4x^3 - 8x^2 + 5$, and our remainder is -7. ∎

Example 3 Suppose $p(x) = 3x^4 - 10x^3 + 8x - 4$. Use synthetic division to find $p(4)$.

Solution From the Remainder Theorem we know that $p(4)$ is the remainder when $3x^4 - 10x^3 + 8x - 4$ is divided by $x - 4$. Using synthetic division, we obtain

$$
\begin{array}{r|rrrrr}
4 & 3 & -10 & 0 & 8 & -4 \\
 & & 12 & 8 & 32 & 160 \\
\hline
 & 3 & 2 & 8 & 40 & \boxed{156} \leftarrow \text{remainder}
\end{array}
$$

Hence $p(4) = 156$ is the remainder. ∎

Example 4 Is $x + 3$ a factor of $3x^3 + 7x^2 - 11x - 15$?

Solution $x + 3 = x - (-3)$. Our division takes the form

$$
\begin{array}{r|rrrr}
-3 & 3 & 7 & -11 & -15 \\
 & & -9 & 6 & 15 \\
\hline
 & 3 & -2 & -5 & \boxed{0} \leftarrow \text{remainder}
\end{array}
$$

Since the remainder is 0, $(x + 3)$ is a factor of $3x^3 + 7x^2 - 11x - 15$. ∎

Note: Suppose a polynomial $p(x)$ is divided by $(x - c)$. Then the following statements are equivalent:

1. The remainder is 0.
2. $(x - c)$ is a factor of $p(x)$.
3. $p(c) = 0$.
4. c is a zero of $p(x)$.
5. c is a root of the equation $p(x) = 0$.

Exercises 5.2 In Exercises 1–11 use synthetic division to find the quotient and the remainder when the first expression is divided by the second.
1. $3x^3 - 7x^2 + 8x - 15,\ x - 2$
2. $3x^4 - 4x^3 + 2x^2 + 2x - 5,\ x - 1$

3. $x^4 + x^3 - 5x + 10, x + 2$
4. $2x^5 + 50x^2 - 4x + 10, x + 3$
5. $2x^3 + 7x^2 - 6x + 4, x - \frac{1}{2}$
6. $x^3 + 1, x + 1$
7. $x^3 + 8, x + 2$
8. $3x^7 - 2x^6 + 4x^4 - 2x - 8, x - 1$
9. $4x^7 - 20x^5 + 34x^2 - 20, x - 2$
10. $x^n - 1$ (where n is any positive integer), $x - 1$
11. $18x^5 - 2x^2 - x + 1, x + \frac{1}{3}$

In Exercises 12–17 use synthetic division to find the required value of $p(x)$.
12. $p(x) = 4x^3 - 3x^2 + 6x - 5; p(2), p(-2)$
13. $p(x) = 4x^4 - 2x^2 + 5x + 6; p(\frac{1}{2}), p(-\frac{1}{2})$
14. $p(x) = x^3 - x^2 - 4x - 12; p(3), p(-2)$
C 15. $p(x) = x^4 - 3x^2 + 5; p(0.1), p(2)$
16. $p(x) = 2x^4 - 3x^3 + 2x^2 - 2x + 1; p(-1), p(5)$
17. $p(x) = x^5 - 20x^3 - 100x - 10; p(5), p(-2)$

In Exercises 18–22 use synthetic division.
18. Show that $x - 2$ and $x + 2$ are factors of $x^4 - 5x^2 + 4$.
19. Show that $x - 1$ is a factor of $2x^4 + x^3 - x^2 + 2x - 4$.
20. Show that $x + 3$ is a factor of $3x^3 + 2x^2 - x + 60$.
21. Show that $x + \frac{1}{2}$ is a factor of $8x^3 + 4x^2 + 6x + 3$.
22. Show that $x - \frac{1}{3}$ is a factor of $9x^3 - 10x + 3$.

In Exercises 23–26 use synthetic division.
23. Show that $\frac{1}{2}$ is a root of $8x^3 + 6x^2 - x - 2 = 0$.
24. Show that $\frac{1}{3}$ is a root of $3x^4 + 14x^3 + 19x^2 + 4x - 4 = 0$.
25. Show that -2 is a root of $x^5 - 3x^4 - 5x^3 + 15x^2 + 4x - 12 = 0$.
26. Show that -2 is a root of $3x^4 + 14x^3 + 19x^2 + 4x - 4 = 0$.

5.3   The Theory of Equations

One might wonder, Does every polynomial have a zero? The following theorem answers this question.

THEOREM 4   *The Fundamental Theorem of Algebra.*
If $p(x)$ is a polynomial of degree $n \geq 1$, there is at least one number c (possibly complex) such that $p(c) = 0$. That is, the equation $p(x) = 0$ has at least one root.

The proof of this theorem is beyond the scope of this book.

From the Fundamental Theorem of Algebra and from the Factor Theorem it follows that if $p(x)$ is a polynomial of degree $n \geq 1$, then there is at least one number c_1 and a polynomial $q_1(x)$ such that

$$p(x) = (x - c_1)q_1(x)$$

If we repeat the process on the polynomial $q_1(x)$, we obtain

$$q_1(x) = (x - c_2)q_2(x)$$

for some number c_2. Hence we see that

$$p(x) = (x - c_1)(x - c_2)q_2(x)$$

If we repeat this process n times, we get the following theorem.

THEOREM 5 If $p(x)$ is a polynomial of degree $n \geq 1$ with leading coefficient a_n, then there exist n numbers $c_1, c_2, \ldots, c_n$ (not necessarily distinct) such that

$$p(x) = a_n(x - c_1)(x - c_2) \cdot \cdot \cdot (x - c_n)$$

Theorem 5 states that every polynomial of degree $n \geq 1$ can be completely factored into n linear factors. Each of the numbers $c_1, c_2, \ldots, c_n$ given in Theorem 5 is a zero of $p(x)$ or a root of the equation $p(x) = 0$. Further, if $p(d) = a_n(d - c_1)(d - c_2) \cdot \cdot \cdot (d - c_n) = 0$, then d must be one of the numbers $c_1, c_2, \ldots, c_n$. Consequently it follows from Theorem 5 that a polynomial $p(x)$ of degree $n \geq 1$ has at most n distinct zeros. We now have the following theorem.

THEOREM 6 A polynomial $p(x)$ of degree $n \geq 1$ has at most n distinct zeros. Equivalently, the polynomial equation $p(x) = 0$ has at most n distinct roots.

Notice that some of the zeros $c_1, c_2, \ldots, c_n$ given in Theorem 5 may be the same.

DEFINITION If a factor $(x - c)$ occurs m times in the factorization guaranteed by Theorem 5, then c is called a zero of **multiplicity** m.

Example 1 The polynomial $p(x) = x^3 - x^2 - 8x + 12$ can be factored as $p(x) = (x + 3)(x - 2)(x - 2)$. Hence we see that -3 is a zero of multiplicity 1, and 2 is a zero of multiplicity 2. ∎

Example 2 Suppose $p(x) = (2x - 1)^2(x + 4)^3(x^2 - 5)^2$. Then

$$p(x) = [2(x - \tfrac{1}{2})]^2(x + 4)^3[(x + \sqrt{5})(x - \sqrt{5})]^2$$
$$= 4(x - \tfrac{1}{2})^2(x + 4)^3(x + \sqrt{5})^2(x - \sqrt{5})^2$$

Thus we see that $\tfrac{1}{2}$ is a zero of multiplicity 2, -4 is a zero of multiplicity 3, and $\pm\sqrt{5}$ are two zeros each having multiplicity 2. ∎

Some of the zeros guaranteed by Theorem 5 may be complex numbers, that is, zeros of the form $a + bi$ where a and b are real numbers and $b \neq 0$. An interesting fact about complex zeros is illustrated in the following theorem.

THEOREM 7 Let $p(x) = a_nx^n + a_{n-1}x^{n-1} + \cdots + a_1x + a_0$ be a polynomial of degree $n \geq 1$ having real coefficients. If $a + bi$ is a complex zero of $p(x)$, then $a - bi$ is also a zero of $p(x)$. Equivalently, if $a + bi$ is a root of the equation $p(x) = 0$ then $a - bi$ is also a root.

Proof: We state some preliminary facts. If $Z = a + bi$, then $a - bi$ is called the **conjugate** of Z and is denoted by $\overline{Z}$ ($\overline{Z} = \overline{a + bi} = a - bi$). The following facts can be proved. If Z_1 and Z_2 are complex numbers and C is a real number, then

(1) $\overline{(Z_1 + Z_2)} = \overline{Z_1} + \overline{Z_2}$

(2) $\overline{(CZ_1)} = C\overline{Z_1}$

(3) $\overline{Z_1^n} = (\overline{Z_1})^n$

Now suppose Z is a complex zero of $p(x)$. Then

$$a_nZ^n + a_{n-1}Z^{n-1} + \cdots + a_1Z + a_0 = 0$$

Hence

$$\overline{a_nZ^n + a_{n-1}Z^{n-1} + \cdots + a_1Z + a_0} = \overline{0} = 0$$

By property (1)

$$\overline{a_nZ^n} + \overline{a_{n-1}Z^{n-1}} + \cdots + \overline{a_1Z} + \overline{a_0} = 0$$

By property (2)

$$a_n\overline{Z^n} + a_{n-1}\overline{Z^{n-1}} + \cdots + a_1\overline{Z} + a_0 = 0$$

Finally, by property (3)

$$\underbrace{a_n(\overline{Z})^n + a_{n-1}(\overline{Z})^{n-1} + \cdots + a_1\overline{Z} + a_0}_{p(\overline{Z})} = 0$$

Hence $p(\overline{Z}) = 0$. This completes the proof. $\square$

The requirement that $p(x)$ have real coefficients is essential. For example, $p(x) = x - (2 + 3i)$ has $2 + 3i$ as a zero; however, $2 - 3i$ is not a zero.

Notice that a polynomial with real coefficients of odd degree must have at least one real zero.

Example 3   Given that $x = 2$ and $x = -1$ are zeros of $p(x) = x^4 + 3x^3 - x^2 - 13x - 10$, find all other zeros of $p(x)$.

Solution   We use synthetic division twice to factor $p(x)$:

$$
\begin{array}{r|rrrrr}
2 & 1 & 3 & -1 & -13 & -10 \\
  &   & 2 & 10 & 18 & 10 \\
\hline
  & 1 & 5 & 9 & 5   & \boxed{0}
\end{array}
$$

$$
\begin{array}{r|rrrr}
-1 & 1 & 5 & 9 & 5 \\
   &   & -1 & -4 & -5 \\
\hline
   & 1 & 4 & 5 & \boxed{0}
\end{array}
$$

Hence $x^4 + 3x^3 - x^2 - 13x - 10 = (x - 2)(x + 1)(x^2 + 4x + 5)$. By the quadratic formula, the zeros of $x^2 + 4x + 5$ are

$$x = \frac{-4 \pm \sqrt{4^2 - 4(1)(5)}}{2(1)} = \frac{-4 \pm 2i}{2} = -2 \pm i$$

Thus all the zeros of the original polynomial are 2, -1, $-2 + i$, and $-2 - i$. ∎

Example 4   Find a polynomial $p(x)$ with real coefficients having $-3i$ and $1 + 2i$ as zeros.

Solution   By Theorem 7, $p(x)$ must also have zeros $3i$ and $1 - 2i$. Hence by the Factor Theorem, $p(x)$ has factors $[x - (-3i)]$, $[x - (3i)]$, $[x - (1 + 2i)]$, and $[x - (1 - 2i)]$. Consequently a polynomial of lowest degree is

$$p(x) = (x + 3i)(x - 3i)[x - (1 + 2i)][x - (1 - 2i)]$$
$$= (x^2 + 9)(x^2 - 2x + 5)$$
$$= x^4 - 2x^3 + 14x^2 - 18x + 45 \quad \blacksquare$$

We can always use the quadratic formula to find the zeros of a second-degree polynomial. However, it is generally very difficult to find zeros of higher-degree polynomials. The next theorem is extremely useful for finding rational zeros, if they exist.

THEOREM 8   Let $p(x) = a_n x^n + a_{n-1} x^{n-1} + \cdots + a_1 x + a_0$ be a polynomial with integer coefficients. If r/s is a rational number reduced to lowest terms such that $p\left(\dfrac{r}{s}\right) = 0$, then r must be a factor of a_0 and s must be a factor of a_n.

Proof: Suppose $p(r/s) = 0$ and r/s is in lowest terms. Then

$$a_n \left(\frac{r}{s}\right)^n + a_{n-1} \left(\frac{r}{s}\right)^{n-1} + \cdots + a_1 \left(\frac{r}{s}\right) + a_0 = 0$$

Multiplying both sides by s^n, we get

$$a_n r^n + a_{n-1} r^{n-1} s + \cdots + a_1 r s^{n-1} + a_0 s^n = 0$$

or

$$a_n r^n + a_{n-1} r^{n-1} s + \cdots + a_1 r s^{n-1} = -a_0 s^n$$

or

$$r(a_n r^{n-1} + a_{n-1} r^{n-2} s + \cdots + a_1 s^{n-1}) = -a_0 s^n$$

By the last equation, r is a factor of $-a_0 s^n$. Since r/s is in lowest terms, r and s have no common factor other than 1. Consequently r is not a factor of s^n. Since r is a factor of $-a_0 s^n$ and not a factor of s^n, it follows that r is a factor of a_0. In the same manner we can show that s is a factor of a_n. $\square$

Notice that a polynomial with integer coefficients need not have any rational zeros. However, if rational zeros exist, they must be of the form (factor of a_0)/(factor of a_n).

Example 5   Find all roots of the equation

$$6x^4 - 7x^3 - 28x^2 + 35x - 10 = 0$$

Solution   If r/s is a root, then r must be a factor of -10 and s must be factor of 6. Hence r must be one of the numbers ± 1, ± 2, ± 5, ± 10, and s must be one of the numbers ± 1, ± 2, ± 3, ± 6. By Theorem 8 the only candidates for rational roots are ± 1, $\pm\frac{1}{2}$, $\pm\frac{1}{3}$, $\pm\frac{1}{6}$, ± 2, $\pm\frac{2}{3}$, ± 5, $\pm\frac{5}{2}$, $\pm\frac{5}{3}$, $\pm\frac{5}{6}$, ± 10, and $\pm\frac{10}{3}$. We check by synthetic division to see which, if any, are roots. We have

$$
\begin{array}{r|rrrrr}
\frac{1}{2} & 6 & -7 & -28 & 35 & -10 \\
& & 3 & -2 & -15 & 10 \\
\hline
& 6 & -4 & -30 & 20 & \boxed{0}
\end{array}
$$

Hence $\frac{1}{2}$ is a root and $6x^3 - 4x^2 - 30x + 20$ is the quotient in the division by $(x - \frac{1}{2})$. The remaining roots must be roots of $6x^3 - 4x^2 - 30x + 20 = 0$ or the equivalent equation $3x^3 - 2x^2 - 15x + 10 = 0$. Hence the list of candidates for the remaining rational roots may be cut down to ± 1, ± 2, ± 5, ± 10, $\pm\frac{1}{3}$, $\pm\frac{2}{3}$, $\pm\frac{5}{3}$, $\pm\frac{10}{3}$. Now we check to see which numbers from this reduced list are roots of $3x^3 - 2x^2 - 15x + 10 = 0$. We have

$$
\begin{array}{r|rrrr}
\frac{2}{3} & 3 & -2 & -15 & 10 \\
& & 2 & 0 & -10 \\
\hline
& 3 & 0 & -15 & \boxed{0}
\end{array}
$$

Hence $\frac{2}{3}$ is a root. The remaining roots must be roots of the equation $3x^2 - 15 = 0$. But $3x^2 - 15 = 0$ if and only if $x = \sqrt{5}$ or $x = -\sqrt{5}$. Hence the roots of the original equation are $\frac{1}{2}$, $\frac{2}{3}$, $\sqrt{5}$, and $-\sqrt{5}$. ∎

Exercises 5.3

1. Find a polynomial of degree 4 such that both -1 and 2 are zeros of multiplicity 2.
2. Find a polynomial of degree 6 such that 0 is a zero of multiplicity 4 and -1 is a zero of multiplicity 2.
3. Find a polynomial of degree 6 such that 0 is a zero of multiplicity 2, -2 is a zero of multiplicity 3, and 1 is a zero of multiplicity 1.
4. Find a polynomial with real coefficients of degree 5 such that 0 is a zero of multiplicity 1, -3 is a zero of multiplicity 2, and $2i$ is a zero of multiplicity 1.

In Exercises 5–10 list the zeros of $p(x)$ and state the multiplicity of each. (See Example 2.)

5. $p(x) = (x - 2)^3(x + 1)^5$
6. $p(x) = (2x - 3)^2(3x + 5)^3(x + 5)$
7. $p(x) = x^4(x^2 - 4)^2(x - 1)$

8. $p(x) = x^5 - 4x^3$
9. $p(x) = (x^2 + 25)(2x - 3)^3$
10. $p(x) = (x^2 + 2x - 3)^3(x^2 - 2)^5$

In Exercises 11–17 find a polynomial with real coefficients of lowest degree having the given zeros. (See Example 4.)

11. $2, 3, -1$ 12. $2i, 2 - i$
13. $1 + 2i, 1 - 3i$ 14. $0, -2, 2 + 3i$
15. $\sqrt{3}, -\sqrt{3}, -3i$ 16. $2 + \sqrt{3}, 2 - \sqrt{3}, i$
17. $0, 2 - 3i, 1 + 4i$
18. Given that 1 is a zero of $p(x) = x^3 - x^2 - 4x + 4$, find all other zeros.
19. Given that -1 is a zero of $p(x) = x^3 + 5x^2 + 5x + 1$, find all other zeros.
20. Given that 2 and -3 are zeros of $p(x) = x^4 + x^3 + 3x^2 + 9x - 54$, find all other zeros.
21. Given that -2 and 5 are zeros of $p(x) = x^4 - 5x^3 + x^2 + 5x - 50$, find all other zeros.

In Exercises 22–30 find all roots of the equation. (See Example 5.)

22. $x^3 - 7x + 6 = 0$
23. $2x^3 + 5x^2 + x - 2 = 0$
24. $8x^3 + 2x^2 - 7x - 3 = 0$
25. $12x^3 - 4x^2 - 3x + 1 = 0$
26. $x^4 + x^3 + 7x^2 + 9x - 18 = 0$
27. $2x^4 + 11x^3 + 20x^2 + 7x - 10 = 0$
28. $x^4 - 2x^3 - 3x^2 + 8x - 4 = 0$
29. $3x^4 + 14x^3 + 19x^2 + 4x - 4 = 0$
30. $x^5 - 3x^4 - 5x^3 + 15x^2 + 4x - 12 = 0$
31. Show that $p(x) = x^3 + 3x^2 + 2x + 3$ has no rational zeros.
32. Show that $p(x) = 3x^3 + x^2 - 2x + 5$ has no rational zeros.

5.4 Graphing a Polynomial

In previous chapters we have graphed linear and quadratic functions. In this section we graph polynomial functions of degree $n \geq 3$.

It can be proved that the graph of a polynomial is continuous. That is, the graph contains no gaps or holes. The following theorem, which follows from the fact that a polynomial function is continuous, is useful in graphing a polynomial and in locating zeros.

THEOREM 9 Suppose that $p(x)$ is a polynomial with real coefficients and that x_1 and x_2 are real numbers such that $p(x_1)$ and $p(x_2)$ differ in sign. Then there is at least one real number x_0 between x_1 and x_2 such that $p(x_0) = 0$.

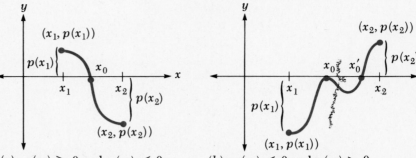

(a) $p(x_1) > 0$ and $p(x_2) < 0$. The graph crosses the x-axis at x_0 between x_1 and x_2.

(b) $p(x_1) < 0$ and $p(x_2) > 0$. There are two numbers x_0 and x_0' between x_1 and x_2 with $p(x_0) = 0$ and $p(x_0') = 0$.

FIGURE 1

The conclusion of Theorem 9 is better understood by considering the graphs in Figure 1. Keep in mind that the statement $p(x_0) = 0$ is equivalent to saying that the graph of $p(x)$ crosses the x-axis at x_0 or possibly touches the x-axis at x_0.

As we see from the graphs in Figure 1, if $p(x_1)$ and $p(x_2)$ differ in sign, there may be several numbers between x_1 and x_2 which are zeros of $p(x)$. In other words, there may be several places between x_1 and x_2 where the graph either touches or crosses the x-axis.

The following theorem is very useful in determining the number of real zeros of a polynomial $p(x)$.

DESCARTES'S RULE OF SIGNS

Let $p(x)$ be a polynomial with real coefficients.
(a) The number of positive zeros of $p(x)$ either is equal to the number of variations in sign in the coefficients of $p(x)$ or is less than that number by an even integer.
(b) The number of negative zeros of $p(x)$ either is equal to the number of variations in sign in the coefficients of $p(-x)$ or is less than that number by an even integer.

Example 1 Determine the possible number of positive, negative, and complex roots of $2x^4 - 7x^3 - 2x^2 + 7x - 3 = 0$.

Solution Let $p(x) = 2x^4 - 7x^3 - 2x^2 + 7x - 3$. Then the roots of $2x^4 - 7x^3 -$

$2x^2 + 7x - 3 = 0$ are the zeros of $p(x)$. As we read from left to right, we have three variations in sign in the coefficients of $p(x)$.

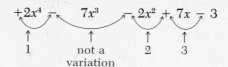

$$+2x^4 \quad -7x^3 \quad -2x^2 \quad +7x \quad -3$$

1   not a   2   3
variation

Hence $p(x) = 0$ has either three positive roots or one. Now

$$p(-x) = 2(-x)^4 - 7(-x)^3 - 2(-x)^2 + 7(-x) - 3$$
$$= 2x^4 + 7x^3 - 2x^2 - 7x - 3$$

has one variation in sign. Hence $p(x) = 0$ has one negative root. Since the equation is of degree 4, we must have four roots. We have the following two possibilities:

1. $p(x) = 0$ has three positive roots and one negative root or
2. $p(x) = 0$ has one positive root, one negative root, and two complex roots. ∎

Example 2   Graph $p(x) = x^3 - 3x^2 - 2x + 6$.

Solution   Now $p(x)$ has two variations in sign and $p(-x) = (-x)^3 - 3(-x)^2 - 2(-x) + 6 = -x^3 - 3x^2 + 2x + 6$ has one variation in sign. Hence by Descartes's Rule of Signs, $p(x)$ has two or zero positive zeros and one negative zero. We will sketch the graph of $p(x)$ by constructing a table of values. We may use synthetic division and the Remainder Theorem to evaluate $p(x)$ for particular choices of x. This procedure is often a convenient shortcut and is illustrated below.

x	-2	-1	0	1	2	3	4
$y = p(x)$	-10	4	6	2	-2	0	14

$$
\begin{array}{r|rrrr}
-2 & 1 & -3 & -2 & 6 \\
   &   & -2 & 10 & -16 \\
\hline
   & 1 & -5 & 8 & \boxed{-10} = p(-2)
\end{array}
$$

$$
\begin{array}{r|rrrr}
4 & 1 & -3 & -2 & 6 \\
  &   & 4 & 4 & 8 \\
\hline
  & 1 & 1 & 2 & \boxed{14} = p(4)
\end{array}
$$

By Theorem 9 there is a zero between -2 and -1 and another between 1 and 2. From our table, 3 is a zero. We plot the points and

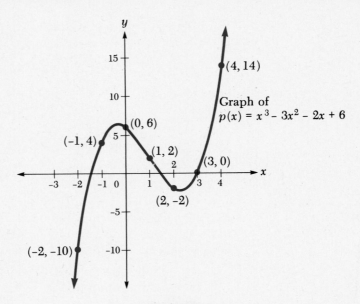

FIGURE 2

connect them with a continuous curve. In order to make the graph manageable, we use an appropriate scale for the y-axis (Figure 2). (*Note:* $(0, 6)$ and $(2, -2)$ are close to but not necessarily the turning points of the graph; therefore our graph is only an approximation of the actual curve.)

Notice that $p(3) = 0$; we may use synthetic division to divide $p(x)$ by $(x - 3)$:

$$p(x) = (x - 3)(x^2 - 2)$$
$$= (x - 3)(x - \sqrt{2})(x + \sqrt{2})$$

Hence $x = 3$, $x = \sqrt{2}$, and $x = -\sqrt{2}$ are zeros of $p(x)$, and the graph of $p(x)$ crosses the x-axis at 3, $\sqrt{2}$, and $-\sqrt{2}$. ∎

Example 3 Graph $p(x) = x^4 - 4x^2 - 4$.

Solution Note that $p(x)$ has one variation in sign and that $p(-x) = (-x)^4 - 4(-x)^2 - 4 = x^4 - 4x^2 - 4$ also has one variation. Hence $p(x)$ has one positive zero and one negative zero. We construct the following table of values using either synthetic division or direct substitution:

x	-3	-2	-1	0	1	2	3
$y = p(x)$	41	-4	-7	-4	-7	-4	41

By Theorem 9 there is a zero between -3 and -2 and another between 2 and 3. For a more accurate sketch, it might be helpful to include several more points in our table of values. Notice that since the exponents of x in $p(x)$ are even, $p(x) = p(-x)$ for any number x.

x	-1.5	-0.5	0.5	1.5
$y = p(x)$	-7.94	-4.94	-4.94	-7.94

The points in both tables are plotted to obtain the graph in Figure 3. ■

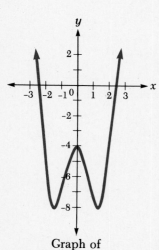

Graph of
$p(x) = x^4 - 4x^2 - 4$

FIGURE 3

From the graph of $p(x) = x^4 - 4x^2 - 4$ we see that there is a zero of $p(x)$ between 2 and 3. If we divide the interval in half, we get

$$p(2.5) \approx 10.06$$

└─approximately equal to

Since $p(2) < 0$ and $p(2.5) > 0$, the zero is between 2 and 2.5. Also

$$p(2.25) \approx 1.38$$

Since $p(2) < 0$ and $p(2.25) > 0$, the zero is between 2 and 2.25. This process can continue until we reach a desired degree of accuracy.

In general, if $p(a)$ and $p(b)$ differ in sign, then by Theorem 9, $p(x)$ has a zero between a and b. Let c be the number exactly halfway between a and b; that is, $c = (a + b)/2$. If $p(c)$ and $p(a)$ differ in sign, the zero is between a and c. If $p(c)$ and $p(b)$ differ in sign, the zero is between c and b. This procedure is called the **half-interval search** and can be continued until a desired degree of accuracy is obtained.

Exercises 5.4

1. Show that $p(x) = x^3 - 4x^2 + 10x - 1$ has a zero between 0 and 1.
2. Show that $p(x) = 2x^3 - 11x^2 + 2x + 15$ has a zero between 1 and 2.
3. Show that $p(x) = x^5 + x^4 - 7x^3 - 7x^2 + 10x + 10$ has a zero between 1 and 2.
4. Show that $p(x) = x^5 + x^4 - 7x^3 - 7x^2 + 10x + 9$ has a zero between -1 and -2.
5. Show that $p(x) = 3x^3 - 4x^2 - 15x + 20$ has a zero between 2 and 3.
6. Show that $p(x) = 3x^3 - 4x^2 - 6x + 6$ has a zero between $\frac{3}{2}$ and 2.

In Exercises 7–19 use Descartes's Rule of Signs to determine the possible number of positive, negative, and complex roots.
7. $2x^4 - 3x^3 + 5x^2 + 3x - 2 = 0$
8. $x^3 + 3x^2 - 2x + 1 = 0$

9. $-3x^3 - 2x^2 + 5 = 0$
10. $2x^6 + 3x^4 + 2x^2 + 1 = 0$
11. $x^5 + 4x^4 + 4x^2 + 3x + 1 = 0$
12. $4x^3 - 5x^2 - 10x + 50 = 0$
13. $x^5 - 4x^4 - 5x^3 - 2x^2 + 3x + 4 = 0$
14. $x^4 + 2x - 5 = 0$
15. $2x^5 - 3x^4 + x^3 + 2x^2 + x - 5 = 0$
16. $x^5 + x^3 + x + 1 = 0$
17. $2x^4 + 3x^2 + 1 = 0$
18. $-x^5 - x^4 + 1 = 0$
19. $x^6 - x^5 + x^4 - x^3 + x^2 - x + 1 = 0$

In Exercises 20–31 sketch the graph of the polynomial.

20. $p(x) = x^3 - 1$
21. $p(x) = x^3 + x$
22. $p(x) = x^3 + 5x^2 + 2x - 8$
23. $p(x) = x^3 - 3x^2 - 9x + 5$
24. $p(x) = 2x^3 - 12x + 1$
25. $p(x) = x^3 - 12x + 10$
26. $p(x) = x^3 - 3x^2 - x + 3$
C 27. $p(x) = x^4 - 2x^3 - 13x^2 + 14x + 24$
C 28. $p(x) = -x^4 + 4x^2 + 2$
C 29. $p(x) = x^4 - 4x^3 - 2x^2 + 12x + 1$
30. $p(x) = x(x + 2)(x - 2)(x - 4)$
31. $p(x) = x^3 - 5x^2 + 7x - 3$

5.5 Graphing a Rational Function

DEFINITION A function of the form $\dfrac{p(x)}{q(x)}$ where $p(x)$ and $q(x)$ are polynomials is called a **rational function.**

If $f(x) = p(x)/q(x)$, then $f(x)$ is not defined at any value of x where $q(x) = 0$. If x_0 is a number such that $q(x_0) = 0$ and $p(x_0) \neq 0$, then the line $x = x_0$ is called a **vertical asymptote.** If x_0 is a vertical asymptote of $f(x) = p(x)/q(x)$, then the value of $|f(x)|$ becomes larger and larger as x gets closer and closer to x_0. If the value of $f(x)$ approaches some value c as $|x|$ gets larger and larger, then the line $y = c$ is called a **horizontal asymptote.**

Example 1 Sketch the graph of $f(x) = \dfrac{1}{x-2}$.

Solution We see that the line $x = 2$ is a vertical asymptote, since the denominator of $1/(x-2)$ is equal to 0 at $x = 2$ and the numerator is not. Also as $|x|$ gets larger and larger, $1/(x-2)$ gets closer and closer to 0. Hence the line $y = 0$ (the x-axis) is a horizontal asymptote. We construct the following table:

x	-2	0	1	1.5	1.8	2.2	2.5	3	4
$y = f(x)$	-0.25	-0.5	-1	-2	-5	5	2	1	0.5

Plotting these points, we obtain the graph in Figure 4. ■

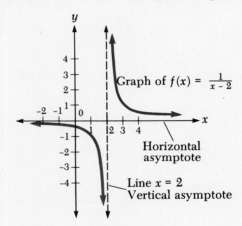

FIGURE 4

Example 2 Graph $f(x) = \dfrac{x-1}{(x-2)(x+1)}$.

Solution The lines $x = 2$ and $x = -1$ are vertical asymptotes. As $|x|$ gets larger and larger, it is hard to tell what happens to $f(x)$, since both the denominator and the numerator of $f(x)$ get large as $|x|$ gets large. To get a better idea of what happens to $f(x)$, we make a table of values:

x	-10	-4	-1.5	0	1	1.8	2.2	4	10
$y = f(x)$	-0.10	-0.28	-1.43	0.5	0	-1.43	1.88	0.3	0.10

Plotting these points, we see that the line $y = 0$ is a horizontal asymptote. The graph appears in Figure 5. ■

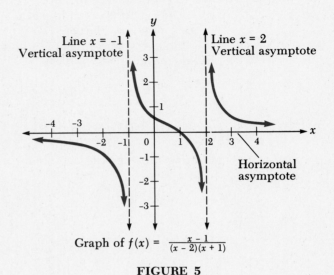

Graph of $f(x) = \dfrac{x-1}{(x-2)(x+1)}$

FIGURE 5

It can be shown that if $f(x) = p(x)/q(x)$ and if the degree of the polynomial $p(x)$ is less than the degree of the polynomial $q(x)$, then $y = 0$ (the x-axis) is a horizontal asymptote. This idea was illustrated in Example 2. Furthermore, if the degree of $p(x)$ is equal to the degree of $q(x)$, then the line $y = c$, where c is the ratio of the leading coefficients of $p(x)$ and $q(x)$, is a horizontal asymptote. This idea will be illustrated in Example 3.

Example 3   Graph $f(x) = \dfrac{6x - 1}{2x + 4}$.

Solution   The line $x = -2$ is a vertical asymptote, since the denominator is 0 at $x = -2$ and the numerator is not. Notice that

$$f(x) = \frac{6x - 1}{2x + 4} = \frac{6 - \dfrac{1}{x}}{2 + \dfrac{4}{x}} \quad \text{if } x \neq 0$$

Hence we see that the value of $f(x)$ gets closer and closer to $(6 - 0)/(2 + 0) = 3$ as $|x|$ gets larger and larger. Therefore the line $y = 3$ is a horizontal asymptote. Plotting several points $(x, f(x))$, we obtain the graph in Figure 6. ∎

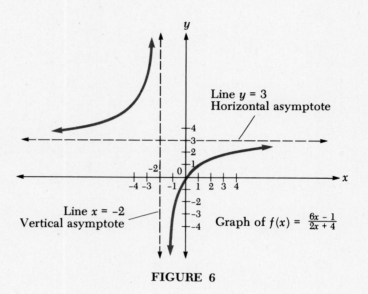

Line $y = 3$
Horizontal asymptote

Line $x = -2$
Vertical asymptote

Graph of $f(x) = \frac{6x - 1}{2x + 4}$

FIGURE 6

We will encounter another type of asymptote, called an **oblique asymptote** or **slant asymptote**, in Example 4.

Example 4   Graph $f(x) = \dfrac{x^2 - x - 1}{x - 2}$.

Solution   We use synthetic division to divide $x^2 - x - 1$ by $x - 2$:

$$
\begin{array}{r|rrr}
2 & 1 & -1 & -1 \\
  &   & 2 & 2 \\
\hline
  & 1 & 1 & \underline{|1} \\
\end{array}
$$

Hence $f(x) = (x^2 - x - 1)/(x - 2) = x + 1 + 1/(x - 2)$. Furthermore, as $|x|$ gets larger and larger, $1/(x - 2)$ gets closer and closer to 0, and so we see that the graph of $f(x)$ approaches the line $y = x + 1$. The line $y = x + 1$ is called an oblique asymptote or a slant asymptote. Notice also that the line $x = 2$ is a vertical asymptote. Plotting several points $(x, f(x))$, we obtain the graph shown below. ∎

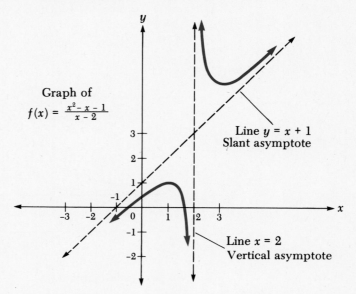

Graph of
$$f(x) = \frac{x^2 - x - 1}{x - 2}$$

Line $y = x + 1$
Slant asymptote

Line $x = 2$
Vertical asymptote

It can be shown that if $f(x) = p(x)/q(x)$ in reduced form, and if

$$\text{degree of } p(x) = \text{degree of } q(x) + 1$$

then $f(x)$ has a slant asymptote. This idea was illustrated in Example 4. Furthermore, if the degree of $p(x)$ is *any* value greater than the degree of $q(x)$, then there are no horizontal asymptotes.

SUMMARY OF ASYMPTOTES

Suppose $f(x) = \dfrac{a_n x^n + a_{n-1} x^{n-1} + \cdots + a_1 x + a_0}{b_m x^m + b_{m-1} x^{m-1} + \cdots + b_1 x + b_0}$.

1. If c is a zero of the denominator polynomial but not a zero of the numerator polynomial, then $x = c$ is a vertical asymptote (see Examples 1, 2, 3, and 4).
2. If $n = m$, then the line $y = a_n/b_m$ is a horizontal asymptote (see Example 3).
3. If $m > n$, then the line $y = 0$ (the x-axis) is a horizontal asymptote (see Examples 1 and 2).
4. If $n > m$, there is no horizontal asymptote (see Example 4).
5. If $f(x)$ is in reduced form and $n = m + 1$, then $f(x)$ has a slant asymptote (see Example 4).

Exercises 5.5

In Exercises 1–21 graph the function and label all asymptotes.

1. $f(x) = \dfrac{1}{x}$

2. $f(x) = \dfrac{1}{x + 2}$

3. $f(x) = \dfrac{3}{x - 4}$

4. $f(x) = \dfrac{1}{2x + 6}$

5. $f(x) = \dfrac{x}{(x + 1)(x - 1)}$

6. $f(x) = \dfrac{x + 2}{(x - 2)(x + 4)}$

7. $f(x) = \dfrac{x}{x^2 - 4}$

8. $f(x) = \dfrac{1}{(x - 1)^2}$

9. $f(x) = \dfrac{1}{(x - 3)^2}$

10. $f(x) = \dfrac{2x - 3}{(x - 1)(x - 3)}$

11. $f(x) = \dfrac{x^2 - x - 5}{x - 3}$

12. $f(x) = \dfrac{x^2 + 1}{x - 1}$

13. $f(x) = \dfrac{x^2 - 3x + 3}{x - 1}$

14. $f(x) = \dfrac{2x^2 + 3x + 2}{x + 1}$

15. $f(x) = \dfrac{x + 3}{x + 1}$

16. $f(x) = \dfrac{6x + 1}{3x + 2}$

17. $f(x) = \dfrac{2x + 1}{4x + 1}$

18. $f(x) = \dfrac{3x - 2}{x - 1}$

19. $f(x) = 2x^2 - 3x - 2$

20. $f(x) = \dfrac{x^2 - 3x + 2}{x - 2}$

21. $f(x) = \dfrac{x^2 + x - 2}{x - 1}$

Key Terms and Formulas

Polynomial
    coefficients of a polynomial
    leading coefficient of a polynomial
    zero of a polynomial
Root of an equation
Remainder Theorem
Factor Theorem
Synthetic division

Fundamental Theorem of Algebra
Zero of multiplicity m
Complex zero of a polynomial
Rational zero of a polynomial
Descartes's Rule of Signs
Vertical asymptote
Horizontal asymptote
Oblique or slant asymptote

Review Exercises

1. Use the Remainder Theorem to find the remainder when
   (a) $3x^3 - 6x + 1$ is divided by $x - 1$.
   (b) $2x^4 + 3x^3 - 2x^2 + 5x + 1$ is divided by $x + 2$.
   (c) $x^7 + 3x^4 + 3x$ is divided by $x + 1$.
   (d) $2x^3 - x^2 - 8x - 7$ is divided by $x - 3$.
   C (e) $2x^3 - 3x^2 + 5x + 3$ is divided by $x - 3.2187$.
2. Use the Factor Theorem to show that
   (a) $x - 1$ is a factor of $4x^3 - 3x^2 + 6x - 7$.
   (b) $x + 2$ is a factor of $x^3 - x^2 - 4x + 4$.
   (c) $x - 2$ is a factor of $x^3 - x^2 - 4x + 4$.
   C (d) $x - \sqrt{2}$ is *not* a factor of $x^5 - 3x^3 - 5x + 9$.
3. Find a polynomial of degree 3 having zeros 2, -1, and 1.
4. Find a polynomial of degree 5 having zeros $\sqrt{3}$, $-\sqrt{3}$, 0, 1, and 2.

In Exercises 5–10 use synthetic division to find the quotient and the remainder when the first expression is divided by the second.

5. $2x^3 - x^2 - 5x + 10$, $x - 3$
6. $3x^5 - 2x^2 + 5$, $x + 1$
7. $4x^7 - 3x^2 + 2x - 5$, $x - 1$
8. $4x^3 - 3x^2 + 4x - 5$, $x + 2$
9. $2x^3 + 7x^2 - 8x + 6$, $x - \frac{1}{2}$
10. $2x^3 + 7x^2 + 7x - 10$, $x + \frac{1}{2}$

11. Use synthetic division to find $p(3)$ if $p(x) = 3x^3 - 2x^2 + 4x - 3$.
12. Use synthetic division to show that $(x - 2)$ and $(x + \frac{1}{2})$ are factors of $4x^3 - 2x^2 - 10x - 4$.
13. Find a polynomial of degree 5 such that 0 is a zero of multiplicity 3 and -1 is a zero of multiplicity 2.
14. Let $p(x) = x(x + 3)^4(x - 1)^3$. List the zeros of $p(x)$ and state the multiplicity of each.
15. Find a polynomial with real coefficients of lowest degree having $1 + i$ and $-2i$ as zeros.
16. Find all roots of $4x^3 - 2x^2 - 10x - 4 = 0$.
17. Find all roots of $x^5 - 3x^4 - 5x^3 + 15x^2 + 4x - 12 = 0$.
18. Find all roots of $24x^3 - 10x^2 - 3x + 1 = 0$.
19. Show that $p(x) = x^3 + 2x^2 - 5$ has a zero between 1 and 2.
20. Determine the possible number of positive, negative, and complex roots of
    (a) $3x^4 - 2x^3 + 5x^2 + x - 1 = 0$
    (b) $3x^4 + 2x^3 + 3x^2 + x - 1 = 0$
21. Sketch the graph of $p(x) = x^3 - 2x^2 - 5x + 6$.
C 22. Sketch the graph of $p(x) = x^4 - 8x^2 + 8$.
23. Sketch the graph of $p(x) = x(x - 1)(x + 1)(x + 3)$.
24. Sketch the graph of $p(x) = 1 - x^3$.
C 25. Sketch the graph of $p(x) = x^3 + 5x^2 + 2x + 2$.
26. Sketch the graph of $p(x) = x^4$.
27. Sketch the graph of $p(x) = x^3$.

28. Sketch the graph of $f(x) = \dfrac{1}{x - 3}$.

29. Sketch the graph of $f(x) = \dfrac{x - 2}{(x - 3)(x + 1)}$.

30. Sketch the graph of $f(x) = \dfrac{2x - 1}{x + 3}$.

31. Sketch the graph of $f(x) = \dfrac{x^2 - 2x + 3}{x - 1}$.

32. State the Factor Theorem.
33. State the Remainder Theorem.
34. Suppose $p(x)$ has the graph shown below.

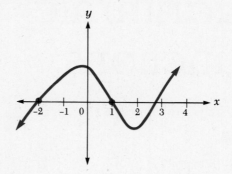

What are two zeros of $p(x)$? Does $p(x)$ have a zero between 2 and 3?
35. Show that $3x^3 + 2x^2 - 3x + 1 = 0$ has no rational roots.

In Exercises 36–39 classify as true or false.
36. A polynomial of degree n has n distinct zeros.
37. If $p(x)$ is a polynomial of degree $n \geq 1$, then there is at least one number c (possibly complex) such that $p(c) = 0$.
38. If c is a zero of $p(x)$, then $(x - c)$ is a factor of $p(x)$.
39. If $p(x)$ is a polynomial and 2 is a zero of $p(x)$, then there must exist points x_1 and x_2 with $x_1 < 2 < x_2$ such that $p(x_1)$ and $p(x_2)$ differ in sign.

CHAPTER 6

Exponential and Logarithmic Functions

6.1   Graphing Exponential and Logarithmic Functions

An exponential function (first introduced in Section 2.5) is a function of the form $y = a^x$, where a is a positive constant. The graph in Figure 1 is from Example 2 of Section 2.5.

Functions of the form $f(x) = b \cdot a^{g(x)} + c$, where b and c are constants and $g(x)$ is a function of x, are also considered exponential functions.

The key to graphing an exponential function is to select values of x that produce both positive and negative exponents.

The graphs shown in Figure 2 illustrate some of the varieties of exponential functions that are frequently found in applications of mathematics.

Looking again at Figure 1 and applying the horizontal line test, we can observe that each horizontal line will cross the graph of $y = 2^x$ in

(c) $3 = \log_4 64$

(d) $\dfrac{1}{2} = \log_{81} 9$ ■

The easiest way to graph logarithmic functions is to use the equivalent exponential form and substitute in values of y (as opposed to our usual way of using x). We should also note that the domain of $f(x) = \log_a x$ is the set of positive real numbers, since $x = a^y$ yields only positive values of x.

Example 2   Graph $y = \log_3 x$.

Solution $y = \log_3 x$ is equivalent to $x = 3^y$. We prepare a table of points using $x = 3^y$:

x	y
$3^2 = 9$	2
$3^1 = 3$	1
$3^0 = 1$	0
$3^{-1} = \dfrac{1}{3}$	-1
$3^{-2} = \dfrac{1}{3^2} = \dfrac{1}{9}$	-2

We plot the points and connect them to generate the graph in Figure 4. ■

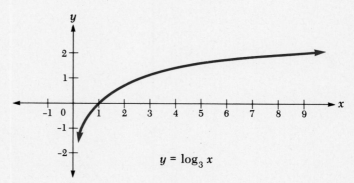

$$y = \log_3 x$$

FIGURE 4

The best way to become proficient at graphing logarithmic functions is to see several examples, then practice for yourself on a variety of exercises.

Example 3   (a) Graph $y = 5 \log_2 x$.   (b) Graph $y = \log_3 (x - 1)$.
(c) Graph $y = 2 + \log_2 x$.   (d) Graph $y = \log_{1/3} x$.

Solution   (a) $y = 5 \log_2 x$ is equivalent to $y/5 = \log_2 x$, or

$$x = 2^{y/5}$$

The graph appears in Figure 5.

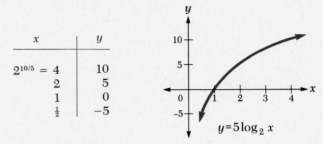

x		y
$2^{10/5} = 4$		10
2		5
1		0
$\frac{1}{2}$		-5

FIGURE 5

(b) $y = \log_3 (x - 1)$ is equivalent to $x - 1 = 3^y$, or

$$x = 3^y + 1$$

The graph appears in Figure 6.

x		y
$3^2 + 1 = 10$		2
4		1
2		0
$1\frac{1}{3}$		-1
$1\frac{1}{9}$		-2

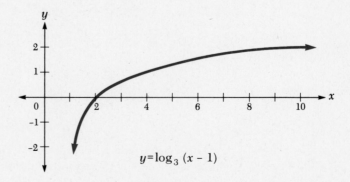

FIGURE 6

(c) $y = 2 + \log_2 x$ is equivalent to $y - 2 = \log_2 x$, or

$$x = 2^{y-2}$$

The graph appears in Figure 7.

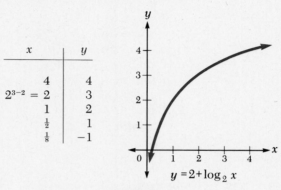

x	y
4	4
$2^{3-2} = 2$	3
1	2
$\frac{1}{2}$	1
$\frac{1}{8}$	-1

$y = 2 + \log_2 x$

FIGURE 7

(d) $y = \log_{1/3} x$ is equivalent to $x = (\frac{1}{3})^y$. The graph appears in Figure 8. ■

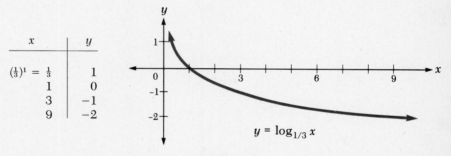

x	y
$(\frac{1}{3})^1 = \frac{1}{3}$	1
1	0
3	-1
9	-2

$y = \log_{1/3} x$

FIGURE 8

Some observations are in order regarding Examples 2 and 3. First, obtaining the exponential form involves isolating the log part of the expression. Second, after obtaining the exponential form, it is usually best to solve for x as an expression in y to simplify calculations since values of y are substituted in the expression to obtain values of x. Third, generally the graph of an exponential function is above a fixed bounding line ($y = b$), and generally the graph of a logarithmic function is to the right of a fixed bounding line ($x = c$). This is because $a^x > 0$ for all x in exponential functions, or equivalently, the domain of $\log_a x$ consists of positive values of x. The bounding line $y = b$ for exponential functions or $x = c$ for logarithmic functions is called an **asymptote**. The graph of the function approaches the asymptote but does not cross it.

Example 4   What are the asymptotes in Figures 1 and 2 and Examples 2 and 3?

Solution   The asymptotes are the following lines:

Figure 1: $y = 0$
Figure 2(a): $y = 0$
Figure 2(b): $y = 0$
Figure 2(c): $y = 1$ $(y = 2^x + 1 \geq 0 + 1)$
Figure 2(d): $y = 0$
Example 2: $x = 0$
Example 3(a): $x = 0$
Example 3(b): $x = 1$ $(x - 1 \geq 0 \rightarrow x \geq 1)$
Example 3(c): $x = 0$
Example 3(d): $x = 0$ ∎

Exercises 6.1

In Exercises 1–8 express the equation in exponential form.
1. $2 = \log_8 64$
2. $3 = \log_5 125$
3. $-1 = \log_2 \frac{1}{2}$
4. $2 = \log_4 16$
5. $6 = \log_{10} 1{,}000{,}000$
6. $5 = \log_3 243$
7. $3 = \log_{1/2} \frac{1}{8}$
8. $-3 = \log_3 \frac{1}{27}$

In Exercises 9–16 express the equation in logarithmic form.
9. $32 = 2^5$
10. $25 = 5^2$
11. $3 = 27^{1/3}$
12. $4 = \sqrt{16}$
13. $10{,}000 = 10^4$
14. $3 = 9^{0.5}$
15. $16 = \left(\frac{1}{2}\right)^{-4}$
16. $\frac{1}{10} = 100^{-1/2}$

In Exercises 17–32 carefully graph the function.
17. $y = 3^x$
18. $y = 3^{-x}$
19. $y = \frac{1}{3} \cdot 2^x$
20. $y = 3^x - 1$
21. $y = 3^{x-1}$
22. $y = \frac{1}{3} \cdot 2^x + 3$
23. $y = \frac{1}{2} \cdot 3^{x+1}$
24. $y = 5^{x-2} + 2$
25. $y = \log_2 x$
26. $y = 2 \log_3 x$
27. $y = (\log_2 x) - 1$
28. $y = \log_2 (x - 3)$
29. $y = \log_3 (x + 2)$
30. $y = \log_{1/2} x$
31. $y = 2 \log_4 (x - 1)$
32. $y = 2 + \log_3 (x + 2)$
33. What are the asymptotes in Exercises 17, 19, 21, 23, 25, 27, 29, and 31?
34. What are the asymptotes in Exercises 18, 20, 22, 24, 26, 28, 30, and 32?
35. The function $N = 3 \cdot 2^x$ describes the number (N) of a particular strain of bacteria as a function of time in hours (x). Describe in words the relationship. (*Hint:* Compute N for $x = 0, 1, 2,$ and 3 and describe the results.)

36. The population of a small country is 1 million and is expected to quadruple every century. Express population as a function of centuries in symbols.

6.2   Properties of Logarithms

Several useful properties result from the definition of logarithms.

PROPERTIES OF LOGARITHMS

Assume that $a > 0$, $a \neq 1$, $m > 0$, $n > 0$, $x \in R$:
1. $\log_a a^x = x$
2. $a^{\log_a m} = m$
3. $\log_a (mn) = \log_a m + \log_a n$
4. $\log_a \dfrac{m}{n} = \log_a m - \log_a n$
5. $\log_a m^x = x \cdot \log_a m$

Properties 1 and 2 are direct results of the facts that $y = a^x$ and $y = \log_a x$ are inverse functions and that the composition of inverse functions yields an identity function (see Section 2.6).

Property 3 results from the definition of $\log_a x$. Let $c = \log_a m$ and $d = \log_a n$, then $m = a^c$ and $n = a^d$. The following steps establish property 3:

$$
\begin{aligned}
\log_a (mn) &= \log_a (a^c a^d) &&\text{(exponential form)} \\
&= \log_a a^{c+d} &&\text{(property of exponents)} \\
&= c + d &&\text{(property 1 of logarithms)} \\
&= \log_a m + \log_a n &&\text{(definition of } c \text{ and } d\text{)}
\end{aligned}
$$

Properties 4 and 5 are established in a similar fashion. At the heart of each is a property of exponents. This should not be surprising, since logarithms are merely exponents of numbers relative to a specified base.

Example 1   (a) Show that $\log_a a = 1$.   (b) Show that $\log_a 1 = 0$.

Solution   (a) $\log_a a = \log_a a^1 = 1$ (property 1)
(b) $\log_a 1 = \log_a a^0$ (since $a^0 = 1$)
          $= 0$ (property 1) ∎

The properties of logarithms generally allow for logarithms of complicated expressions to be rewritten as expressions involving logarithms of simpler expressions. This fact is useful for computations involving logarithms (see Section 6.4).

Example 2   (a) Rewrite $\log_2 \dfrac{8\sqrt{x}(x + 1)}{5}$.   (b) Rewrite $\log_{10} (3.14)^{1.41}$.

(c) Rewrite $\log_3 9$.

Solution   (a) $\log_2 \dfrac{8\sqrt{x}(x + 1)}{5} = \log_2 [8\sqrt{x}(x + 1)] - \log_2 5$ (property 4)

$$= \log_2 8 + \log_2 \sqrt{x} + \log_2 (x + 1) - \log_2 5$$

(property 3)

$$= \log_2 2^3 + \log_2 x^{1/2} + \log_2 (x + 1) - \log_2 5$$

(since $8 = 2^3$ and $\sqrt{x} = x^{1/2}$)

$$= 3 + \frac{1}{2}\log_2 x + \log_2 (x + 1) - \log_2 5$$

(properties 1 and 5)

(b) $\log_{10} (3.14)^{1.41} = 1.41 \log_{10} 3.14$ (property 5)

(c) $\log_3 9 = \log_3 3^2$ (since $9 = 3^2$)

$= 2$ (property 1) ■

The properties of logarithms may also allow us to combine terms involving logarithms into a single expression. This is useful in solving certain types of equations (see Section 6.6).

Example 3   (a) Combine $\log_2 14 - \log_2 7$.   (b) Combine $\log_{10} 3 + 2 \log_{10} x$.
(c) Combine $1 - \log_3 \pi$.

Solution   (a) $\log_2 14 - \log_2 7 = \log_2 \dfrac{14}{7}$ (property 4)

$$= \log_2 2 \qquad \left(\frac{14}{7} = 2\right)$$

$$= 1 \qquad \text{(see Example 1)}$$

(b) $\log_{10} 3 + 2 \log_{10} x = \log_{10} 3 + \log_{10} x^2$ (property 5)

$\qquad\qquad\qquad = \log_{10} 3x^2$ (property 3)

(c) $1 - \log_3 \pi = \log_3 3 - \log_3 \pi$ (see Example 1)

$\qquad\qquad\quad = \log_3 \dfrac{3}{\pi}$ (property 4) ∎

Exercises 6.2

In Exercises 1–16 rewrite the expression in a manner similar to that used in Example 2.

1. $\log_{10} 2x$
2. $\log_2 4z$
3. $\log_3 \dfrac{a}{7}$
4. $\log_{10} \dfrac{10}{c}$
5. $\log_2 x^5$
6. $\log_7 4^w$
7. $\log_3 9^2$
8. $\log_5 5^{1.23}$
9. $\log_2 xy^2z$
10. $\log_{10} \dfrac{2x}{\sqrt{y}}$
11. $\log_3 \sqrt{11ab^2}$
12. $\log_2 \sqrt{\dfrac{3}{2x}}$
13. $\log_2 32$
14. $\log_3 81$
15. $\log_{10} \sqrt{2}(x + y + z)$
16. $\log_5 x^{\sqrt{2}}$

In Exercises 17–32 combine the expressions in a manner similar to that used in Example 3.

17. $\log_2 9 + \log_2 x$
18. $\log_{10} x + \log_{10} y + \log_{10} z$
19. $\log_3 2 - \log_3 w$
20. $\log_7 x - \log_7 16$
21. $2 \log_{10} x + \log_{10} z$
22. $\log_3 4 + 5 \log_3 x$
23. $\frac{1}{2}(\log_7 x - \log_7 y)$
24. $3 \log_2 x - 5 \log_2 y$
25. $\log_{10} 600 - \log_{10} 6$
26. $\log_{10} 25 + \log_{10} 4$
27. $2 - \log_3 x$
28. $3 + \log_2 y$
29. $\log_2 x - 3 \log_2 y - \log_2 5$
30. $\frac{1}{2} \log_3 x + \frac{1}{3} \log_3 y - \log_3 z$
31. $1 - \log_{10} x + 2 \log_{10} c$
32. $2 \log_7 x - \log_7 y - \log_7 x$
33. Establish property 4 of logarithms.
34. Establish property 5 of logarithms.

6.3   Scientific Notation

Since scientists frequently deal with very large and very small numbers, they use a special way of representing numbers, called **scientific notation.** Scientific notation is also useful in connection with tables of logarithms and computations with logarithms (see Section 6.4).

A number other than 0 is written in scientific notation when it has the form: a nonzero digit (1 through 9) followed by a decimal point followed by the remaining digits in the number, then multiplied by an appropriate power of 10. Another way of expressing this form is $a \times 10^b$, where a is a real number with $1 \le a < 10$ and b is an integer.

Example 1 Write the following numbers in scientific notation:
(a) 327 (b) 1,308,000
(c) 51.9 (d) 0.00336

Solution (a) 3.27×10^2
(b) 1.308×10^6
(c) 5.19×10^1
(d) 3.36×10^{-3} ∎

Note in Example 1 that multiplication by 10 moves the decimal point one place to the right and multiplication by 10^{-1} (division by 10) moves the decimal point one place to the left. Thus multiplication by 10^6 moves the decimal point 6 places to the right, and multiplication by 10^{-3} moves the decimal point 3 places to the left.

It is also useful to represent numbers in standard decimal form when they are given in scientific notation.

Example 2 Write the following numbers in standard decimal form:
(a) 1.0×10^4 (b) 7.69×10^{-2}
(c) 2.71×10^0 (d) 4.475×10^8

Solution (a) 10,000
(b) 0.0769
(c) 2.71
(d) 447,500,000 ∎

Numbers arising in real-world applications are usually approximate or measured only to a certain degree of accuracy. This results in the concept of **significant digits.** What is meant is the number of digits that are valid or accurate.

Example 3 Express 203.9 in scientific notation with:
(a) two digits of accuracy. (b) three digits of accuracy.
(c) four digits of accuracy.

Solution   (a) 2.0×10^2
              (b) 2.04×10^2
              (c) 2.039×10^2 ■

The answers in Example 3 illustrate **rounding** to the lowest digit of accuracy. If the next digit is 5 or larger, add 1 to the last digit retained; otherwise, simply drop the extra digits.

A common-sense rule related to significant digits is that the result of computations can be no more accurate than the least accurate of the quantities involved.

Example 4   Multiply 3.19×10^4 by 8.235×10^2.

Solution

$$(3.19 \times 10^4) \cdot (8.235 \times 10^2) = 26.26965 \times 10^6$$

(multiply numbers and add exponents of 10)

$$= 2.626965 \times 10^7$$

(adjust to scientific notation)

$$= 2.63 \times 10^7$$

(round to three digits of accuracy, since 3.19 has only three digits of accuracy)

■

It is interesting to note that computers store decimal numbers in a form of exponential notation. Roughly speaking, a number like 23.71 would be stored as $+0.2371E+02$. There are four pieces in this number: the sign of the number, the digits from the number, the sign after E (E takes the place of $\times 10$) to indicate the sign of the exponent, and the digits after E, which indicate the exponent of 10.*

Exercises 6.3   In Exercises 1–10 write the number in scientific notation.

1.	1920	2.	370,100
3.	36.9	4.	186.23
5.	0.21	6.	0.00021
7.	5.29	8.	0.0667
9.	0.0000000118	10.	91,002,000,000

* The form of internal storage in a computer is actually in binary (base 2) as opposed to our familiar decimal (base 10). The form given is sometimes used for input and/or output with some computer programming languages.

In Exercises 11–20 write the number in standard decimal form.

11. 7.2×10^1

12. 8.04×10^{-1}

13. 9.98×10^3

14. 1.824×10^0

15. 5.0437×10^{-5}

16. 3.10×10^2

17. 6.55×10^5

18. 1.7×10^{-2}

19. 3.96×10^{-12}

20. 4.223×10^{11}

In Exercises 21–26 express the number to the following degree of accuracy: (a) 2 digits; (b) 3 digits; (c) 4 digits.

21. 5.362×10^4

22. 1.8×10^3

23. 7.49×10^1

24. 6.8172×10^{-2}

25. 2.345×10^{-6}

26. 9.9938×10^7

C In Exercises 27–34 perform the indicated computations and express your answer in scientific notation with the appropriate number of significant digits.

27. $(1.6 \times 10^4) \cdot (2.36 \times 10^2)$

28. $(3.11 \times 10^1) \cdot (6.04 \times 10^{-3})$

29. $(9.26 \times 10^{-3}) \cdot (8.305 \times 10^{-2})$

30. $(4.18 \times 10^3) \div (6.4 \times 10^8)$

31. $(7.221 \times 10^9) \div (8.138 \times 10^5)$

32. $(8.016 \times 10^4) \div (2.6 \times 10^7)$

33. $(3.99 \times 10^0) - (1.3 \times 10^0)$

34. $(1.478 \times 10^{-2}) + (8.87 \times 10^{-2})$

6.4   Common Logarithms

Computations with logarithms have traditionally made use of base 10 logarithms.* Since our number system is base 10, this has generally been a good approach. Base 10 logarithms are called **common logarithms** and written log with base 10 understood.

Now that a base is fixed, the computations make use of a table of common logarithms. We enter the table and exit the table by means of scientific notation (see Section 6.3). To handle three digits of accuracy, the table runs from 1.00 to 9.99. To find the common log of $a.bc$, we look for the row with $a.b$ at the left and go across that row until we reach the entry in the column labeled c. This entry is log $a.bc$. (See Figure 9 for log 1.13; here $a = 1$, $b = 1$, $c = 3$.) More generally, log $(a.bc \times 10^d)$ is log $(a.bc) + d$ by the properties of logarithms:

$$\log (a.bc \times 10^d) = \log (a.bc) + \log (10^d) = \log (a.bc) + d$$

The table of common logarithms is given at the back of the book.

* See "A Brief History of Logarithms," by R. C. Pierce, Jr., *Two Year College Mathematics Journal*, **8** (1977), pp. 22–26.

Table of Common Logarithms

	0	1	2	3	4	5	6	7	8	9
1.0	0.0000	0.0043	0.0086	0.0128	0.0170	0.0212	0.0253	0.0294	0.0334	0.0374
→ 1.1	0.0414	0.0453	0.0492	0.0531	0.0569	0.0607	0.0645	0.0682	0.0719	0.0755

FIGURE 9   Using the table to get log 1.13 = 0.0531.

Example 1   Determine each of the following:
(a) log 1.29   (b) log 318
(c) log 0.00725

Solution   (a) log 1.29 =   0.1106   (direct from table)

(b) log 318 = log (3.18×10^2)
$\qquad\qquad = \log (3.18) + 2$

$\qquad\qquad = 0.5024 + 2$

(c) log 0.00725 = log (7.25×10^{-3})
$\qquad\qquad\qquad = (\log 7.25) - 3$

$\qquad\qquad\qquad = 0.8603 - 3$ ■

Note that the logarithms we are using have four digits of accuracy. Any level of accuracy is possible, but four digits are adequate for our purposes. The four-digit part of the logarithm is called the **mantissa,** and the integer exponent of 10 is called the **characteristic.** The two can be combined, but many applications are worked more easily if they are separate. Note that the mantissa is *always* nonnegative (greater than or equal to 0).

It is also useful to be able to reconstruct a number from its logarithm.

Example 2   Find x in each of the following:
(a) log x = 0.4669   (b) log x = 5.7528
(c) log x = 0.9036 − 2

Solution   (a) log x = 0.4669

x = 2.93   (direct from table)

(b) $\log x = 0.7528 + 5$ (usual form)

$$x = 5.66 \times 10^5 \quad \text{or} \quad 566{,}000$$

(c) $\log x = 0.9036 - 2$

$$x = 8.01 \times 10^{-2} \quad \text{or} \quad 0.0801 \quad \blacksquare$$

The value of x for each part of Example 2 is called the **antilogarithm** of the given number; for instance, 2.93 is the antilogarithm of 0.4669 (equivalently, $10^{0.4669} = 2.93$). The antilogarithm of the four-digit decimal is found by reversing our earlier process; that is, find the four-digit decimal in the body of the table and locate $a.b$ at the left and c above to give $a.bc$ (refer again to Figure 9). The integer becomes the exponent of 10 in scientific notation.

The process of using logarithms in computations usually involves taking logarithms of a complicated expression, applying the properties of logarithms, performing the arithmetic, then finding the antilogarithm to obtain the desired answer.

Example 3 (a) Compute $\dfrac{(3.14)(9.72)}{(7.11)^2}$ using logarithms.

(b) Compute $\sqrt[5]{20}$ using logarithms.

(c) Compute $(0.0361)^{2/3}$

Solution (a) Let $x = \dfrac{(3.14)(9.72)}{(7.11)^2}$. Taking the logarithm of each side and using the properties of logarithms gives

$$\log x = \log 3.14 + \log 9.72 - 2 \log 7.11$$

$$\log x = 0.4969 + 0.9877 - 2(0.8519) \quad \text{(from the table)}$$

$$\log x = 1.4846 - 1.7038$$

$$\log x = -0.2192$$

$$\log x = 0.7808 - 1 \quad \text{(usual form: add } +1 \text{ to } -0.2192 \text{ to obtain a positive mantissa; subtract 1 to compensate for adding 1)}$$

$$x = 6.04 \times 10^{-1} \text{ or } \boxed{0.604} \quad \text{(antilogarithm)}$$

(b) Let $x = \sqrt[5]{20} = 20^{1/5}$. Then

$$\log x = \frac{1}{5} \log 20 \qquad \text{(take logarithms; use property 5)}$$

$$= \frac{1}{5} \log (2.00 \times 10^1)$$

$$= \frac{1}{5} (\log 2.00 + 1)$$

$$= \frac{1}{5} (0.3010 + 1)$$

$$= \frac{1}{5} (1.3010)$$

$$\log x = 0.2602$$

$$x = \boxed{1.82} \qquad \text{(antilogarithm)}$$

(c) Let $x = (0.0361)^{2/3}$. Then

$$\log x = \frac{2}{3} \log 0.0361 \qquad \text{(take logarithms; use property 5)}$$

$$= \frac{2}{3} \log (3.61 \times 10^{-2})$$

$$= \frac{2}{3} (\log 3.61 - 2)$$

$$= \frac{2}{3} (0.5575 - 2)$$

$$= \frac{2}{3} (1.5575 - 3) \qquad \text{(add 1 to mantissa and } -1 \text{ to characteristic so new integer is divisible by 3)}$$

$$\log x = 1.0383 - 2$$

$$\log x = 0.0383 - 1 \qquad \text{(usual form)}$$

$$x = 1.09 \times 10^{-1} \text{ or } \boxed{0.109} \qquad \text{(antilogarithm)} \quad \blacksquare$$

Calculations such as those in Example 3 often result in mantissas that are not in the table. Normally, we take the value that is closest. In case of a tie, pick either of the values that is closest.

We can estimate the logarithm of a four-digit number by a process called **interpolation.** If the accuracy is justified, we can also find an antilogarithm to four digits by the same process.

Example 4   (a) Estimate log 3.926.

(b) Find x (to four digits) if log $x = 0.8150 + 2$.

Solution   (a)

$$10\left(6\left(\begin{matrix}\log 3.92 = .5933 \\ \log 3.926 = ? \\ \log 3.93 = .5944\end{matrix}\right) x\right)11$$

We form a proportion to find x. The units must be consistent (on the left they are each hundredths; on the right they are each ten-thousandths).

$$\frac{6}{10} = \frac{x}{11}$$

$$10x = 66$$

$$x = 6.6 \quad \text{round to 7}$$

Thus log $3.926 = 0.5933 + 0.0007$

$$= \boxed{0.5940}$$

(b)

$$10\left(y\left(\begin{matrix}\log 6.53 = .8149 \\ \log ? = .8150 \\ \log 6.54 = .8156\end{matrix}\right) 1\right)7$$

We form a proportion to find y. The units must be consistent (on the left they are each hundredths; on the right they are each ten-thousandths).

$$\frac{y}{10} = \frac{1}{7}$$

$$7y = 10$$

$$y = 1.4 \quad \text{round to 1}$$

Thus log $(6.53 + 0.001) = \log 6.531 = 0.8150$, so that

$$\log x = 0.8150 + 2$$

$$\boxed{x = 6.531 \times 10^2 \quad \text{or} \quad 653.1} \quad \blacksquare$$

Inexpensive calculators have made life much easier with respect to logarithms. For instance, the results of Examples 1, 2, 3, and 4 can all be found quickly with many calculators.

Most calculators with a logarithmic function either are algebraic in nature or use the "Reverse Polish Notation" (RPN). The next example shows the button sequences for both. The $\boxed{\text{C}}$ is the clear button; $\boxed{\text{ENTER}}$ denotes the enter button for RPN; $\boxed{\text{LOG}}$ denotes the common log button; $\boxed{\text{INV}}$ indicates the inverse function button; and $\boxed{y^x}$ indicates the exponential button, for which you specify the base and the exponent. All other buttons are self-explanatory. (*Note:* Your calculator may label the buttons in a slightly different way; consult the user's guide.) The interpolation example is not a problem, since the calculator can provide more accuracy than a log table.

Example 5 Use a calculator to find the following:
(a) log 318   (b) log 0.00725
(c) x if log $x = 5.7528$ (d) x if log $x = 0.9036 - 2$
(e) $\sqrt[5]{20}$ (f) $(0.0361)^{2/3}$

Solution (a) *Algebraic:* [C] [3] [1] [8] [LOG] [=]

Display is [2.5024271]

RPN: [C] [3] [1] [8] [ENTER] [LOG]

Display is [2.502427]

(b) *Algebraic:* [C] [.] [0] [0] [7] [2] [5] [LOG] [=]

Display is [−2.139662]

RPN: [C] [.] [0] [0] [7] [2] [5] [ENTER] [LOG]

Display is [−2.139662]

(c) *Algebraic:* [C] [5] [.] [7] [5] [2] [8] [INV]

[LOG] [=]

Display is [565978.58]

RPN: [C] [5] [.] [7] [5] [2] [8] [ENTER] [INV]

[LOG]

Display is [565977.2]

(d) *Algebraic:* [C] [.] [9] [0] [3] [6] [−] [2] [=]

[INV] [LOG] [=]

Display is [0.080094]

RPN: [C] [.] [9] [0] [3] [6] [ENTER] [2] [−]

[INV] [LOG]

Display is [.080094]

(e) (Use .2 for $\frac{1}{5}$ power, $\sqrt[5]{}$.)

Algebraic: [C] [2] [0] [y^x] [.] [2] [=]

Display is [1.8205642]

RPN: [C] [2] [0] [ENTER] [.] [2] [y^x]

Display is [1.820563]

(f) (Use .66666667 for $\frac{2}{3}$ power.)

Algebraic: [C] [.] [0] [3] [6] [1] [y^x] [.] [6]

[6] [6] [6] [6] [6] [6] [7] [=]

Display is [.10922904]

RPN: [C] [.] [0] [3] [6] [1] [ENTER] [.] [6]

[6] [6] [6] [6] [6] [6] [7] [y^x]

Display is [.1092291] ■

Some observations should be made about Example 5. First, the calculator can give us more digits of accuracy than we may be justified in using, so the answers should be rounded to the appropriate degree of accuracy. Second, the mantissa-characteristic separation is not needed or used with calculators. Third, if the [INV] button is not present, we can use the [y^x] button with 10 as the base and the logarithm as the exponent.

Exercises 6.4

In Exercises 1–10 use the table to find the logarithm of the given number.

1. 1
2. 38,100
3. 16.3
4. 709
5. 0.25
6. 0.0632
7. 0.000114
8. 0.00553
9. 18,500,000,000
10. 0.0000000694

In Exercises 11–20 use the table of logarithms to find the antilogarithm of the given number.

11. 0.8910 + 2
12. 0.7076 + 5
13. 0.4232
14. 0.2967 + 1
15. 0.3365 − 2
16. 0.5551 − 1
17. 0.6415 − 4
18. 0.9952 − 3
19. 4.1173
20. − 1.5421

In Exercises 21–30 use logarithms to compute the value.

21. $(80.1)^2(6.54)$

22. $\dfrac{497}{(2.23)(16.5)}$

23. $\sqrt{90.2}$

24. $\sqrt{0.00468}$

25. $(571)^{3/4}$

26. $(0.293)^{0.3}$

27. $(5.62)^4(10.3)^2$

28. $\dfrac{\sqrt{728}}{6.87}$

29. $\sqrt{\dfrac{728}{6.87}}$

30. $[(1.31)(6.77)(4.92)]^3$

In Exercises 31–36 use the technique of Example 4 to interpolate.
31. Estimate log 9.016.
32. Estimate log 245.3.
33. Estimate log 0.05558.
34. Find x (to four digits) if log $x = 0.5709$.
35. Find x (to four digits) if log $x = 0.7377 + 2$.
36. Find x (to four digits) if log $x = 0.9461 - 1$.

C In Exercises 37–44 use a calculator to work the problem.
37. Rework exercises 1, 3, 5, 7, and 9.
38. Rework exercises 2, 4, 6, 8, and 10.
39. Rework exercises 11, 13, 15, 17, and 19.
40. Rework exercises 12, 14, 16, 18, and 20.
41. Rework exercises 31, 33, and 35.
42. Rework exercises 32, 34, and 36.
43. Rework exercises 21, 23, 25, 27, and 29.
44. Rework exercises 22, 24, 26, 28, and 30.

6.5   Bases Other Than 10

Since tables and calculators are available for common logarithms, it is convenient to have a way of converting between such a tabulated base and any other base that may be needed.

CHANGE OF BASE FORMULAS

$$\log_a n = \frac{\log_b n}{\log_b a}$$

$$\log_a b = \frac{1}{\log_b a}$$

The following sequence establishes the first change of base formula:

$$\log_b n = \log_b (a^{\log_a n}) \qquad \text{(since } a^{\log_a n} = n)$$

$$\log_b n = (\log_a n)(\log_b a) \qquad \text{(property 5 of logarithms)}$$

$$\frac{\log_b n}{\log_b a} = \log_a n \qquad \text{(divide by } \log_b a)$$

$$\log_a n = \frac{\log_b n}{\log_b a} \qquad \text{(interchange sides)}$$

The second change of base formula is a special case of the first formula for $n = b$:

$$\log_a b = \frac{\log_b b}{\log_b a} = \frac{1}{\log_b a}$$

Example 1 (a) Compute $\log_2 7$.
(b) Compute $\log_3 10$.
(c) Express $\log_5 x$ with respect to base 10.

Solution (a) $\log_2 7 = \dfrac{\log_{10} 7}{\log_{10} 2} = \dfrac{0.8451}{0.3010} = \boxed{2.808}$

(b) $\log_3 10 = \dfrac{1}{\log_{10} 3} = \dfrac{1}{0.4771} = \boxed{2.096}$

(c) $\log_5 x = \dfrac{\log_{10} x}{\log_{10} 5} = \dfrac{\log x}{0.6990} = \boxed{1.431 \log x}$ ■

The sequence of numbers $(1 + \frac{1}{2})^2$, $(1 + \frac{1}{3})^3$, $(1 + \frac{1}{4})^4$, and so on, gets closer and closer to a natural constant designated e. The constant e appears in business settings (interest compounded continuously), in engineering settings (damped oscillations), and in many other applications of mathematics. The following numbers give a feel for the value of e:

$$\left(1 + \frac{1}{2}\right)^2 = 2.25$$

$$\left(1 + \frac{1}{3}\right)^3 = 2.3704$$

$$\left(1 + \frac{1}{4}\right)^4 = 2.4414$$

$$\left(1 + \frac{1}{10}\right)^{10} = 2.5937$$

$$\left(1 + \frac{1}{100}\right)^{100} = 2.7048$$

The actual value is $\boxed{e = 2.7183}$ (rounded to four decimal places).

The constant e is an irrational number (an infinite nonrepeating decimal). Because e occurs so frequently in applications, logarithms with a base of e are called **natural logarithms** and written $\log_e x$ or simply $\ln x$ (to distinguish from "log" for common logarithms). This is the first time we have encountered logarithms with a base that is not a rational number. The change of base formulas and calculators are particularly useful here.

Example 2   (a) Calculate $\ln 3.62$.   (b) Find x if $\ln x = 0.7514$.

Solution   (a) *Table method:*

$$\ln 3.62 = \frac{\log_{10} 3.62}{\log_{10} e}$$

$$= \frac{0.5587}{0.4346} \quad \longleftarrow \log 2.72$$

$$\boxed{= 1.286}$$

Calculator:

(Algebraic) $\boxed{C}$ $\boxed{3}$ $\boxed{.}$ $\boxed{6}$ $\boxed{2}$ $\boxed{LN}$ $\boxed{=}$

Display is $\boxed{1.286474}$

(RPN) $\boxed{C}$ $\boxed{3}$ $\boxed{.}$ $\boxed{6}$ $\boxed{2}$ $\boxed{ENTER}$ $\boxed{LN}$

Display is $\boxed{1.286474}$

(b) *Table method:*

$$\ln x = \frac{\log_{10} x}{\log_{10} e} = \frac{\log x}{0.4346}$$

Thus,

$$\frac{\log x}{0.4346} = 0.7514$$

$$\log x = (0.4346)(0.7514)$$

$$\log x = 0.3266$$

$$\boxed{x = 2.12} \qquad \text{(antilog)}$$

Calculator:

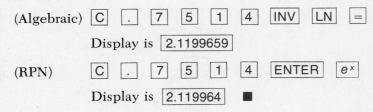

(Algebraic) [C] [.] [7] [5] [1] [4] [INV] [LN] [=]

Display is [2.1199659]

(RPN) [C] [.] [7] [5] [1] [4] [ENTER] [e^x]

Display is [2.119964] ∎

Again, you should check the user's guide for your calculator if the indicated buttons are not available or not labeled in the same way. The accuracy of calculators is quite good, but the methods used by the calculator in response to [LOG], [LN], [e^x], and [y^x] are approximate, and thus the actual accuracy of the display may be a digit or two less than what it appears to be. Since most calculators with these buttons have at least eight digits in the display, there is still greater accuracy than with tables. Do not forget that you cannot get more significant digits out of a calculator than you punch in.

Exercises 6.5

In Exercises 1–10 compute the logarithm.

C 1. $\log_5 10$ 2. $\log_{12} 10$
3. $\log_7 5$ 4. $\log_5 7$
5. $\log_2 32$ 6. $\log_3 2$
7. $\log_9 18$ 8. $\log_2 11$
9. $\log_3 568$ 10. $\log_{11} 0.00321$

C In Exercises 11–16 express the logarithm with respect to base 10.
11. $\log_2 x$ 12. $\log_3 x$
13. $\log_7 x^2$ 14. $\log_4 2x$
15. $\log_5 \sqrt{x}$ 16. $\log_2 x + \log_3 x$

C In Exercises 17–24 calculate the natural logarithm.
17. $\ln 1.03$ 18. $\ln 7.91$
19. $\ln 82.6$ 20. $\ln 109$
21. $\ln 0.132$ 22. $\ln 0.0527$
23. $\ln 0.00416$ 24. $\ln 10$

C In Exercises 25–30 find x.
25. $\ln x = 0.1062$ 26. $\ln x = 0.3452$

27. $\ln x = 0.8417$
29. $\ln x = 1.637$

28. $\ln x = 2.840$
30. $\ln x = -1.715$

6.6 Solving Logarithmic and Exponential Equations

Many useful applications of logarithms involve solving logarithmic and exponential equations. A key to solving many logarithmic equations is the following property of logarithms:

$$\text{If } \log_a m = \log_a n, \quad \text{then} \quad m = n$$

This is developed as follows:

$\log_a m = \log_a n$ (hypothesis)

$a^{\log_a m} = a^{\log_a n}$ (same base with equal exponents)

$m = n$ (property of logarithms)

Example 1 (a) Solve for x: $3 \log x = \log 8$.
      (b) Solve for x: $\log x + \log (x + 1) = \log 6$.

Solution (a) $3 \log x = \log 8$ (given)

$\log x^3 = \log 8$ (property of logarithms)

$x^3 = 8$ (property of logarithms)

$x = \sqrt[3]{8}$

$x = 2$

(b) $\log x + \log(x - 1) = \log 6$ (given)

$\log [x(x - 1)] = \log 6$ (property of logarithms)

$x(x - 1) = 6$ (property of logarithms)

$x^2 - x - 6 = 0$

$(x - 3)(x + 2) = 0$

$x = 3$ or $x = -2$

($x = -2$ is discarded, since $\log x$ is undefined for negative values of x.) ∎

An exponential equation is usually solved by applying logarithms to each side of the equation.

Example 2   (a) Solve for x: $2^{x-1} = 9$.   (b) Solve for x: $2^x = 3^{2x+1}$.
                (c) Solve for x: $x^{\sqrt{2}} = 7.11$.   (d) Solve for x: $e^x = 4$.

Solution   (a) $2^{x-1} = 9$ (given)

$$\log 2^{x-1} = \log 9 \qquad \text{(apply logarithms to each side)}$$

$$(x - 1) \log 2 = \log 9 \qquad \text{(property of logarithms)}$$

$$x - 1 = \frac{\log 9}{\log 2}$$

$$x = \frac{\log 9}{\log 2} + 1$$

$$= \frac{0.9542}{0.3010} + 1$$

$$x = 4.17 \qquad \text{(from a calculator)}$$

(b) $2^x = 3^{2x+1}$ (given)

$$\log 2^x = \log 3^{2x+1} \qquad \text{(apply logarithms to each side)}$$

$$x \log 2 = (2x + 1) \log 3 \qquad \text{(property of logarithms)}$$

$$x \log 2 = 2x \log 3 + \log 3$$

$$x \log 2 - 2x \log 3 = \log 3$$

$$x(\log 2 - 2 \log 3) = \log 3$$

$$x = \frac{\log 3}{\log 2 - 2 \log 3}$$

$$= \frac{0.4771}{0.3010 - 2(0.4771)}$$

$$x = -0.73 \qquad \text{(from a calculator)}$$

(c) $x^{\sqrt{2}} = 7.11$ (given)

$$\log x^{\sqrt{2}} = \log 7.11 \qquad \text{(apply logarithms to each side)}$$

$$\sqrt{2} \log x = \log 7.11 \qquad \text{(property of logarithms)}$$

$$\log x = \frac{\log 7.11}{\sqrt{2}}$$

$$\log x = \frac{0.8519}{1.414}$$

$$\log x = 0.6025$$

$$x = 4 \qquad \text{(antilog)}$$

(d) $\qquad e^x = 4 \qquad$ (given)

$$\log e^x = \log 4 \qquad \text{(apply logarithms to each side)}$$

$$x \log e = \log 4 \qquad \text{(property of logarithms)}$$

$$x = \frac{\log 4}{\log e}$$

$$= \frac{0.6020}{0.4346}$$

$$x = 1.39$$

If we have a table for ln or a calculator with an $\boxed{\text{LN}}$ button, we can apply natural logarithms rather than common logarithms:

$$e^x = 4 \qquad \text{(given)}$$

$$\ln e^x = \ln 4 \qquad \text{(apply ln)}$$

$$x = \ln 4 \qquad (\ln x \text{ and } e^x \text{ are inverse functions)}$$

$$x = 1.39 \qquad \blacksquare$$

We now consider two applications of the preceding techniques.

Example 3 A certain strain of bacteria is modeled by the function $N = 3 \cdot 2^x$, where N is the number of colonies of the bacteria and x is the number of hours after introduction of the bacteria into a test culture. When will there be 100 colonies?

Solution We need to solve the equation

$$100 = 3 \cdot 2^x$$

$$\log 100 = \log (3 \cdot 2^x) \qquad \text{(apply logarithms to each side)}$$

$$2 = \log 3 + x \log 2 \qquad \text{(property of logarithms; } 100 = 10^2)$$

$$x = \frac{2 - \log 3}{\log 2} \qquad \text{(solve for } x)$$

$$x = \frac{2 - 0.4771}{0.3010}$$

$$x = \boxed{5.06 \text{ hours after introduction}} \quad \blacksquare$$

Example 4   The amount, A, accumulated when a principal, P, is invested at a nominal rate of r for t years, where interest is compounded continuously, is $A = Pe^{rt}$. How long does it take for an investment to double if the nominal rate is 12%? (Note that the number of dollars invested is irrelevant to solving the problem.)

Solution

$A = Pe^{rt}$ (given)

$2P = Pe^{0.12t}$ (2P represents a doubled principal, and 0.12 represents 12%)

$2 = e^{0.12t}$ (divide out P)

$\log 2 = \log e^{0.12t}$ (apply logarithms)

$\log 2 = 0.12t \log e$ (property of logarithms)

$t = \dfrac{\log 2}{0.12 \log e}$

$ = \dfrac{0.3010}{(0.12)(0.4346)}$

$t = \boxed{5.78 \text{ years}}$

With a calculator we could use the following sequence;

$$2 = e^{0.12t}$$

$$\ln 2 = \ln e^{0.12t} \quad \text{(apply ln)}$$

$$0.6931 = 0.12t$$

$$t = \boxed{5.78 \text{ years}} \quad \blacksquare$$

Exercises 6.6   In Exercises 1–24 solve the equation.

1. $4 \log x = \log 81$
2. $\log 2x = \log (4x - 7)$
3. $\log x - \log 3 = \log (x - 4)$
4. $\log 5 + \log x = 2$
5. $1 + \log x = \log 9$
6. $\log (x + 4) - \log x = \log 2$
7. $\log 2 + \log (x - 1) = \log 3$

8. $\log (x - 1) + \log (x - 2) = \log (2x - 4)$

9. $\log (2x + 3) - \log (2x - 3) = \log x$

10. $\frac{1}{2} \log 4 + \log x = \log 8$

11. $2 \log x = 2$

12. $\frac{3}{4} \log 16 + \frac{1}{2} \log x = \log 6 + \log 4$

C 13. $5^x = 10$ C 14. $7^{x+1} = 19$

C 15. $3^{2x} = 4$ C 16. $3^x = 5^{x-1}$

17. $2^{5x} = 8^{x+3}$ C 18. $10^{x-2} = 7^{4x+1}$

C 19. $x^{1.29} = 83.5$ C 20. $x^{3.01} = 142$

C 21. $x^{11.7} = 2.68$ C 22. $e^{2x} = 7$

C 23. $e^{x-1} = 3.14$ C 24. $e^{2x+3} = 781$

C 25. Rework Example 3 for $N = 1200 \cdot 2^{x-1}$ if we want to know when there will be 2000 colonies.

C 26. The population of a certain small country is modeled by $N = N_0 \cdot 10^{0.0146t}$, where N is the population after t years and N_0 is the initial population. In how many years will the population be half again as large as the initial population?

C 27. Rework Example 4 to determine how long it takes for an investment to triple if the nominal rate is 10%.

C 28. The formula for compound interest is $A = P(1 + i)^n$, where A is the accumulated amount, P is the original principal, i is the interest rate per period, and n is the number of periods. If the interest rate is 15% and the interest period is one year, how long will it take for an investment to double in value?

C 29. Radioactive decay is modeled by $x = x_0 e^{-kt}$, where x is the amount of the material present at time t, x_0 is the initial amount, and k is the constant of decay for the material. What is the decay constant if 5% of the material has decayed after 24 units of time?

C 30. In Exercise 29, what is the half-life of the substance (the time when $x = \frac{1}{2}x_0$) if k is 0.06?

Key Terms and Formulas

Exponential function

Inverse function

Logarithm

Base

$y = \log_a x$ if and only if $x = a^y$

Logarithmic function

Asymptote

Properties of logarithms:

$\log_a a^x = x$

$a^{\log_a x} = x$

$\log_a (mn) = \log_a m + \log_a n$

$\log_a \dfrac{m}{n} = \log_a m - \log_a n$

$\log_a m^x = x \cdot \log_a m$

Scientific notation

Significant digits

Rounding

Common logarithm (log)

Mantissa

Characteristic

Antilogarithm

Interpolation

Change of base formula:

$$\log_a n = \frac{\log_b n}{\log_b a}$$

e

Natural logarithm (ln)

Logarithmic equation

Exponential equation

Review Exercises

1. Express in exponential form: $2 = \log 100$.
2. Express in exponential form: $4 = \log_3 81$.
3. Express in logarithmic form: $5 = \sqrt[3]{125}$.
4. Express in logarithmic form: $1{,}000{,}000 = 10^6$.
5. Carefully graph $y = 2^{x-1}$ and identify any asymptotes.
6. Carefully graph $y = \log_2 (x - 1)$ and identify any asymptotes.

7. Rewrite $\log \left(\dfrac{x + y}{z}\right)^2$ by using properties of logarithms.

8. Rewrite $\ln e^2$ by using properties of logarithms.
9. Combine $2 + \frac{1}{2} \log_3 x$ by using properties of logarithms.
10. Combine $\log x - \log z + 3 \log w$ by using properties of logarithms.
11. Write 37,608 in scientific notation.
12. Write 0.0037608 in scientific notation.
13. Write 1.992×10^{-5} in standard decimal form.
14. Write 2.1×10^2 in standard decimal form.
15. Express 9.145×10^3 with two digits of accuracy.
16. Express 8.3×10^{-18} with three digits of accuracy.

C 17. Perform the indicated computation and express your answer in scientific notation with the appropriate number of significant digits:

$$(3.978 \times 10^8) \div (9.6 \times 10^5)$$

C 18. Perform the indicated computation and express your answer in scientific notation with the appropriate number of significant digits:

$$(1.92 \times 10^3) \cdot (6.771 \times 10^{-3})$$

19. Use the table of common logarithms to find log 137.

C 20. Use a calculator to find log 335.4.

21. Use the table of common logarithms to find x if $\log x = 0.4065 - 2$.

C 22. Use a calculator to find x if $\log x = -3.1182$.

23. Use logarithms to compute $\sqrt{2}$.

24. Use logarithms to compute $\dfrac{(41.6)(2.22)}{37.2}$.

25. Use the table of common logarithms and interpolation to estimate log 1042.

26. Use the table of common logarithms and interpolation to estimate x if log $x = 0.9044 + 3$.

C 27. Compute $\log_2 10$.

C 28. Compute $\log_7 13$.

29. Express $\log_9 x$ with respect to base 10.

30. Express $\log_2 (1/x)$ with respect to base 10.

C 31. Calculate ln 33.7.

C 32. Calculate ln 0.000337.

C 33. Find x if ln $x = 2.103$.

C 34. Find x if ln $x = 0.3714$.

35. Solve $\log x - \log (x - 1) = \log 2$.

36. Solve $\dfrac{1}{3} \log x = \log 3$.

C 37. Solve $x^9 = 2.17$.

C 38. Solve $3^x = 2^{3x-1}$.

C 39. The radioactive decay of a certain element is modeled by $x = x_0 e^{-37t}$, where t is time in years and x_0 is the initial amount of material. What is the half-life of the element, that is, when will one half of the material have decayed?

C 40. If $300 is invested at 18% compounded continuously, when will the investment double in value?

CHAPTER 7

Matrix Algebra

7.1 Introduction

You probably began your study of mathematics with arithmetic. In arithmetic you learned to perform mathematical operations (such as addition) using numerals. A typical problem is $2 + 3 = 5$.

After mastering arithmetic, you progressed to algebra, which is sometimes defined as "generalized arithmetic." In algebra you learned to perform mathematical operations (such as addition) using literal numbers. A typical problem is $2x + 3x = 5x$.

In matrix algebra you will learn to perform mathematical operations using rectangular arrays of elements. Each horizontal line of elements in a matrix is called a **row**, and each vertical line of elements is called a **column**.

DEFINITION   A **matrix** is a rectangular array of elements consisting of m rows and n columns. The elements may be variables or numerals or both.

The array $\begin{pmatrix} 2 & 1 & y \\ 1 & x & -2 \end{pmatrix}$ is a matrix containing 2 rows and 3 columns.

The array $\begin{pmatrix} 1 \\ 2 \\ 3 \end{pmatrix}$ is a matrix containing 3 rows and 1 column. A matrix consisting of a single column (or a single row) is also called a **vector**.

The array $\begin{pmatrix} a & 1 & -2 \\ 0 & x & 5 \\ 1 & 1 & 1 \end{pmatrix}$ is a matrix containing 3 rows and 3 columns. A matrix consisting of the same number of rows and columns is called a **square matrix.**

The development of matrix algebra is similar to the development of other mathematical systems that you have studied. The development may be summarized in four steps:

1. A set of elements is defined. In matrix algebra the elements are rectangular arrays.
2. Certain operations are explained. In matrix algebra we discuss
   addition (Section 7.2)
   subtraction (Section 7.3)
   scalar multiplication (Section 7.3)
   matrix multiplication (Section 7.4)
   row operations (Section 7.5)
   inversion (Section 7.6)
   determinants (Section 7.8)

   The operations of addition, subtraction, and multiplication are binary operations—that is, two matrices are required. The remaining operations are performed on a single matrix. (In arithmetic, addition is binary; two numbers are required. The operation "square root" is not binary; only one number is required.)
3. Properties of the operations are determined. It will be noted that matrix addition is commutative and associative; matrix multiplication is not commutative.
4. The elements and operations are used to solve realistic problems. Two types of applications will be made. The first application of matrices capitalizes on the fact that a matrix is basically a table that can be used to organize information (Sections 7.2, 7.3, and 7.4). The second application is to use matrices to solve systems of linear equations. The matrix methods for solving linear systems (Sections 7.5, 7.7, and 7.8) are called **numerical methods.** The systems will be solved using only arrays of numerals.

7.2   Matrices and Addition

The study of matrix algebra requires a more detailed definition of a matrix than the one given in Section 7.1. In particular, we need a definition that will provide a way for us to refer to any particular entry in the matrix. This will help us understand operations involving two or more matrices.

DEFINITION

A **matrix** is a rectangular array of entries denoted by

$$A = \begin{pmatrix} a_{11} & a_{12} & a_{13} & \cdots & a_{1n} \\ a_{21} & a_{22} & a_{23} & \cdots & a_{2n} \\ a_{31} & a_{32} & a_{33} & \cdots & a_{3n} \\ \cdot & \cdot & \cdot & & \cdot \\ \cdot & \cdot & \cdot & & \cdot \\ \cdot & \cdot & \cdot & & \cdot \\ a_{m1} & a_{m2} & a_{m3} & \cdots & a_{mn} \end{pmatrix} = (a_{ij})_{m \times n}$$

Each entry is called an **element** of the matrix. The second notation given is a commonly used abbreviation.

The location of each element in the matrix is expressed by a unique pair of subscripts. The first subscript indicates the row in which the element is located; the second subscript indicates the column. For instance, a_{31} is in row 3 and column 1.

Example 1   Find the a_{11}, a_{23}, and a_{32} elements in the matrix

$$A = \begin{pmatrix} 0 & 1 & 2 \\ -2 & 1 & \frac{1}{2} \\ 7 & 0 & 10 \end{pmatrix}$$

What is the location of the element 10?

Solution   The element $a_{11} = 0$, $a_{23} = \frac{1}{2}$, and $a_{32} = 0$. The element 10 is in the a_{33} position.   ∎

Order of a Matrix

The order of a matrix is an (ordered) pair of numbers. The *first* mentioned number indicates the number of *rows* in the matrix. The *sec-*

ond mentioned number indicates the number of *columns* in the matrix.

If a matrix has m rows and n columns, it is said to be of **order** (or size) m by n, denoted $m \times n$.

(*Note:* $m \times n$ does *not* indicate multiplication.)

A matrix that has 6 rows and 2 columns is of order 6×2. A matrix of order 3×5 has 3 rows and 5 columns.

Example 2 Determine the order of each of the following:

$$A = \begin{pmatrix} 1 & 2 & 3 & 4 \\ 5 & 6 & 7 & 8 \\ 9 & 7 & 4 & 1 \end{pmatrix}; \qquad B = \begin{pmatrix} 1 & 2 \\ 3 & 4 \end{pmatrix};$$

$$C = (1 \quad 2 \quad 3); \qquad D = \begin{pmatrix} 1 \\ 2 \\ 3 \end{pmatrix};$$

$$E = \begin{pmatrix} 0 & 0 & 0 \\ 0 & 0 & 0 \\ 0 & 0 & 0 \end{pmatrix}; \qquad F = (1).$$

Solution The matrix A is of order 3×4, B is of order 2×2, C is of order 1×3, D is of order 3×1, E is of order 3×3, and F is of order 1×1. ∎

SPECIAL-ORDER MATRICES

1. A matrix that has only one column ($n = 1$) is called a **column matrix.**
2. A matrix that has only one row ($m = 1$) is called a **row matrix.**
3. A matrix that has the same number of rows and columns ($m = n$) is called a **square matrix.**

In Example 2, C is a row matrix and D is a column matrix. Matrices B and E are square matrices. Matrix F is a square matrix consisting of a single row and a single column.

The first consideration given to relationships between two matrices involves the concept of matrix equality.

DEFINITION OF EQUAL MATRICES

The matrix $A = (a_{ij})_{m \times n}$ is **equal** to the matrix $B = (b_{ij})_{m \times n}$, denoted $A = B$, provided that

1. they are the same size (both $m \times n$), and
2. their corresponding elements are equal, that is, each $a_{ij} = b_{ij}$ for all i and j.

Example 3 Let

$$A = \begin{pmatrix} 1 & 2 \\ 3 & 4 \end{pmatrix}; \qquad B = \begin{pmatrix} 1 & 2 & 3 \\ 4 & 5 & 6 \end{pmatrix};$$

$$C = \begin{pmatrix} 1 & 2 \\ 3 & 5 \end{pmatrix}; \qquad D = \begin{pmatrix} 1 \\ 2 \\ 3 \end{pmatrix};$$

$$E = \begin{pmatrix} \frac{3}{3} & 2 \\ \frac{12}{4} & \frac{-8}{-2} \end{pmatrix}.$$

Which matrices are equal?

Solution Matrix A = matrix E, since they are the same size (2×2) and their corresponding elements are equal, that is, $a_{11} = 1 = \frac{3}{3}$, $a_{12} = 2$ (for both), $a_{21} = 3 = \frac{12}{4}$, and $a_{22} = 4 = -8/(-2)$.

No other pair of (different) matrices satisfies the definition. (Of course, each matrix is equal to itself.) ∎

DEFINITION A **matrix equation** is a statement of equality between two matrices.

Matrix equations may be classified as follows:

1. An **identity** such as $\begin{pmatrix} 0 & 1 \\ 2 & 3 \end{pmatrix} = \begin{pmatrix} 0 & 1 \\ 2 & 3 \end{pmatrix}$,

2. A **conditional equation** such as $\begin{pmatrix} x \\ y \end{pmatrix} = \begin{pmatrix} 1 \\ 2 \end{pmatrix}$, or

3. An **impossible equation** such as $\begin{pmatrix} 1 \\ 2 \end{pmatrix} = \begin{pmatrix} 1 & 2 \\ 3 & 4 \end{pmatrix}$,

The solution of a conditional equation is the matrix that when put in place of the unknown makes the equation become an identity. The numbers in the matrix are determined by setting corresponding elements equal and solving.

Example 4 Determine the values of the unknowns in the matrix equation

$$\begin{pmatrix} a & b \\ c & d \end{pmatrix} = \begin{pmatrix} 1 & 2 \\ 3 & 4 \end{pmatrix}$$

Solution By the definition of equal matrices, $a = 1$, $b = 2$, $c = 3$, and $d = 4$. ∎

Addition of Matrices

The next step in the development of **matrix algebra** is that of defining operations. In this section, addition is defined.

DEFINITION OF MATRIX ADDITION

If $A = (a_{ij})_{m \times n}$ and $B = (b_{ij})_{m \times n}$, then

$$A + B = (a_{ij} + b_{ij})_{m \times n}.$$

Notice that *the matrices to be added must be of the same order.* A matrix of order 2×3 can be added *only* to another matrix of order 2×3. If the matrices are of the same order, *their sum is found by adding their corresponding elements.* That is, the a_{11} element is added to the b_{11} element and their sum $a_{11} + b_{11}$ is put in the first row and first column of the $A + B$ matrix.

Example 5 Add

$$A = \begin{pmatrix} -1 & 2 & 3 \\ 4 & 5 & 6 \end{pmatrix} \quad \text{and} \quad B = \begin{pmatrix} 1 & 2 & 0 \\ 1 & 2 & 4 \end{pmatrix}$$

Solution
$$A + B = \begin{pmatrix} -1 & 2 & 3 \\ 4 & 5 & 6 \end{pmatrix} + \begin{pmatrix} 1 & 2 & 0 \\ 1 & 2 & 4 \end{pmatrix}$$
$$= \begin{pmatrix} -1+1 & 2+2 & 3+0 \\ 4+1 & 5+2 & 6+4 \end{pmatrix} = \begin{pmatrix} 0 & 4 & 3 \\ 5 & 7 & 10 \end{pmatrix} \quad \blacksquare$$

DEFINITION   Two matrices of the *same order* are said to be **conformable** (or **compatible**) for addition. The matrices being added are called **summands**. The answer in addition is called the **sum**.

Two matrices of different orders cannot be added. If $C = \begin{pmatrix} 1 & 2 \\ 3 & 4 \end{pmatrix}$ and $D = \begin{pmatrix} 1 \\ 0 \end{pmatrix}$, then C and D cannot be added. They are *not conformable for addition.* This seems reasonable, since there are elements in C that have *no corresponding elements* in D.

Two or more matrices of the *same order* may be added by simply adding all their corresponding elements and putting the total in the appropriate position in the sum matrix.

Example 6   Find the sum of

$$A = \begin{pmatrix} -1 & 2 \\ 0 & 3 \end{pmatrix}, \qquad B = \begin{pmatrix} 2 & 0 \\ \frac{1}{2} & 5 \end{pmatrix}, \quad \text{and} \quad C = \begin{pmatrix} 3 & \frac{1}{4} \\ 6 & 2.7 \end{pmatrix}.$$

Solution $A + B + C = \begin{pmatrix} -1 + 2 + 3 & 2 + 0 + \frac{1}{4} \\ 0 + \frac{1}{2} + 6 & 3 + 5 + 2.7 \end{pmatrix}$

$$= \begin{pmatrix} 4 & \frac{9}{4} \\ \frac{13}{2} & 10.7 \end{pmatrix} \quad \blacksquare$$

In arithmetic, the number 0 is called the additive identity because $a + 0 = a$ for any number a. In other words, adding 0 to a number does not change the number. The situation is somewhat different in matrix algebra. There is not just one additive identity matrix. *There is an additive identity matrix for each different order.*

ZERO MATRIX (ADDITIVE IDENTITY)

A matrix in which every element is 0 is called a **zero matrix**, denoted O.

Example 7   Find the sum of

$$A = \begin{pmatrix} 1 & 2 & 3 \\ 4 & 5 & 6 \end{pmatrix} \text{ and } O = \begin{pmatrix} 0 & 0 & 0 \\ 0 & 0 & 0 \end{pmatrix}.$$

Solution   Since A and O are of the same order (2×3), they are conformable. Hence

$$A + O = \begin{pmatrix} 1 + 0 & 2 + 0 & 3 + 0 \\ 4 + 0 & 5 + 0 & 6 + 0 \end{pmatrix} = \begin{pmatrix} 1 & 2 & 3 \\ 4 & 5 & 6 \end{pmatrix} = A \quad \blacksquare$$

Matrix addition is both commutative and associative.

PROPERTIES OF ADDITION

Commutative Property of Matrix Addition:

$$A + B = B + A \text{ if } A \text{ and } B \text{ are conformable.}$$

Associative Property of Matrix Addition:

$$(A + B) + C = A + (B + C) \text{ if } A, B, \text{ and } C \text{ are conformable.}$$

The associative and commutative properties of matrix addition are direct consequences of these properties in arithmetic.

Example 8   Let $A = \begin{pmatrix} 1 \\ 2 \end{pmatrix}$, $B = \begin{pmatrix} 3 \\ 5 \end{pmatrix}$, and $C = \begin{pmatrix} 7 \\ 0 \end{pmatrix}$. Find the sum $A + B + C$ using both possible associations. Are the sums equal?

Solution

$$(A + B) + C = \begin{pmatrix} 1 + 3 \\ 2 + 5 \end{pmatrix} + \begin{pmatrix} 7 \\ 0 \end{pmatrix} \qquad A + (B + C) = \begin{pmatrix} 1 \\ 2 \end{pmatrix} + \begin{pmatrix} 3 + 7 \\ 5 + 0 \end{pmatrix}$$

$$= \begin{pmatrix} 4 \\ 7 \end{pmatrix} + \begin{pmatrix} 7 \\ 0 \end{pmatrix} \qquad\qquad = \begin{pmatrix} 1 \\ 2 \end{pmatrix} + \begin{pmatrix} 10 \\ 5 \end{pmatrix}$$

$$= \begin{pmatrix} 11 \\ 7 \end{pmatrix} \qquad\qquad\qquad = \begin{pmatrix} 11 \\ 7 \end{pmatrix}$$

The sums are equal, that is, $\begin{pmatrix} 11 \\ 7 \end{pmatrix} = \begin{pmatrix} 11 \\ 7 \end{pmatrix}$. $\blacksquare$

An Application of Matrix Addition

Matrices are frequently used to organize and display data in a clear, concise manner. After the data have been expressed in matrix notation, they may be combined with other data using matrix operations such as addition. The next example shows how this may be done.

Example 9 An automobile company makes the following weekly shipments during the month of June:

First week.   To Dallas: 75 sports cars, 150 sedans, 50 compacts
              To Los Angeles: 175 sports cars, 50 sedans, 400 compacts
              To Atlanta: 60 sports cars, 300 sedans, 150 compacts
Second week. To Dallas: 125 sports cars, 200 sedans, 100 compacts
              To Los Angeles: 100 sports cars, 100 sedans, 200 compacts
              To Atlanta: 120 sports cars, 50 sedans, 100 compacts

Make a matrix representation of the weekly shipments. Use matrix addition to find the total number of each type car shipped to each city for the two weeks.

Solution Let column 1 represent sports cars, column 2 sedans, and column 3 compacts. Matrix A will represent the first week, matrix B the second week. Let row 1 represent Dallas, row 2 represent Los Angeles, and row 3 represent Atlanta. Then

$$A = \begin{pmatrix} 75 & 150 & 50 \\ 175 & 50 & 400 \\ 60 & 300 & 150 \end{pmatrix}, \qquad B = \begin{pmatrix} 125 & 200 & 100 \\ 100 & 100 & 200 \\ 120 & 50 & 100 \end{pmatrix}$$

The sum of the individual shipments for two weeks is

$$A + B = \begin{matrix} \text{Sports} \\ \text{cars} & \text{Sedans} & \text{Compacts} \\ \begin{pmatrix} 200 & 350 & 150 \\ 275 & 150 & 600 \\ 180 & 350 & 250 \end{pmatrix} & \begin{matrix} \text{Dallas} \\ \text{Los Angeles} \\ \text{Atlanta} \end{matrix} \end{matrix}$$

The 275 in the final matrix means that 275 sports cars were shipped to Los Angeles in two weeks. ∎

In arithmetic, additive inverses were defined as two numbers whose sum is zero (the additive identity).

ADDITIVE INVERSES (MATRICES)

Two matrices are called **additive inverses** if their sum is a zero matrix. That is, if

$$A + (-A) = O$$

then $(-A)$ is called the additive inverse of A.

The additive inverse of a matrix can be determined (or solved for) by using the definition of matrix equality.

Example 10 Find the additive inverse of $A = \begin{pmatrix} 1 & -2 & 3 \\ 4 & 5 & -6 \end{pmatrix}$.

Solution By definition, the additive inverse of A is a matrix $-A$ such that $A + (-A) = 0$. Therefore,

$$\begin{pmatrix} 1 & -2 & 3 \\ 4 & 5 & -6 \end{pmatrix} + \begin{pmatrix} b_{11} & b_{12} & b_{13} \\ b_{21} & b_{22} & b_{23} \end{pmatrix} = \begin{pmatrix} 0 & 0 & 0 \\ 0 & 0 & 0 \end{pmatrix}$$

By the definition of matrix addition,

$$\begin{pmatrix} 1 + b_{11} & -2 + b_{12} & 3 + b_{13} \\ 4 + b_{21} & 5 + b_{22} & -6 + b_{23} \end{pmatrix} = \begin{pmatrix} 0 & 0 & 0 \\ 0 & 0 & 0 \end{pmatrix}$$

By the definition of equal matrices,

$$1 + b_{11} = 0; \qquad -2 + b_{12} = 0; \qquad 3 + b_{13} = 0$$
$$4 + b_{21} = 0; \qquad 5 + b_{22} = 0; \qquad -6 + b_{23} = 0$$

By solving each equation we find that

$$b_{11} = -1; \quad b_{12} = +2; \quad b_{13} = -3; \quad b_{21} = -4, \quad b_{22} = -5; \quad b_{23} = +6$$

Hence

$$-A = \begin{pmatrix} -1 & +2 & -3 \\ -4 & -5 & +6 \end{pmatrix} \quad \blacksquare$$

The reason for denoting the additive inverse of A as $(-A)$ is clear. The additive inverse of a matrix is another matrix all of whose elements are the negatives of those in the given matrix.

Exercises 7.2 In Exercises 1–9 let

$$A = \begin{pmatrix} 1 & 2 \\ -1 & 0 \end{pmatrix}; \qquad B = (1 \quad 2 \quad 3); \qquad C = \begin{pmatrix} 1 \\ 2 \\ 3 \end{pmatrix};$$

$$D = \begin{pmatrix} 1 & 2 & 1 \\ 1 & 0 & 1 \\ 1 & 3 & 5 \end{pmatrix}; \qquad E = \begin{pmatrix} a \\ b \\ c \end{pmatrix}; \qquad F = \begin{pmatrix} \frac{3}{3} & 2 \\ -1 & \frac{0}{4} \end{pmatrix}.$$

1. Give the order of each matrix.
2. Which of the matrices are square?

3. In matrix D, name the d_{21} element, the d_{11} element, and the d_{33} element.
4. Which of the matrices are equal?
5. Why is C not equal to B?
6. If $C = E$, then $a =$ _____, $b =$ _____, and $c =$ _____.
7. Write an array of numbers which is *not a matrix*.
8. Which of A through F is a row matrix?
9. Which of A through F are column matrices?

10. In abbreviated notation $B = (b_{ij})_{3 \times 3}$, where $b_{ij} = 5$ for $i = j$ and $b_{ij} = 1$ for $i \neq j$. What is the matrix B in ordinary notation?

In Exercises 11–14 determine the values of the *unknowns* in the matrix equation.

11. $\begin{pmatrix} x \\ y \end{pmatrix} = \begin{pmatrix} -1 \\ 2 \end{pmatrix}$

12. $\begin{pmatrix} x & y \\ 2 & 3 \end{pmatrix} = \begin{pmatrix} -4 & 6 \\ 2 & 3 \end{pmatrix}$

13. $\begin{pmatrix} x + 3 \\ y + 2 \end{pmatrix} = \begin{pmatrix} 7 \\ 3 \end{pmatrix}$

14. $\begin{pmatrix} x + y \\ x - y \end{pmatrix} = \begin{pmatrix} 3 \\ 1 \end{pmatrix}$

15. Mr. Jones invested $1000 in stocks, $2000 in bonds, and $5000 in real estate. Mr. Smith invested $6000 in stocks, $3000 in bonds, and $8000 in real estate. Mrs. Brown invested no money in stocks, $3000 in bonds, and $10,000 in real estate. Represent their investments in the form of a matrix where each person's investments form one row of the matrix.

16. A company manufactures three different machines for digging ditches. During July the company sold 30 heavy-duty machines, 15 regular machines, and 45 lightweight machines. In August sales were 20, 12, and 60 machines, respectively, and in September sales were 40, 30, and 25, respectively. Organize these data in a matrix with the columns as the different types of machines and the rows as the different months.

In Exercises 17–31 let

$$A = \begin{pmatrix} 1 & 2 \\ 3 & 4 \end{pmatrix}; \qquad B = \begin{pmatrix} 1 \\ 2 \end{pmatrix}; \qquad C = (3 \quad 4);$$

$$D = \begin{pmatrix} -1 \\ -2 \end{pmatrix}; \qquad E = (+3 \quad -4); \qquad F = \begin{pmatrix} 1 & 0 \\ 0 & 2 \end{pmatrix};$$

$$G = \begin{pmatrix} 7 \\ 8 \end{pmatrix}; \qquad H = \begin{pmatrix} 3 & 1 \\ 0 & 5 \end{pmatrix}; \qquad \text{and } O = \begin{pmatrix} 0 & 0 \\ 0 & 0 \end{pmatrix}.$$

In Exercises 17–31 find the sum whenever possible. If addition of matrices is not possible, explain why.

17. $A + F$
18. $B + G$
19. $A + G$
20. $A + O$
21. $B + D$
22. $F + A$
23. $G + B$
24. $(A + H) + F$
25. $A + (H + F)$
26. $O + C$
27. $E + C$.
28. Which matrix is an additive identity?

29. What is the additive inverse of A? of E?
30. Is matrix $B + G$ equal to $G + B$? Why?
31. Compute $A + (-A)$.

32. An appliance company shipped 30 dishwashers, 20 refrigerators, and no freezers to Albany during the month of June. They shipped 15 dishwashers, 30 refrigerators, and 12 freezers to Memphis during the month of June. In July the company shipped 20 dishwashers, 12 refrigerators, and 10 freezers to Albany, and they shipped 6 dishwashers, 14 refrigerators, and 18 freezers to Memphis. Form a matrix for each month and use matrix addition to determine the total number of dishwashers, refrigerators, and freezers shipped to each city.

33. Consider the following matrices:

$$J = \begin{pmatrix} 235 & 126 & 217 \\ 120 & 150 & 206 \\ 30 & 75 & 26 \\ 100 & 80 & 120 \end{pmatrix}, \quad F = \begin{pmatrix} 175 & 204 & 319 \\ 106 & 140 & 182 \\ 20 & 50 & 40 \\ 120 & 90 & 140 \end{pmatrix},$$

$$M = \begin{pmatrix} 85 & 75 & 36 \\ 100 & 75 & 80 \\ 10 & 40 & 12 \\ 75 & 30 & 40 \end{pmatrix}$$

Let the rows represent the cities to which an item has been shipped (row one Chicago, row two Denver, row three Salt Lake City, and row four Portland). Let the columns represent the three types of bicycles being shipped to those cities (column one ten-speeds, column two regular bicycles, and column three children's bicycles). Matrix J represents shipments made in January, matrix F represents shipments made in February, and matrix M represents shipments made in March.
   (a) How many ten-speeds were shipped to Portland during March?
   (b) How many regular bicycles were shipped to Denver during January?
   (c) How many children's bicycles were shipped to Salt Lake City during February?

34. By using the matrices in Exercise 33, find
   (a) $J + F + M$.
   (b) the number of ten-speeds shipped to Chicago during this three-month period.

7.3 Matrix Subtraction; Scalar Multiplication

In arithmetic, the number being subtracted is called the **subtrahend.** The number being subtracted from is called the **minuend.** The answer is called the **difference.** The same terminology is used for matrix subtraction.

When subtracting signed numbers (positive and negatives) you learned to do subtraction by the process of *reversing the sign of the subtrahend and adding.* This is the process we want to use for subtraction of matrices. But how does one change the sign of a matrix?

SCALAR MULTIPLICATION

A **scalar** is a real number. **Scalar multiplication** is the process of multiplying a scalar (number) by a matrix as follows:

$$c \cdot A = (ca_{ij})_{m \times n}$$

To multiply a scalar times a matrix, multiply the *scalar times each element* in the matrix.

Note: The multiplication dot may be used to indicate multiplication or it may be omitted; that is, $c \cdot A$ means cA.

Example 1 Multiply the scalar 3 by the matrix

$$B = \begin{pmatrix} 1 & 0 & -3 \\ 0.5 & 2 & \frac{1}{3} \end{pmatrix}$$

Solution $3 \cdot B = \begin{pmatrix} 3 \times 1 & 3 \times 0 & 3 \times (-3) \\ 3 \times 0.5 & 3 \times 2 & 3 \times \frac{1}{3} \end{pmatrix} = \begin{pmatrix} 3 & 0 & -9 \\ 1.5 & 6 & 1 \end{pmatrix}$ ∎

Example 2 The matrix $\begin{bmatrix} \$8 \\ \$10 \\ \$40 \end{bmatrix}$ represents the selling prices of the Tri-Product Company. If the company wants to reduce their prices to 80% of their current level, find the new price matrix.

Solution By scalar multiplication,

$$80\% \begin{bmatrix} \$8 \\ \$10 \\ \$40 \end{bmatrix} = 0.80 \begin{bmatrix} \$8 \\ \$10 \\ \$40 \end{bmatrix} = \begin{bmatrix} \$6.40 \\ \$8.00 \\ \$32.00 \end{bmatrix}$$ ∎

Matrix Subtraction

To do matrix subtraction, we want to "change the signs" of the subtrahend and add. To change the signs of a matrix, multiply by the scalar

(-1). This will change the sign of every (nonzero) element in the matrix. Hence, subtraction can be defined as follows:

MATRIX SUBTRACTION

Let matrix A be the minuend and B be the subtrahend. Then

$$A - B = A + (-1)B$$

In other words, to subtract a matrix, change the sign of the subtrahend (by multiplying by -1) and add.

Note: Since subtraction is performed through the use of addition, the matrices must be of the same order (that is, conformable for addition).

Example 3   Find $A - B$ where $A = \begin{pmatrix} -4 & 2 \\ 0 & 4 \\ -3 & 1 \end{pmatrix}$ and $B = \begin{pmatrix} 2 & -1 \\ 3 & 0 \\ -2 & 5 \end{pmatrix}$.

Solution $A - B = A + (-1)B = \begin{pmatrix} -4 & 2 \\ 0 & 4 \\ -3 & 1 \end{pmatrix} + (-1)\begin{pmatrix} 2 & -1 \\ 3 & 0 \\ -2 & 5 \end{pmatrix}$

$$= \begin{pmatrix} -4 & 2 \\ 0 & 4 \\ -3 & 1 \end{pmatrix} + \begin{pmatrix} -2 & 1 \\ -3 & 0 \\ 2 & -5 \end{pmatrix}$$

$$= \begin{pmatrix} -6 & 3 \\ -3 & 4 \\ -1 & -4 \end{pmatrix} \quad \blacksquare$$

Many common business problems require the use of subtraction. Some of these are shown in the following example.

Example 4   The Tri-Product Company's total revenue from the sale of its products is $R = \begin{pmatrix} \$800 \\ \$1000 \\ \$4000 \end{pmatrix}$. The cost of producing these items was $C = \begin{pmatrix} \$600 \\ \$700 \\ \$3000 \end{pmatrix}$.
Find the matrix representing the profit yielded from the production and sale of the items.

Solution   Profit = revenue − cost. Hence

$$P = \begin{pmatrix} \$800 \\ \$1000 \\ \$4000 \end{pmatrix} - \begin{pmatrix} \$600 \\ \$700 \\ \$3000 \end{pmatrix} = \begin{pmatrix} \$200 \\ \$300 \\ \$1000 \end{pmatrix}$$

The profit produced by the first item is $200, by the second item is $300, and by the third item is $1000.   ■

Example 5   The selling prices of a company's products are given by

$$S = (\$8 \quad \$10 \quad \$12 \quad \$6)$$

If the company reduces all prices by 10%, find the matrix of new selling prices.

Solution   To get the amount of the price decreases or discounts, compute

$$\begin{aligned} D &= 10\%(\$8 \quad \$10 \quad \$12 \quad \$6) \\ &= 0.10(\$8 \quad \$10 \quad \$12 \quad \$6) \\ &= (\$0.80 \quad \$1.00 \quad \$1.20 \quad \$0.60) \end{aligned}$$

The new selling prices are the original prices (S) minus the discounts (D). Hence the new selling prices are given by

$$\begin{aligned} S - D &= (\$8 \quad \$10 \quad \$12 \quad \$6) - (\$0.80 \quad \$1.00 \quad \$1.20 \quad \$0.60) \\ &= (\$7.20 \quad \$9.00 \quad \$10.80 \quad \$5.40) \quad ■ \end{aligned}$$

Exercises 7.3   In Exercises 1–9 perform the operations if possible.

1. $5 \begin{pmatrix} -1 & 3 \\ 4 & 2 \\ 1 & 6 \end{pmatrix}$

2. $\dfrac{1}{2} \begin{pmatrix} 2 & 6 & 4 \\ -3 & 1 & 0 \end{pmatrix}$

3. $20\% \begin{pmatrix} 2 & 7 \\ 8 & 6 \end{pmatrix}$

4. $\begin{pmatrix} 7 & 8 \\ 9 & 10 \end{pmatrix} - \begin{pmatrix} 1 & 2 \\ 3 & 4 \end{pmatrix}$

5. $\begin{pmatrix} 1 & 0 & 0 \\ 0 & 1 & 0 \\ 0 & 0 & 1 \end{pmatrix} - \begin{pmatrix} 1 \\ 2 \\ 3 \end{pmatrix}$

6. $\begin{pmatrix} 10 \\ -2 \\ 4 \end{pmatrix} - \begin{pmatrix} 8 \\ 2 \\ -6 \end{pmatrix}$

7. $3 \begin{pmatrix} 4 \\ 7 \\ 8 \end{pmatrix} - \begin{pmatrix} 7 \\ 2 \\ 10 \end{pmatrix}$

8. $10\% \begin{pmatrix} \$2 \\ \$7 \\ \$8 \end{pmatrix} + 5\% \begin{pmatrix} \$10 \\ \$8 \\ \$1 \end{pmatrix}$

9. $4 \begin{pmatrix} 2 & 0 & 0 \\ 0 & -3 & 0 \\ 0 & 0 & 7 \end{pmatrix}$

10. This year three stores had payrolls of

$$P = (\$100{,}000 \quad \$150{,}000 \quad \$80{,}000)$$

A 5% increase in payroll is expected next year. What will be next year's payrolls?

11. If the revenues produced by the sales of four items are given by $\begin{pmatrix} \$800 \\ \$600 \\ \$500 \\ \$700 \end{pmatrix}$

and their respective costs are $\begin{pmatrix} \$600 \\ \$300 \\ \$100 \\ \$100 \end{pmatrix}$, find the profit produced by each item.

12. Four items having selling prices $S = \begin{pmatrix} \$6 \\ \$2 \\ \$8 \\ \$10 \end{pmatrix}$ are put on sale at 25% off the regular price. Find the matrix of sale prices.

13. Using the matrices in Exercise 11, find the profit produced by each item if revenue increases 10% and cost increases 20%.

14. Solve the following matrix equations. (*Hint:* Perform the indicated operations, first scalar multiplication then addition; next, use equality of matrices to establish a system of linear equations.)

(a) $x \begin{pmatrix} 2 \\ 1 \end{pmatrix} - y \begin{pmatrix} 1 \\ -1 \end{pmatrix} = \begin{pmatrix} 3 \\ 2 \end{pmatrix}$

(b) $x \begin{pmatrix} 2 \\ 1 \end{pmatrix} - y \begin{pmatrix} -3 \\ 1 \end{pmatrix} = \begin{pmatrix} 6 \\ 2 \end{pmatrix}$

(c) $x \begin{pmatrix} 1 \\ 2 \\ 0 \end{pmatrix} + y \begin{pmatrix} -1 \\ 0 \\ 1 \end{pmatrix} - z \begin{pmatrix} -1 \\ 1 \\ -1 \end{pmatrix} = \begin{pmatrix} 1 \\ 2 \\ 3 \end{pmatrix}$

7.4 Matrix Multiplication

The next operation to be considered is matrix multiplication. As usual, quantities that are multiplied together are called **factors**. The answer is called the **product**. First we will consider the special case of multiplying a row matrix (written on the left) times a column matrix (written on the right). Later the procedure will be generalized to other matrices, but the requirement of the left factor being a row and the right factor being a column will be retained.

MULTIPLICATION OF A ROW MATRIX BY A COLUMN MATRIX

Let $\quad R = (r_{11} \quad r_{12} \quad r_{13} \cdots r_{1n})_{1 \times n}$ and $C = \begin{pmatrix} c_{11} \\ c_{21} \\ c_{31} \\ \cdot \\ \cdot \\ \cdot \\ c_{n1} \end{pmatrix}_{n \times 1}$

Then $\quad RC = (r_{11} \quad r_{12} \cdots r_{1n}) \begin{pmatrix} c_{11} \\ c_{21} \\ c_{31} \\ \cdot \\ \cdot \\ \cdot \\ c_{n1} \end{pmatrix}$

$$= (r_{11}c_{11} + r_{12}c_{21} + r_{13}c_{31} + \cdots + r_{1n}c_{n1})_{1 \times 1}$$

Note: The number of columns in R must be *equal* to the number of rows in C. The product will be a *single number* when all of the terms are added together, that is, a 1×1 matrix.

Example 1 Find the product of $R = (1 \quad 2 \quad 3)$ and $C = \begin{pmatrix} 2 \\ 0 \\ 4 \end{pmatrix}$.

Solution $RC = (1 \quad 2 \quad 3) \begin{pmatrix} 2 \\ 0 \\ 4 \end{pmatrix} = (1 \times 2 + 2 \times 0 + 3 \times 4) = (2 + 0 + 12) = 14$

It is crucial that the number of columns in the left factor be the same as the number of rows in the right factor. In a sense, you are multiplying corresponding elements and summing all of the products to get a single element in the product matrix.

Example 2 Suppose that the Work Saver Appliance Company ships 30 dishwashers, 20 refrigerators, and 6 freezers to its outlet in Albany at prices of

$200, $400, and $300, respectively. What is the total price of the shipment? Formulate the problem using matrices.

Solution   The items shipped can be represented by a row vector (30 20 6).

The price of each can be represented by the column matrix $\begin{pmatrix} \$200 \\ \$400 \\ \$300 \end{pmatrix}$.

The product is

$$(30 \quad 20 \quad 6) \begin{pmatrix} \$200 \\ \$400 \\ \$300 \end{pmatrix} = (30 \times \$200 + 20 \times \$400 + 6 \times \$300)$$

$$= (\$6000 + \$8000 + \$1800)$$

$$= (\$15{,}800)$$

The total price of the shipment is $15,800. ∎

To multiply matrices that have more than one row and column, a general definition is needed.

MATRIX MULTIPLICATION

Let A be an $m \times p$ matrix and B be an $p \times n$ matrix. The product AB is an $m \times n$ matrix where each element is obtained by multiplying the ith row of A by the jth column of B (as given in the previous definition) and putting the result in the ith row and jth column of the product matrix AB.

Schematic of matrix multiplication:

$$\text{First row} \rightarrow \begin{pmatrix} 1 & 2 & 3 \\ . & . & . \\ . & . & . \end{pmatrix} \begin{pmatrix} . & 1 & . \\ . & 2 & . \\ . & 3 & . \end{pmatrix} = \begin{pmatrix} . & 14 & . \\ . & . & . \\ . & . & . \end{pmatrix}$$

First row, second column (top)

Second column (↑)

$$\text{Second row} \rightarrow \begin{pmatrix} . & . & . \\ 1 & 2 & 3 \\ . & . & . \end{pmatrix} \begin{pmatrix} . & . & 1 \\ . & . & 2 \\ . & . & 3 \end{pmatrix} = \begin{pmatrix} . & . & . \\ . & . & 14 \\ . & . & . \end{pmatrix} \leftarrow \text{Second row, third column}$$

Third column (↑)

This process is continued until all the entries in the product matrix are determined.

The following illustration shows how to determine whether multiplication is possible, and if so, how to find the dimension of the product AB. Suppose A is 2×3 and B is 3×4.

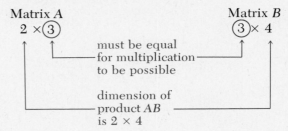

Example 3 Let $A = \begin{pmatrix} 1 & 2 \\ 3 & 4 \end{pmatrix}$ and $B = \begin{pmatrix} 2 & 0 & 4 \\ 3 & 1 & 5 \end{pmatrix}$. Find AB.

Solution Since the number of columns of A is equal to the number of rows of B, the matrices can be multiplied. (The order of AB will be 2×3.)

$$AB = \begin{pmatrix} 1 & 2 \\ 3 & 4 \end{pmatrix}_{2 \times 2} \begin{pmatrix} 2 & 0 & 4 \\ 3 & 1 & 5 \end{pmatrix}_{2 \times 3}$$

$$\underbrace{\qquad}_{\text{match}}$$

$$= \begin{pmatrix} 1 \cdot 2 + 2 \cdot 3 & 1 \cdot 0 + 2 \cdot 1 & 1 \cdot 4 + 2 \cdot 5 \\ 3 \cdot 2 + 4 \cdot 3 & 3 \cdot 0 + 4 \cdot 1 & 3 \cdot 4 + 4 \cdot 5 \end{pmatrix}$$

$$= \begin{pmatrix} 8 & 2 & 14 \\ 18 & 4 & 32 \end{pmatrix}_{2 \times 3} \quad \blacksquare$$

Note that if the order of A and B were reversed, then we could *not* multiply. The product BA cannot be computed because B has three columns and A has two rows. *Matrix multiplication is not commutative.*

Example 4 Let $C = \begin{pmatrix} 1 & 0 & -1 \\ 2 & 3 & 0 \\ 0 & 1 & 2 \end{pmatrix}$ and $D = \begin{pmatrix} 1 \\ 2 \\ 3 \end{pmatrix}$. Find CD.

Solution The product CD can be computed, and the order of the product matrix will be 3×1. We have

$$CD = \begin{pmatrix} 1 \cdot 1 + 0 \cdot 2 + (-1)(3) \\ 2 \cdot 1 + 3 \cdot 2 + 0 \cdot 3 \\ 0 \cdot 1 + 1 \cdot 2 + 2 \cdot 3 \end{pmatrix} = \begin{pmatrix} -2 \\ 8 \\ 8 \end{pmatrix} \quad \blacksquare$$

In arithmetic, the number 1 is called the multiplicative identity, since $1 \cdot a = a \cdot 1 = a$ for all numbers a. Is there a (unique) multiplicative identity for matrices, that is, is there an identity matrix I such that $IA = AI = A$ for any matrix A?

MULTIPLICATIVE IDENTITIES FOR SQUARE MATRICES

A square matrix $I = (a_{ij})_{n \times n}$ in which $a_{ij} = 1$ for $i = j$ and $a_{ij} = 0$ for all $i \neq j$ is called a **multiplicative identity**. The a_{ij} elements where $i = j$ form the **main diagonal** of a square matrix.

Multiplicative identities are matrices of the form

$$(1), \quad \begin{pmatrix} 1 & 0 \\ 0 & 1 \end{pmatrix}, \quad \begin{pmatrix} 1 & 0 & 0 \\ 0 & 1 & 0 \\ 0 & 0 & 1 \end{pmatrix}, \quad \begin{pmatrix} 1 & 0 & 0 & 0 \\ 0 & 1 & 0 & 0 \\ 0 & 0 & 1 & 0 \\ 0 & 0 & 0 & 1 \end{pmatrix}, \text{ and so on}$$

All multiplicative identities are denoted I. The particular order will be clear from the context.

Example 5 Find the products:

(a) $\begin{pmatrix} 1 & 0 \\ 0 & 1 \end{pmatrix}\begin{pmatrix} 1 & 2 \\ 3 & 4 \end{pmatrix}$

(b) $\begin{pmatrix} 1 & 0 & 0 \\ 0 & 1 & 0 \\ 0 & 0 & 1 \end{pmatrix}\begin{pmatrix} 1 & 2 & 3 \\ 4 & 5 & 6 \\ 7 & 8 & 9 \end{pmatrix}$

(c) $\begin{pmatrix} 1 & 0 \\ 0 & 1 \end{pmatrix}\begin{pmatrix} 2 \\ 3 \end{pmatrix}$

Solution (a) $\begin{pmatrix} 1 & 0 \\ 0 & 1 \end{pmatrix}\begin{pmatrix} 1 & 2 \\ 3 & 4 \end{pmatrix} = \begin{pmatrix} 1 & 2 \\ 3 & 4 \end{pmatrix}$

(b) $\begin{pmatrix} 1 & 0 & 0 \\ 0 & 1 & 0 \\ 0 & 0 & 1 \end{pmatrix}\begin{pmatrix} 1 & 2 & 3 \\ 4 & 5 & 6 \\ 7 & 8 & 9 \end{pmatrix} = \begin{pmatrix} 1 & 2 & 3 \\ 4 & 5 & 6 \\ 7 & 8 & 9 \end{pmatrix}$

(c) $\begin{pmatrix} 1 & 0 \\ 0 & 1 \end{pmatrix}\begin{pmatrix} 2 \\ 3 \end{pmatrix} = \begin{pmatrix} 2 \\ 3 \end{pmatrix}$ ∎

An Application of Matrix Multiplication in Business

Example 6 Suppose the automobile company in Example 9 of Section 7.2 charges the receiving point $7000 for each sports car, $5000 for each sedan, and $3000 for each compact. Use matrix multiplication to determine the cost to each city.

Solution   By taking the matrix formed by the summation of the two matrices and then multiplying (using matrix multiplication) by a column matrix that involves cost, we can obtain the bill sent to each receiving point in matrix form.

$$\begin{pmatrix} 200 & 350 & 150 \\ 275 & 150 & 600 \\ 180 & 350 & 250 \end{pmatrix} \begin{pmatrix} \$7000 \\ \$5000 \\ \$3000 \end{pmatrix} = \begin{pmatrix} \$3{,}600{,}000 \\ \$3{,}925{,}000 \\ \$3{,}760{,}000 \end{pmatrix} \begin{matrix} \text{Dallas} \\ \text{Los Angeles} \\ \text{Atlanta} \end{matrix}$$

The cost to each city is reflected in the product matrix by the value in each row. For example, the cost to Atlanta is $3,760,000.   ∎

An Application of Matrix Multiplication in Linear Systems

The definitions of matrix equality, matrix equation, and matrix multiplication can be applied to express a linear system as a single matrix equation.

Consider the linear system

$$2x + y = 3$$
$$x + y = 2$$

The system consists of three matrices:

1. The matrix of *coefficients* of the variables, which is generally denoted A:

$$A = \begin{pmatrix} 2 & 1 \\ 1 & 1 \end{pmatrix}$$

2. The (column) matrix of *unknowns*, which is generally denoted X:

$$X = \begin{pmatrix} x \\ y \end{pmatrix}$$

3. The (column) matrix of *constants*, which is generally denoted B:

$$B = \begin{pmatrix} 3 \\ 2 \end{pmatrix}$$

The linear system can be expressed as the matrix equation $AX = B$:

$$\begin{pmatrix} 2 & 1 \\ 1 & 1 \end{pmatrix} \begin{pmatrix} x \\ y \end{pmatrix} = \begin{pmatrix} 3 \\ 2 \end{pmatrix}$$

The equivalence of the matrix equation $AX = B$ to the original system in ordinary algebraic notation can be demonstrated by performing the indicated operations as follows:

$$\begin{pmatrix} 2 & 1 \\ 1 & 1 \end{pmatrix} \begin{pmatrix} x \\ y \end{pmatrix} = \begin{pmatrix} 3 \\ 2 \end{pmatrix} \qquad \text{(matrix equation)}$$

$$\begin{pmatrix} 2x + y \\ x + y \end{pmatrix} = \begin{pmatrix} 3 \\ 2 \end{pmatrix} \qquad \text{(matrix multiplication on the left)}$$

original system $\longrightarrow$ $\begin{cases} 2x + y = 3 \\ x + y = 2 \end{cases}$ (equal matrices have equal elements)

Any linear system of n equations in n unknowns can be expressed in the form $AX = B$.

DEFINITION OF A MATRIX EQUATION

Let

$$\begin{aligned}
a_{11}x_1 + a_{12}x_2 + \cdots + a_{1n}x_n &= b_1 \\
a_{21}x_1 + a_{22}x_2 + \cdots + a_{2n}x_n &= b_2 \\
&\ \ \vdots \\
a_{n1}x_1 + a_{n2}x_2 + \cdots + a_{nn}x_n &= b_n
\end{aligned}$$

represent an algebraic system of n linear equations in n unknowns. This may be expressed in **matrix equation** notation as

$$\begin{pmatrix} a_{11} & a_{12} & \cdots & a_{1n} \\ a_{21} & a_{22} & \cdots & a_{2n} \\ \vdots & \vdots & & \vdots \\ a_{n1} & a_{n2} & \cdots & a_{nn} \end{pmatrix} \begin{pmatrix} x_1 \\ x_2 \\ \vdots \\ x_n \end{pmatrix} = \begin{pmatrix} b_1 \\ b_2 \\ \vdots \\ b_n \end{pmatrix}$$

That is, $AX = B$.

Since there can be a large number of unknowns, they are expressed as $x_1, x_2 \cdots x_n$ rather than with different letters. For example, if the system of equations contained fifty variables, we would not have enough different letters in the alphabet to represent each one differently.

Example 7 Write the following system of linear equations in matrix equation notation:

$$x_1 - x_2 + x_3 = 1$$
$$2x_1 \qquad - x_3 = 2$$
$$x_2 + x_3 = 3$$

Solution

$$\begin{pmatrix} 1 & -1 & 1 \\ 2 & 0 & -1 \\ 0 & 1 & 1 \end{pmatrix} \begin{pmatrix} x_1 \\ x_2 \\ x_3 \end{pmatrix} = \begin{pmatrix} 1 \\ 2 \\ 3 \end{pmatrix}$$

$$AX = B$$

Remember that $2x_1 - x_3 = 2$ is the same as $2x_1 + 0x_2 - x_3 = 2$. The zero coefficient *must be included* in the coefficient matrix. ∎

Exercises 7.4

In Exercises 1–16 find the product if possible.

1. $(1 \quad 2) \begin{pmatrix} 3 \\ 4 \end{pmatrix}$

2. $(1 \quad 2 \quad 3) \begin{pmatrix} 1 \\ 4 \\ -2 \end{pmatrix}$

3. $\begin{pmatrix} 3 \\ 4 \end{pmatrix} (1 \quad 2)$

4. $\begin{pmatrix} 1 & -1 & 0 \\ 0 & 2 & 3 \end{pmatrix} \begin{pmatrix} 3 \\ 1 \\ 4 \end{pmatrix}$

5. $(0 \quad 0) \begin{pmatrix} 1 \\ 2 \end{pmatrix}$

6. $\begin{pmatrix} 1 & 2 \\ 2 & 1 \end{pmatrix} \begin{pmatrix} 2 & 0 \\ 0 & 2 \end{pmatrix}$

7. $\begin{pmatrix} 1 & 2 & 3 & -1 \\ 2 & 1 & 0 & 1 \end{pmatrix} \begin{pmatrix} 1 & 2 \\ 0 & 3 \\ -1 & 5 \\ 2 & -5 \end{pmatrix}$

8. $(1 \quad 2 \quad 3 \quad 4 \quad 5) \begin{pmatrix} -5 \\ -4 \\ 3 \\ 2 \\ 1 \end{pmatrix}$

9. $\begin{pmatrix} 1 & 2 \\ -1 & 3 \end{pmatrix} \begin{pmatrix} x \\ y \end{pmatrix}$

10. $(a \quad b \quad c) \begin{pmatrix} x \\ y \end{pmatrix}$

11. $\begin{pmatrix} 1 & 0 \\ 0 & 1 \end{pmatrix} \begin{pmatrix} 1 & 2 \\ 3 & 4 \end{pmatrix}$

12. $\begin{pmatrix} 1 & 1 \\ 2 & 1 \end{pmatrix} \begin{pmatrix} -1 & 1 \\ 2 & -1 \end{pmatrix}$

13. $\begin{pmatrix} \frac{1}{5} & \frac{2}{5} & \frac{1}{5} \\ -\frac{2}{5} & \frac{1}{5} & \frac{3}{5} \\ \frac{2}{5} & -\frac{1}{5} & \frac{2}{5} \end{pmatrix} \begin{pmatrix} 1 & -1 & 1 \\ 2 & 0 & -1 \\ 0 & 1 & 1 \end{pmatrix}$

14. $\begin{pmatrix} 1 & -1 & 1 \\ 2 & 0 & -1 \\ 0 & 1 & 1 \end{pmatrix} \begin{pmatrix} 1 & 0 & 0 \\ 0 & 1 & 0 \\ 0 & 0 & 1 \end{pmatrix}$

15. $\begin{pmatrix} 0 & 0 \\ 0 & 0 \end{pmatrix} \begin{pmatrix} 1 & 2 \\ 3 & 4 \end{pmatrix}$

16. $\begin{pmatrix} 7 & 0 & 0 \\ 0 & 5 & 0 \\ 0 & 0 & 6 \end{pmatrix} \begin{pmatrix} \$4 \\ \$8 \\ \$3 \end{pmatrix}$

17. A person has investments that yield 3%, 4%, 6%, and 10% annually. The amounts invested at each rate are $1000, $2000, $8000, and $3000, respectively. Express the rates of yield in a row matrix and the amounts in a column matrix. What total amount of interest will be yielded at the end of 1 year?

18. Express the percents in Exercise 17 as a diagonal matrix and determine

the matrix in which the amount of interest from each investment is shown.

19. A company has 100 units of A, 250 units of B, and 500 units of C that retail for \$2, \$3, and \$1.50, respectively. Express the situation in matrix notation and use matrix multiplication to determine the total revenue that would result if all the items were sold.

20. Express each of the following systems in the form $AX = B$.

(a) $x + y = 5$
$\quad 2x - y = 1$

(b) $x_1 + x_2 = 5$
$\quad 2x_1 + 2x_2 = 16$

(c) $x + y + z = 4$
$\quad -y + 2z = 1$
$\quad x + y \quad\;\; = 2$

(d) $x_1 \quad\;\; + x_3 = 5$
$\quad x_2 - 2x_3 = 3$
$\quad 2x_1 + x_2 + x_3 = 7$

(e) $x = 4 - y$
$\quad x + 2 = y$
(*Hint:* Put the system in standard form.)

(f) $2x + y = 3 - z$
$\quad x = 3 + z - y$
$\quad x + z = 3$

7.5 Row Operations; The Gauss–Jordan Method

In Chapter 3 you learned how to solve a linear system by the addition and substitution methods. In this section we want to review and expand the addition method. The addition method as presented in Chapter 3 is a "partial" elimination method; that is, after eliminating one variable from the system, we completed the solution by elementary algebra and replacement. Our goal was to reduce the system from two equations in two unknowns to one equation in one unknown.

We could carry the addition method farther, with *no* reduction in the number of equations. Also, we can now represent the system using a matrix. In particular, the system can be represented by an **augmented matrix**. The augmented matrix consists of the coefficients (A) and the constants (B) in the linear system. In the following example we show the "complete" elimination by addition process and the corresponding matrix representation. Study this example carefully.

Example 1 Solve the system R_1: $2x + y = 3$
$\qquad\qquad\qquad R_2$: $\quad x + y = 2$

Solution For emphasis, all understood coefficients are shown.

Algebraic problem

R_1: $2x + 1y = 3$
R_2: $1x + 1y = 2$

Augmented matrix representation

$$(A \mid B) = \begin{pmatrix} 2 & 1 & | & 3 \\ 1 & 1 & | & 2 \end{pmatrix}$$

constants} matrix B

coefficients of y}
coefficients of x} matrix A

Multiply R_2 by -2 and replace R_2 by the resulting product:

$$R_1: \quad 2x + 1y = 3$$
$$R_2: \quad -2x - 2y = -4$$

$$\begin{pmatrix} 2 & 1 & | & 3 \\ -2 & -2 & | & -4 \end{pmatrix}$$

Add R_1 and R_2; put their sum in place of R_2:

$$R_1: \quad 2x + 1y = 3$$
$$R_2: \quad 0x - 1y = -1$$

$$\begin{pmatrix} 2 & 1 & | & 3 \\ 0 & -1 & | & -1 \end{pmatrix}$$

Add R_1 and R_2; replace R_1 by their sum:

$$R_1: \quad 2x + 0y = 2$$
$$R_2: \quad 0x - 1y = -1$$

$$\begin{pmatrix} 2 & 0 & | & 2 \\ 0 & -1 & | & -1 \end{pmatrix}$$

Now multiply R_1 by $\frac{1}{2}$ and R_2 by -1:

$$R_1: \quad 1x + 0y = 1$$
$$R_2: \quad 0x + 1y = 1$$

$$\begin{pmatrix} 1 & 0 & | & 1 \\ 0 & 1 & | & 1 \end{pmatrix}$$

From equation R_1 we have $x = 1$, and from equation R_2 we have $y = 1$. Therefore $x = 1$, $y = 1$ is the solution of the system. ∎

The repeated addition process leads to the solution using only arithmetic operations of multiplication and addition. We could have performed the arithmetic operations on the rows of matrix $(A \mid B)$ until we achieved the matrix

$$(I \mid C) = \begin{pmatrix} 1 & 0 & | & a \\ 0 & 1 & | & b \end{pmatrix}$$

which gives $x = a$ and $y = b$. This process is called the **Gauss–Jordan method.**

The arithmetic operations preformed on the rows of a matrix are called **elementary row operations.** There are three of them:

1. Interchanging two rows.
2. Multiplication of any row by a nonzero number.
3. Addition of two rows and replacement of either one by their sum.

These operations will be expressed symbolically as follows:

1. Interchange rows 1 and 2 ($\rightarrow$ means replace):

$$R_1 \rightleftarrows R_2$$

2. Multiply row 3 by $\frac{1}{2}$ and replace row 3 by the product:

$$\frac{1}{2} R_3 \rightarrow R_3$$

3. Add row 1 and row 3; replace row 3 by their sum:

$$R_1 + R_3 \rightarrow R_3$$

4. Multiply row 2 by 5, add the product to row 3, replace row 3 by their sum.

$$5R_2 + R_3 \rightarrow R_3$$

The following samples show the effect of particular row operations on the given matrices.

ROW OPERATION 1

Any two rows of a matrix may be interchanged.

Sample:
$$\begin{pmatrix} 1 & -1 & 1 \\ 2 & 0 & -1 \\ 0 & 1 & 1 \end{pmatrix}$$
Given matrix. $R_1 \rightleftarrows R_2$

$$\begin{pmatrix} 2 & 0 & -1 \\ 1 & -1 & 1 \\ 0 & 1 & 1 \end{pmatrix}$$
New matrix with first and second rows interchanged

ROW OPERATION 2

We may multiply any row of a matrix by a nonzero number.

Sample:
$$\begin{pmatrix} 1 & -1 & 1 \\ 2 & 0 & -1 \\ 0 & 1 & 1 \end{pmatrix}$$
Given matrix. $-2R_3 \rightarrow R_3$

$$\begin{pmatrix} 1 & -1 & 1 \\ 2 & 0 & -1 \\ 0 & -2 & -2 \end{pmatrix}$$
New matrix obtained by multiplying row 3 by -2 and replacing row 3 by the product

---

ROW OPERATION 3

Two rows of a matrix may be added together and either row replaced by their sum. The row *not* replaced by the sum must be returned to its original form and replaced in its original position.

Sample:
$$\begin{pmatrix} 1 & -1 & 1 \\ 2 & 0 & -1 \\ 0 & 1 & 1 \end{pmatrix}$$
Given matrix. $R_2 + R_3 \rightarrow R_3$

$$\begin{pmatrix} 1 & -1 & 1 \\ 2 & 0 & -1 \\ 2 & 1 & 0 \end{pmatrix}$$
New matrix with rows 2 and 3 added and row 3 replaced by the sum; row 2 is returned to its original form and position.

Note: Row operations 2 and 3 are frequently used together, i.e., one row will be multiplied by a number and the product added to another row.

Sample:

$$\begin{pmatrix} 1 & -1 & 1 \\ 2 & 0 & -1 \\ 0 & 1 & 1 \end{pmatrix} \qquad -3R_1 + R_3 \to R_3$$

$$\begin{pmatrix} 1 & -1 & 1 \\ 2 & 0 & -1 \\ -3 & 4 & -2 \end{pmatrix} \qquad$$ New matrix with -3 times row 1 added to row 3 and row 3 replaced by their sum; row 1 is returned to its original form and position.

Justification of these operations results directly from the properties of equality. Any sequence of row operations that achieves the desired final form of the matrix representation of the system is permissible. You might want to try a different sequence from the one used to solve Example 1.

The Gauss–Jordan method is demonstrated again using the system in Example 2 of Section 3.1.

Example 2 Use matrices to solve the system

$$R_1: \quad 2x + 3y = 6$$
$$R_2: \quad x - y = 2$$

Solution Express the system using matrices.

Matrix	**Operation Leading to Next Matrix**
$(A \mid B) = \begin{pmatrix} 2 & 3 & \mid & 6 \\ 1 & -1 & \mid & 2 \end{pmatrix}$	$\frac{1}{2}R_1 \to R_1$ (to obtain 1 in the a_{11} position)
$\begin{pmatrix} 1 & \frac{3}{2} & \mid & 3 \\ 1 & -1 & \mid & 2 \end{pmatrix}$	$-1R_1 + R_2 \to R_2$ (to obtain 0 in the a_{21} position)
$\begin{pmatrix} 1 & \frac{3}{2} & \mid & 3 \\ 0 & -\frac{5}{2} & \mid & -1 \end{pmatrix}$	$-\frac{2}{5}R_2 \to R_2$ (to obtain 1 in the a_{22} position)
$\begin{pmatrix} 1 & \frac{3}{2} & \mid & 3 \\ 0 & 1 & \mid & \frac{2}{5} \end{pmatrix}$	$-\frac{3}{2}R_2 + R_1 \to R_1$ (to obtain 0 in the a_{12} position)
$\begin{pmatrix} 1 & 0 & \mid & \frac{24}{10} \\ 0 & 1 & \mid & \frac{2}{5} \end{pmatrix}$	Simplify
$\begin{pmatrix} 1 & 0 & \mid & 2\frac{2}{5} \\ 0 & 1 & \mid & \frac{2}{5} \end{pmatrix}$	$= (I \mid C)$

Hence the solution is

$$x = 2\frac{2}{5}, \qquad y = \frac{2}{5} \quad \blacksquare$$

The Gauss–Jordan method is particularly useful in solving larger systems containing three or more equations and variables using only arithmetic. For three equations and three variables x, y, and z, you must achieve the form

$$\begin{pmatrix} 1 & 0 & 0 & a \\ 0 & 1 & 0 & b \\ 0 & 0 & 1 & c \end{pmatrix}$$

which implies that $x = a$, $y = b$, and $z = c$. (Here the first column represents the coefficients of x, the second column the coefficients of y, the third column the coefficients of z, and the rightmost column consists of constants.)

Example 3   Solve the system
$$\begin{aligned} R_1: & \quad x - y + z = 1 \\ R_2: & \quad 2x \phantom{{}- y} - z = 2 \\ R_3: & \quad y + z = 3 \end{aligned}$$

Solution

Matrix	Comment

$$(A\,|\,B) = \begin{pmatrix} 1 & -1 & 1 & 1 \\ 2 & 0 & -1 & 2 \\ 0 & 1 & 1 & 3 \end{pmatrix}$$
$-2R_1 + R_2 \rightarrow R_2$ puts the first column in the proper form

$$\begin{pmatrix} 1 & -1 & 1 & 1 \\ 0 & 2 & -3 & 0 \\ 0 & 1 & 1 & 3 \end{pmatrix}$$
$R_2 \rightleftarrows R_3$

$$\begin{pmatrix} 1 & -1 & 1 & 1 \\ 0 & 1 & 1 & 3 \\ 0 & 2 & -3 & 0 \end{pmatrix}$$
$R_1 + R_2 \rightarrow R_1$

$$\begin{pmatrix} 1 & 0 & 2 & 4 \\ 0 & 1 & 1 & 3 \\ 0 & 2 & -3 & 0 \end{pmatrix}$$
$-2R_2 + R_3 \rightarrow R_3$

$$\begin{pmatrix} 1 & 0 & 2 & 4 \\ 0 & 1 & 1 & 3 \\ 0 & 0 & -5 & -6 \end{pmatrix}$$
$-\frac{1}{5}R_3 \rightarrow R_3$

$$\begin{pmatrix} 1 & 0 & 2 & | & 4 \\ 0 & 1 & 1 & | & 3 \\ 0 & 0 & 1 & | & \frac{6}{5} \end{pmatrix} \qquad -1R_3 + R_2 \rightarrow R_2$$

$$\begin{pmatrix} 1 & 0 & 2 & | & 4 \\ 0 & 1 & 0 & | & \frac{9}{5} \\ 0 & 0 & 1 & | & \frac{6}{5} \end{pmatrix} \qquad -2R_3 + R_1 \rightarrow R_1$$

$$\begin{pmatrix} 1 & 0 & 0 & | & \frac{8}{5} \\ 0 & 1 & 0 & | & \frac{9}{5} \\ 0 & 0 & 1 & | & \frac{6}{5} \end{pmatrix} = (I \,|\, C) \qquad \text{final matrix}$$

Therefore $x = \frac{8}{5}$, $y = \frac{9}{5}$, $z = \frac{6}{5}$. ■

SYSTEM CLASSIFICATION

If the resulting matrix has any row of the form $(0 \quad 0 \ldots 0 \,|\, b)$, $b \neq 0$, the system is *inconsistent*.

If the resulting matrix has any row of the form $(0 \quad 0 \quad 0 \ldots 0 \,|\, 0)$, the system is *dependent*.

The matrix method is probably the most commonly used method of solving linear systems because of its adaptability to computers.

Notice that the matrix method is applied to a system in which all equations are in *standard form* $ax + by + cz = d$. That is, the system

$$\begin{cases} x + z = 1 + y \\ 2x \quad = z + 2 \\ 0 = 3 - y - z \end{cases} \quad \text{must be expressed as} \quad \begin{cases} x - y + z = 1 \\ 2x \quad - z = 2 \\ y + z = 3 \end{cases}$$

before attempting the solution by the matrix method.

Example 4 A chemical company wants to mix nitrogen, phosphoric acid, and potash to make 3600 pounds of fertilizer. The number of pounds of nitrogen is equal to three times the number of pounds of phosphoric acid. The total amount of phosphoric acid and potash is 1200 pounds. How many pounds of each chemical are required?

Solution Let x = the number of pounds of nitrogen

y = the number of pounds of phosphoric acid

z = the number of pounds of potash

The system of equations that describes the problem is

$$x + y + z = 3600$$
$$x - 3y = 0$$
$$ y + z = 1200$$

The matrix notation for the problem is as follows:

$$\begin{pmatrix} 1 & 1 & 1 & | & 3600 \\ 1 & -3 & 0 & | & 0 \\ 0 & 1 & 1 & | & 1200 \end{pmatrix} \quad -1R_1 + R_2 \to R_2$$

$$\begin{pmatrix} 1 & 1 & 1 & | & 3600 \\ 0 & -4 & -1 & | & -3600 \\ 0 & 1 & 1 & | & 1200 \end{pmatrix} \quad -\tfrac{1}{4}R_2 \to R_2$$

$$\begin{pmatrix} 1 & 1 & 1 & | & 3600 \\ 0 & 1 & \tfrac{1}{4} & | & 900 \\ 0 & 1 & 1 & | & 1200 \end{pmatrix} \quad \begin{array}{l} -1R_2 + R_1 \to R_1 \\ \text{and} \\ -1R_2 + R_3 \to R_3 \end{array}$$

$$\begin{pmatrix} 1 & 0 & \tfrac{3}{4} & | & 2700 \\ 0 & 1 & 0 & | & 900 \\ 0 & 0 & \tfrac{3}{4} & | & 300 \end{pmatrix} \quad \tfrac{4}{3}R_3 \to R_3$$

$$\begin{pmatrix} 1 & 0 & \tfrac{3}{4} & | & 2700 \\ 0 & 1 & \tfrac{1}{4} & | & 900 \\ 0 & 0 & 1 & | & 400 \end{pmatrix} \quad \begin{array}{l} -\tfrac{1}{4}R_3 + R_2 \to R_2 \\ \text{and} \\ -\tfrac{3}{4}R_3 + R_1 \to R_1 \end{array}$$

$$\begin{pmatrix} 1 & 0 & 0 & | & 2400 \\ 0 & 1 & 0 & | & 800 \\ 0 & 0 & 1 & | & 400 \end{pmatrix} \quad \text{final matrix}$$

Therefore $x = 2400$ pounds of nitrogen, $y = 800$ pounds of phosphoric acid, and $z = 400$ pounds of potash. ∎

Computational Note: Any sequence of row operations that achieves the form $\begin{pmatrix} 1 & 0 \\ 0 & 1 \end{pmatrix}$ for two equations in two unknowns or $\begin{pmatrix} 1 & 0 & 0 \\ 0 & 1 & 0 \\ 0 & 0 & 1 \end{pmatrix}$ for three equations in three unknowns in the left partition of the matrix is permissible. However, experience has shown that, in general, the most efficient way to proceed is to obtain 1 in the upper left corner, then multiply the first row by the additive inverse of the first element in the second row and add row 1 to row 2, replacing row 2 by their sum. Then multiply row 1 by the additive inverse of the first element of row 3 and add row 1 to row 3, replacing row 3 by their sum. This will put the first column in the desired form.

Next, obtain a 1 in the second row and second column and use multiplication (by additive inverses) and addition to obtain zeros in the remainder of the column. Then obtain a 1 in the third row and third column and proceed to obtain zeros in the remainder of the third column.

For three equations in three unknowns, a maximum of nine steps is required (in the order shown):

$$\begin{pmatrix} 1_1 & 0_5 & 0_9 \\ 0_2 & 1_4 & 0_8 \\ 0_3 & 0_6 & 1_7 \end{pmatrix}$$

Exercises 7.5

In Exercises 1–12 solve the system using the Gauss–Jordan method. (See Examples 1 and 2.)

1. R_1: $x - y = 3$
   R_2: $x + y = 1$

2. R_1: $x + y = -3$
   R_2: $x - y = -5$

3. R_1: $x + y = 5$
   R_2: $2x + 3y = 12$

4. R_1: $x + 2y = 8$
   R_2: $2x + 3y = 11$

5. R_1: $x + y = 1$
   R_2: $x - 3y = 3$

6. R_1: $x - y = 1$
   R_2: $2x + y = 0$

7. R_1: $2x + y = -3$
   R_2: $x - 2y = 11$

8. R_1: $3x + 2y = 3$
   R_2: $3x - 4y = -10$

9. R_1: $x + y + z = 2$
   R_2: $y - z = 2$
   R_3: $x - y = 1$

10. R_1: $x - y + z = 6$
    R_2: $y + z = -1$
    R_3: $x - z = 2$

11. R_1: $x + 2y - z = 6$
    R_2: $2y + z = -1$
    R_3: $x - 2y = 2$

12. R_1: $x - 3z = 1$
    R_2: $4x - y + 12z = 12$
    R_3: $y + 12z = -2$

In Exercises 13–16 use the matrix method to classify the system of equations as dependent or inconsistent.

13. R_1: $x + y = 5$
    R_2: $x + y = -3$

14. R_1: $x + y + z = 3$
    R_2: $x + z = 1$
    R_3: $2x + y + 2z = 2$

15. R_1: $2x - y = 4$
    R_2: $x - \frac{1}{2}y = 2$

16. R_1: $x - y + z = 4$
    R_2: $y - 2z = 3$
    R_3: $x - z = 7$

17. A coin-operated machine which accepts nickels, dimes, and quarters contained 24 coins. The number of dimes and quarters were the same, but there were 6 more nickels than quarters. Find the number of each type of coin.

18. A service station operator sells three types of gasoline: regular, lead-free, and premium. The demand for regular gasoline is three times the demand for lead-free and premium combined. The demand for lead-free is twice the demand for premium. If the dealer desires to order 1500 gallons of gasoline, how many gallons of each type should be ordered to meet the demand?

19. A chemical company wants to mix nitrogen, phosphoric acid, and potash to make 4500 pounds of fertilizer. The number of pounds of nitrogen is equal to three times the number of pounds of phosphoric acid and potash

combined. The total amount of phosphoric acid and potash is 1000 pounds. How many pounds of each chemical are required?

20. A business firm has $5000 to spend on advertising an upcoming sale. The money is to be divided among television, radio, and newspapers. The firm has determined to spend three times as much money on television advertising as it spends on radio advertising. They also decided to spend $500 less on radio advertising than they spend on newspaper advertising. How much advertising money will be spent on television, radio, and newspapers, respectively?

7.6   The Multiplicative Inverse of a Matrix

Thus far, addition, subtraction, scalar and matrix multiplication, and row operations have been discussed. The next operation one would expect to discuss is division, but division (in the usual sense) is not defined for matrices. However, by restricting our attention to square matrices, there is the *potential* for finding the multiplicative inverse of a matrix. To develop this concept further, let us review multiplicative inverses in arithmetic.

In arithmetic, the multiplicative inverse of a (nonzero) number was defined as follows: Two numbers whose product is 1 (the multiplicative identity) are called multiplicative inverses. For example, the multiplicative inverse of 2 is $\frac{1}{2}$ because $(2)(\frac{1}{2}) = 1$ (the multiplicative identity).

An important use of multiplicative inverses is to accomplish division by multiplication. That is, $6 \div 2$ is the same as $6 \times \frac{1}{2}$. This process is the one you learned for dividing by a fraction; to divide by a fraction, you multiplied by its multiplicative inverse (which was also called the reciprocal).

If a matrix has a multiplicative inverse, we can in effect divide by the given matrix through multiplication of its multiplicative inverse. This will be extremely useful in developing another method for solving linear systems using matrices.

MULTIPLICATIVE INVERSES OF MATRICES

Let A be a square matrix. The **multiplicative inverse** of A (if it exists) is a square matrix of the same order, denoted A^{-1}, such that $AA^{-1} = A^{-1}A = I$ (the multiplicative identity).

Note that multiplication of (multiplicative) inverses *is commutative*.

In arithmetic the multiplicative inverse of a number was the reciprocal of the number (if it exists). For matrices the multiplicative inverse is much more difficult to find. Some matrices that are multiplicative inverses are shown in Example 1.

Example 1 Let $A = \begin{pmatrix} 2 & 0 & 1 \\ 3 & 1 & 2 \\ 1 & 0 & 1 \end{pmatrix}$ and $B = \begin{pmatrix} 1 & 0 & -1 \\ -1 & 1 & -1 \\ -1 & 0 & 2 \end{pmatrix}$. Are A and B multiplicative inverses?

Solution $AB = \begin{pmatrix} 2 & 0 & 1 \\ 3 & 1 & 2 \\ 1 & 0 & 1 \end{pmatrix} \begin{pmatrix} 1 & 0 & -1 \\ -1 & 1 & -1 \\ -1 & 0 & 2 \end{pmatrix} = \begin{pmatrix} 1 & 0 & 0 \\ 0 & 1 & 0 \\ 0 & 0 & 1 \end{pmatrix} = I$

Since $AB = I$, $A = B^{-1}$ (or equivalently $B = A^{-1}$). You should also verify that $BA = I$. ■

Some square matrices do not have multiplicative inverses. The matrix $N = \begin{pmatrix} 0 & 0 & 0 \\ 2 & 3 & 7 \\ 5 & 9 & 8 \end{pmatrix}$ is a case in point. No matter what matrix is multiplied by N, the product matrix will have 0 in the first row and first column position, that is, $a_{11} = 0$. This means that the product cannot be the identity (I), since an identity must have a 1 in that position.

A matrix that has a multiplicative inverse is called **nonsingular**. Further, a nonsingular matrix has *one and only one* multiplicative inverse.

Some square matrices have multiplicative inverses, and others do not (as was just shown). Unfortunately, the determination of the multiplicative inverse of a square matrix (if it exists) is not as simple as finding reciprocals. Two methods will be presented. The first method uses the definitions of multiplicative inverses, matrix multiplication, matrix equality, and some algebra from Chapter 3. This method is illustrated in the following example.

Example 2 Find the multiplicative inverse of $A = \begin{pmatrix} 2 & 1 \\ 1 & 1 \end{pmatrix}$.

Solution The multiplicative inverse of A is the matrix A^{-1} such that $AA^{-1} = I$.

That is,

$$\begin{pmatrix} 2 & 1 \\ 1 & 1 \end{pmatrix}\begin{pmatrix} b_{11} & b_{12} \\ b_{21} & b_{22} \end{pmatrix} = \begin{pmatrix} 1 & 0 \\ 0 & 1 \end{pmatrix}$$

By multiplication,

$$\begin{pmatrix} 2b_{11} + b_{21} & 2b_{12} + b_{22} \\ b_{11} + b_{21} & b_{12} + b_{22} \end{pmatrix} = \begin{pmatrix} 1 & 0 \\ 0 & 1 \end{pmatrix}.$$

By the definition of equal matrices,

$$\begin{cases} 2b_{11} + b_{21} = 1 \\ b_{11} + b_{21} = 0 \end{cases} \qquad \begin{cases} 2b_{12} + b_{22} = 0 \\ b_{12} + b_{22} = 1 \end{cases}$$

This gives two systems of two equations in two unknowns. The solution of the systems (using the methods in Chapter 3) yields $b_{11} = 1$, $b_{21} = -1$, $b_{12} = -1$, and $b_{22} = 2$. Therefore

$$A^{-1} = \begin{pmatrix} 1 & -1 \\ -1 & 2 \end{pmatrix}$$

To check,

$$AA^{-1} = \begin{pmatrix} 2 & 1 \\ 1 & 1 \end{pmatrix}\begin{pmatrix} 1 & -1 \\ -1 & 2 \end{pmatrix} = \begin{pmatrix} 1 & 0 \\ 0 & 1 \end{pmatrix} = I \quad \blacksquare$$

If matrix A had been of order 3×3, there would have been *nine* equations and nine unknowns to solve. A matrix of order 4×4 would have *sixteen* equations, and so on. The (first) method becomes long and tedious rather rapidly.

If the given (square) matrix has *no* multiplicative inverse, then the system of equations is dependent or inconsistent.

The second method for determining the multiplicative inverse of a matrix uses an augmented matrix and row operations.

To find the multiplicative inverse of a square matrix A (if it exists), form the augmented matrix $(A|I)$ as shown in the box on page 280.

After the augmented matrix has been formed, row operations are performed on $(A|I)$ until a matrix of the form $(I|B)$ is obtained. If the matrix $(I|B)$ can be obtained using row operations, then the matrix B is the desired multiplicative inverse of A, that is, $B = A^{-1}$.

If it is *not* possible to obtain the matrix $(I|B)$, then A *does not have* a multiplicative inverse.

AUGMENTED MATRIX FOR FINDING THE MULTIPLICATIVE INVERSE

This **augmented matrix** is a rectangular matrix formed from an identity matrix I and a square matrix A as follows: If

$$A = \begin{pmatrix} a_{11} & a_{12} & \cdots & a_{1n} \\ a_{21} & a_{22} & \cdots & a_{2n} \\ \cdot & \cdot & & \cdot \\ \cdot & \cdot & & \cdot \\ \cdot & \cdot & & \cdot \\ a_{n1} & a_{n2} & \cdots & a_{nn} \end{pmatrix}_{n \times n} \quad \text{and } I = \begin{pmatrix} 1 & 0 & \cdots & 0 \\ 0 & 1 & \cdots & 0 \\ \cdot & \cdot & & \cdot \\ \cdot & \cdot & & \cdot \\ \cdot & \cdot & & \cdot \\ 0 & 0 & \cdots & 1 \end{pmatrix}_{n \times n}$$

then

$$(A \mid I) = \left(\begin{array}{cccc|cccc} a_{11} & a_{12} & \cdots & a_{1n} & 1 & 0 & \cdots & 0 \\ a_{21} & a_{22} & \cdots & a_{2n} & 0 & 1 & \cdots & 0 \\ \cdot & \cdot & & \cdot & \cdot & \cdot & & \cdot \\ \cdot & \cdot & & \cdot & \cdot & \cdot & & \cdot \\ \cdot & \cdot & & \cdot & \cdot & \cdot & & \cdot \\ a_{n1} & a_{n2} & \cdots & a_{nn} & 0 & 0 & \cdots & 1 \end{array} \right)$$

The vertical line between A and I is included for clarity, to show the relative location of A and I in the augmented matrix.

Example 3   Let $A = \begin{pmatrix} 1 & -1 & 1 \\ 2 & 0 & -1 \\ 0 & 1 & 1 \end{pmatrix}$. Form the augmented matrix $(A \mid I)$ and find the multiplicative inverse of A.

Solution   First, form the matrix $(A \mid I)$. Remember that the purpose of the following manipulations is to use the elementary row operations to transform $(A \mid I)$ into the form $(I \mid B)$.

$$(A \mid I) = \left(\begin{array}{ccc|ccc} 1 & -1 & 1 & 1 & 0 & 0 \\ 2 & 0 & -1 & 0 & 1 & 0 \\ 0 & 1 & 1 & 0 & 0 & 1 \end{array} \right)$$

Operation performed
$-2R_1 + R_2 \to R_2$
(gives 0 in the
a_{21} position)

$$\left(\begin{array}{ccc|ccc} 1 & -1 & 1 & 1 & 0 & 0 \\ 0 & 2 & -3 & -2 & 1 & 0 \\ 0 & 1 & 1 & 0 & 0 & 1 \end{array} \right)$$

$R_2 \rightleftarrows R_3$
(gives 1 in the
a_{22} position)

$$\begin{pmatrix} 1 & -1 & 1 & | & 1 & 0 & 0 \\ 0 & 1 & 1 & | & 0 & 0 & 1 \\ 0 & 2 & -3 & | & -2 & 1 & 0 \end{pmatrix}$$

$-2R_2 + R_3 \rightarrow R_3$
(gives 0 in the
a_{32} position)

$$\begin{pmatrix} 1 & -1 & 1 & | & 1 & 0 & 0 \\ 0 & 1 & 1 & | & 0 & 0 & 1 \\ 0 & 0 & -5 & | & -2 & 1 & -2 \end{pmatrix}$$

$R_2 + R_1 \rightarrow R_1$
(gives 0 in the
a_{12} position)

$$\begin{pmatrix} 1 & 0 & 2 & | & 1 & 0 & 1 \\ 0 & 1 & 1 & | & 0 & 0 & 1 \\ 0 & 0 & -5 & | & -2 & 1 & -2 \end{pmatrix}$$

$-\frac{1}{5}R_3 \rightarrow R_3$
(gives 1 in the
a_{33} position)

(Note that we did not interchange rows 2 and 3 even though this would have put a 1 in the a_{33} position. That would have eliminated the 1 in the a_{22} position and would have been self-defeating.)

$$\begin{pmatrix} 1 & 0 & 2 & | & 1 & 0 & 1 \\ 0 & 1 & 1 & | & 0 & 0 & 1 \\ 0 & 0 & 1 & | & \frac{2}{5} & -\frac{1}{5} & \frac{2}{5} \end{pmatrix}$$

$-1R_3 + R_2 \rightarrow R_2$
(gives 0 in the
a_{23} position)

$$\begin{pmatrix} 1 & 0 & 2 & | & 1 & 0 & 1 \\ 0 & 1 & 0 & | & -\frac{2}{5} & \frac{1}{5} & \frac{3}{5} \\ 0 & 0 & 1 & | & \frac{2}{5} & -\frac{1}{5} & \frac{2}{5} \end{pmatrix}$$

$-2R_3 + R_1 \rightarrow R_1$
(gives 0 in the
a_{13} position)

$$\begin{pmatrix} 1 & 0 & 0 & | & \frac{1}{5} & \frac{2}{5} & \frac{1}{5} \\ 0 & 1 & 0 & | & -\frac{2}{5} & \frac{1}{5} & \frac{3}{5} \\ 0 & 0 & 1 & | & \frac{2}{5} & -\frac{1}{5} & \frac{2}{5} \end{pmatrix} = (I \,|\, B)$$ Final matrix

Now we have a matrix in the form of $(I \,|\, B)$ where B is the multiplicative inverse of A; $B = A^{-1}$. Therefore, the multiplicative inverse of

$$\begin{pmatrix} 1 & -1 & 1 \\ 2 & 0 & -1 \\ 0 & 1 & 1 \end{pmatrix} \text{ is } \begin{pmatrix} \frac{1}{5} & \frac{2}{5} & \frac{1}{5} \\ -\frac{2}{5} & \frac{1}{5} & \frac{3}{5} \\ \frac{2}{5} & -\frac{1}{5} & \frac{2}{5} \end{pmatrix}$$

This can be checked by using matrix multiplication:

$$\begin{pmatrix} 1 & -1 & 1 \\ 2 & 0 & -1 \\ 0 & 1 & 1 \end{pmatrix}\begin{pmatrix} \frac{1}{5} & \frac{2}{5} & \frac{1}{5} \\ -\frac{2}{5} & \frac{1}{5} & \frac{3}{5} \\ \frac{2}{5} & -\frac{1}{5} & \frac{2}{5} \end{pmatrix} = \begin{pmatrix} 1 & 0 & 0 \\ 0 & 1 & 0 \\ 0 & 0 & 1 \end{pmatrix} \quad \blacksquare$$

The steps performed in Example 3 are not the only steps that would have led to determination of the multiplicative inverse. Another person could have found the inverse of A by the same method but with an entirely different sequence of row operations. All that is important is the goal.

> To find the multiplicative inverse of a matrix A, perform row operations on $(A \,|\, I)$ until a matrix of the form $(I \,|\, B)$ (an augmented matrix with I on the left side) is obtained. Any sequence of row operations that accomplishes the goal is permissible.

Note: For a 3×3 matrix, the nine-step process shown on page 276 is recommended.

You might want to work Example 4 using different steps.

Example 4 Find the multiplicative inverse of the matrix $A = \begin{pmatrix} 3 & 2 \\ 5 & -1 \end{pmatrix}$

Solution First, form $(A \,|\, I)$:

Operation performed

$$(A \,|\, I) = \begin{pmatrix} 3 & 2 & | & 1 & 0 \\ 5 & -1 & | & 0 & 1 \end{pmatrix} \quad \begin{array}{l} \frac{1}{3}R_1 \to R_1 \\ \text{(gives 1 in the} \\ a_{11} \text{ position)} \end{array}$$

$$\begin{pmatrix} 1 & \frac{2}{3} & | & \frac{1}{3} & 0 \\ 5 & -1 & | & 0 & 1 \end{pmatrix} \quad \begin{array}{l} -5R_1 + R_2 \to R_2 \\ \text{(gives 0 in the} \\ a_{21} \text{ position)} \end{array}$$

$$\begin{pmatrix} 1 & \frac{2}{3} & | & \frac{1}{3} & 0 \\ 0 & -\frac{13}{3} & | & -\frac{5}{3} & 1 \end{pmatrix} \quad \begin{array}{l} -\frac{3}{13}R_2 \to R_2 \\ \text{(gives 1 in the} \\ a_{22} \text{ position)} \end{array}$$

$$\begin{pmatrix} 1 & \frac{2}{3} & | & \frac{1}{3} & 0 \\ 0 & 1 & | & \frac{5}{13} & -\frac{3}{13} \end{pmatrix} \quad \begin{array}{l} -\frac{2}{3}R_2 + R_1 \to R_1 \\ \text{(gives 0 in the} \\ a_{12} \text{ position)} \end{array}$$

$$\begin{pmatrix} 1 & 0 & | & \frac{1}{13} & \frac{2}{13} \\ 0 & 1 & | & \frac{5}{13} & -\frac{3}{13} \end{pmatrix} = (I \,|\, B)$$

Therefore the multiplicative inverse of $\begin{pmatrix} 3 & 2 \\ 5 & -1 \end{pmatrix}$ is $\begin{pmatrix} \frac{1}{13} & \frac{2}{13} \\ \frac{5}{13} & -\frac{3}{13} \end{pmatrix}$. This can be checked by multiplication:

$$\begin{pmatrix} 3 & 2 \\ 5 & -1 \end{pmatrix}\begin{pmatrix} \frac{1}{13} & \frac{2}{13} \\ \frac{5}{13} & -\frac{3}{13} \end{pmatrix} = \begin{pmatrix} 1 & 0 \\ 0 & 1 \end{pmatrix} \quad \blacksquare$$

As was previously stated, not all square matrices have a multiplicative inverse. If there is no inverse for a particular square matrix, then the procedure will fail. How do we know that the procedure has failed? The following example will illustrate this point.

Example 5 Find the multiplicative inverse of the matrix $A = \begin{pmatrix} 1 & 1 & 0 \\ 1 & 0 & -1 \\ 2 & 1 & -1 \end{pmatrix}$ if it exists.

Solution The reasons for each step are left to you.

$$\left(\begin{array}{ccc|ccc} 1 & 1 & 0 & 1 & 0 & 0 \\ 1 & 0 & -1 & 0 & 1 & 0 \\ 2 & 1 & -1 & 0 & 0 & 1 \end{array}\right)$$

$$\left(\begin{array}{ccc|ccc} 1 & 1 & 0 & 1 & 0 & 0 \\ 0 & -1 & -1 & -1 & 1 & 0 \\ 2 & 1 & -1 & 0 & 0 & 1 \end{array}\right)$$

$$\left(\begin{array}{ccc|ccc} 1 & 1 & 0 & 1 & 0 & 0 \\ 0 & -1 & -1 & -1 & 1 & 0 \\ 0 & -1 & -1 & -2 & 0 & 1 \end{array}\right)$$

$$\left(\begin{array}{ccc|ccc} 1 & 1 & 0 & 1 & 0 & 0 \\ 0 & 1 & 1 & 1 & -1 & 0 \\ 0 & -1 & -1 & -2 & 0 & 1 \end{array}\right)$$

$$\left(\begin{array}{ccc|ccc} 1 & 1 & 0 & 1 & 0 & 0 \\ 0 & 1 & 1 & 1 & -1 & 0 \\ 0 & 0 & 0 & -1 & -1 & 1 \end{array}\right) \quad \blacksquare$$

Note that the first three columns of row 3 all contain zeros. There is no way to obtain a 1 in the a_{33} position without disrupting some other row. Thus A does not have an inverse.

> When the transformation of $(A \mid I)$ produces a step $(C \mid D)$ where C has a row with all zero entries, then the square matrix A does not have a multiplicative inverse.

Exercises 7.6 In Exercises 1–6 state whether the pair of matrices are multiplicative inverses.

1. $\begin{pmatrix} 1 & 0 \\ 0 & 1 \end{pmatrix}; \begin{pmatrix} 1 & 1 \\ 2 & 1 \end{pmatrix}$

2. $\begin{pmatrix} 1 & 1 \\ 2 & 1 \end{pmatrix}; \begin{pmatrix} -1 & 1 \\ 2 & -1 \end{pmatrix}$

3. $\begin{pmatrix} 1 & -1 & 1 \\ 2 & 0 & 1 \\ 0 & 1 & 1 \end{pmatrix}; \begin{pmatrix} \frac{1}{5} & \frac{2}{5} & \frac{1}{5} \\ -\frac{2}{5} & \frac{1}{5} & \frac{3}{5} \\ \frac{2}{5} & -\frac{1}{5} & \frac{2}{5} \end{pmatrix}$

4. $\begin{pmatrix} 2 & 4 \\ -1 & -3 \end{pmatrix}; \begin{pmatrix} \frac{3}{2} & 2 \\ -\frac{1}{2} & -1 \end{pmatrix}$

5. $\begin{pmatrix} 1 & 2 \\ 1 & 3 \end{pmatrix}; \begin{pmatrix} 1 & -1 \\ -1 & 1 \end{pmatrix}$

6. $\begin{pmatrix} 1 & 2 & -1 \\ 0 & -1 & 3 \end{pmatrix}; \begin{pmatrix} 2 & 0 \\ -1 & 1 \\ 1 & 2 \end{pmatrix}$

In Exercises 7–17 find the multiplicative inverse of the matrix, if possible, by the matrix method.

7. $\begin{pmatrix} 1 & 1 & 1 \\ 0 & -1 & 2 \\ 1 & 1 & 0 \end{pmatrix}$

8. $\begin{pmatrix} -1 & 1 \\ 1 & -2 \end{pmatrix}$

9. $\begin{pmatrix} 2 & 1 & 1 \\ 0 & -1 & 2 \\ 0 & 1 & -1 \end{pmatrix}$

10. $\begin{pmatrix} 1 & 1 \\ 3 & 3 \end{pmatrix}$

11. $\begin{pmatrix} 2 & 1 & -1 \\ 1 & 0 & -1 \\ 1 & 1 & 0 \end{pmatrix}$

12. $\begin{pmatrix} 1 & 0 & 1 \\ 0 & 1 & -2 \\ 2 & 1 & 1 \end{pmatrix}$

13. $\begin{pmatrix} 1 & 1 \\ 2 & -1 \end{pmatrix}$

14. $\begin{pmatrix} 1 & 2 & 1 \\ 2 & 3 & 1 \\ -1 & 1 & 1 \end{pmatrix}$

15. $\begin{pmatrix} 4 & 9 \\ 5 & 6 \end{pmatrix}$

16. $\begin{pmatrix} 1 & 3 & 5 \\ 6 & -4 & 2 \\ 5 & 4 & 11 \end{pmatrix}$

17. $\begin{pmatrix} 1 & 0 & 0 \\ -2 & 1 & 3 \\ 4 & 1 & 2 \end{pmatrix}$

18. Find the inverse of $\begin{pmatrix} 1 & 2 \\ 3 & 4 \end{pmatrix}$ by the method used in Example 2 of this section.

In Exercises 19–20 find the inverse by either method.

19. $\begin{pmatrix} 2 & 0 & 0 \\ 0 & 3 & 0 \\ 0 & 0 & 4 \end{pmatrix}$

20. $\begin{pmatrix} 0 & 0 & 1 \\ 0 & 1 & 0 \\ 1 & 0 & 0 \end{pmatrix}$

7.7   The Inverse Method for Solving a Linear System

In Section 7.4 we expressed a linear system as the matrix equation $AX = B$. We now want to solve the matrix equation for (the matrix of unknowns) X.

The process is similar to that of solving an ordinary equation of the form $ax = b$. Consider the process for solving $\frac{1}{2}x = 6$. For such an algebraic equation, the solution is obtained by multiplying both sides of the equation by the multiplicative inverse of the coefficient $\frac{1}{2}$, which is 2:

$$2\left(\frac{1}{2}\right)x = 2(6)$$

$$x = 12$$

Although the matrix equation $AX = B$ is not an ordinary algebraic equation, it can be solved in a similar manner.

Let $AX = B$. Now multiply both sides of the equation by the multiplicative inverse of A:

$$A^{-1}(AX) = A^{-1}B$$

$$(A^{-1}A)X = A^{-1}B$$

$$IX = A^{-1}B$$

Since the unknown matrix is multiplied by the identity (I), we have

$$X = A^{-1}B$$

Note: A^{-1} is multiplied on the *left* side of each member of the equation. This gives $A^{-1}B$ as the solution (not BA^{-1}). Since matrix multiplication is not commutative, $A^{-1}B \neq BA^{-1}$.

Now, by applying the definition of equal matrices, the value of each variable in X can be determined.

SOLUTION OF A MATRIX EQUATION

The solution of a matrix equation $AX = B$ is $X = A^{-1}B$, provided that A^{-1} exists.

The key to solving $AX = B$ is the existence of A^{-1}. Unless A^{-1} exists, this process cannot be used.

Example 1 Solve $\begin{pmatrix} 2 & 1 \\ 1 & 1 \end{pmatrix}\begin{pmatrix} x \\ y \end{pmatrix} = \begin{pmatrix} 2 \\ 3 \end{pmatrix}$ given that the inverse of the coefficient matrix is $\begin{pmatrix} 1 & -1 \\ -1 & 2 \end{pmatrix}$.

Solution   We are given that $A^{-1} = \begin{pmatrix} 1 & -1 \\ -1 & 2 \end{pmatrix}$. Therefore

$$X = A^{-1}B$$

$$\begin{pmatrix} x \\ y \end{pmatrix} = \begin{pmatrix} 1 & -1 \\ -1 & 2 \end{pmatrix}\begin{pmatrix} 2 \\ 3 \end{pmatrix}$$

$$= \begin{pmatrix} -1 \\ 4 \end{pmatrix}$$

By the definition of equal matrices $x = -1$, $y = 4$.

Check:
Replace x and y by -1 and 4, respectively, in the *original* problem:

$$\begin{pmatrix} 2 & 1 \\ 1 & 1 \end{pmatrix}\begin{pmatrix} -1 \\ 4 \end{pmatrix} \overset{?}{=} \begin{pmatrix} 2 \\ 3 \end{pmatrix} \qquad \text{Multiply}$$

$$\begin{pmatrix} -2 + 4 \\ -1 + 4 \end{pmatrix} \overset{?}{=} \begin{pmatrix} 2 \\ 3 \end{pmatrix}$$

$$\begin{pmatrix} 2 \\ 3 \end{pmatrix} \overset{\checkmark}{=} \begin{pmatrix} 2 \\ 3 \end{pmatrix} \qquad \blacksquare$$

Example 2   Solve the matrix equation $\begin{pmatrix} 1 & -1 & 1 \\ 2 & 0 & -1 \\ 0 & 1 & 1 \end{pmatrix}\begin{pmatrix} x \\ y \\ z \end{pmatrix} = \begin{pmatrix} 1 \\ 2 \\ 3 \end{pmatrix}$ if the inverse of

the coefficient matrix is $\begin{pmatrix} \frac{1}{5} & \frac{2}{5} & \frac{1}{5} \\ -\frac{2}{5} & \frac{1}{5} & \frac{3}{5} \\ \frac{2}{5} & -\frac{1}{5} & \frac{2}{5} \end{pmatrix}$

Solution   To solve, multiply B on the left by A^{-1}. Hence

$$X = A^{-1}B$$

$$\begin{pmatrix} x \\ y \\ z \end{pmatrix} = \begin{pmatrix} \frac{1}{5} & \frac{2}{5} & \frac{1}{5} \\ -\frac{2}{5} & \frac{1}{5} & \frac{3}{5} \\ \frac{2}{5} & -\frac{1}{5} & \frac{2}{5} \end{pmatrix}\begin{pmatrix} 1 \\ 2 \\ 3 \end{pmatrix}$$

$$\begin{pmatrix} x \\ y \\ z \end{pmatrix} = \begin{pmatrix} \frac{8}{5} \\ \frac{9}{5} \\ \frac{6}{5} \end{pmatrix}$$

Therefore $x = \frac{8}{5}$, $y = \frac{9}{5}$, and $z = \frac{6}{5}$. $\blacksquare$

For A^{-1} to exist, the first consideration is that A be square. Since this section is restricted to systems of equations that have the same numbers of equations and unknowns, A will always be square.

However, being square is not sufficient for A^{-1} to exist. We saw square matrices in Section 7.6 that did not have inverses. The existence of A^{-1} depends on the type of linear system. If the linear system is *inconsistent* (has no solution) or *dependent* (with infinitely many solutions), A^{-1} will not exist.

> The inverse of the coefficient matrix (A^{-1}) can be used to solve a matrix equation $AX = B$, provided that A is *square* and that the system of linear equations has a *unique* solution.

When the inverse can be found, the matrix equation method is a practical approach for solving certain systems of equations. The advantage is that the constants can be changed and new solutions can be obtained without starting over. This is not the case with other matrix methods. The next example will illustrate the advantage of using the inverse.

Consider a store manager who is ordering a summer selection of shirts.

Example 3   A store manager wants to stock casual shirts, sport shirts, and dress shirts in its summer selection. The cost per shirt of each type is \$2, \$3, and \$6, respectively. The shirts will retail for \$4, \$5, and \$10, respectively.

(a) How many shirts of each type should be purchased if the company needs a total of 1000 shirts costing \$3200 and selling for \$5600?

(b) How many shirts of each type should be purchased if the company needs 800 shirts costing \$2100 and selling for \$3900?

Let x = the number of casual shirts, y = the number of sport shirts, and z = the number of dress shirts. We find the inverse of the coefficient matrix and use matrix equation notation to solve the problem.

Solution   (a) algebraic equations: matrix equation:

$$
\begin{aligned}
x + y + z &= 1000 \text{ (quantity)} \\
2x + 3y + 6z &= 3200 \text{ (cost)} \\
4x + 5y + 10z &= 5600 \text{ (retail)}
\end{aligned}
\qquad
\begin{pmatrix} 1 & 1 & 1 \\ 2 & 3 & 6 \\ 4 & 5 & 10 \end{pmatrix}
\begin{pmatrix} x \\ y \\ z \end{pmatrix}
=
\begin{pmatrix} 1000 \\ 3200 \\ 5600 \end{pmatrix}
$$

Now we find the inverse by the method given in Section 7.6:

$$
\left(\begin{array}{ccc|ccc} 1 & 1 & 1 & 1 & 0 & 0 \\ 2 & 3 & 6 & 0 & 1 & 0 \\ 4 & 5 & 10 & 0 & 0 & 1 \end{array}\right)
\rightarrow
\left(\begin{array}{ccc|ccc} 1 & 1 & 1 & 1 & 0 & 0 \\ 0 & 1 & 4 & -2 & 1 & 0 \\ 0 & 1 & 6 & -4 & 0 & 1 \end{array}\right)
\rightarrow
$$

$$
\left(\begin{array}{ccc|ccc} 1 & 0 & -3 & 3 & -1 & 0 \\ 0 & 1 & 4 & -2 & 1 & 0 \\ 0 & 0 & 2 & -2 & -1 & 1 \end{array}\right)
\rightarrow
\left(\begin{array}{ccc|ccc} 1 & 0 & 0 & 0 & -\frac{5}{2} & \frac{3}{2} \\ 0 & 1 & 0 & 2 & 3 & -2 \\ 0 & 0 & 1 & -1 & -\frac{1}{2} & \frac{1}{2} \end{array}\right)
$$

Hence $A^{-1} = \begin{pmatrix} 0 & -\frac{5}{2} & \frac{3}{2} \\ 2 & 3 & -2 \\ -1 & -\frac{1}{2} & \frac{1}{2} \end{pmatrix}$. (You should check that $AA^{-1} = I$.)

Now we solve the matrix equation.

$$\begin{matrix} X & = & A^{-1} & & B \end{matrix}$$

$$\begin{pmatrix} x \\ y \\ z \end{pmatrix} = \begin{pmatrix} 0 & -\frac{5}{2} & \frac{3}{2} \\ 2 & 3 & -2 \\ -1 & -\frac{1}{2} & \frac{1}{2} \end{pmatrix} \begin{pmatrix} 1000 \\ 3200 \\ 5600 \end{pmatrix}$$

$$\begin{pmatrix} x \\ y \\ z \end{pmatrix} = \begin{pmatrix} 400 \\ 400 \\ 200 \end{pmatrix}$$

Therefore $x = 400$ casual shirts, $y = 400$ sport shirts, and $z = 200$ dress shirts.

(b) Since nothing has changed but the constants, there is no need to find the inverse again. We simply use the new constants.

$$\begin{pmatrix} x \\ y \\ z \end{pmatrix} = \begin{pmatrix} 0 & -\frac{5}{2} & \frac{3}{2} \\ 2 & 3 & -2 \\ -1 & -\frac{1}{2} & \frac{1}{2} \end{pmatrix} \begin{pmatrix} 800 \\ 2100 \\ 3900 \end{pmatrix}$$

$$\begin{pmatrix} x \\ y \\ z \end{pmatrix} = \begin{pmatrix} 600 \\ 100 \\ 100 \end{pmatrix}$$

Therefore $x = 600$ casual shirts, $y = 100$ sport shirts, and $z = 100$ dress shirts. ■

Exercises 7.7

In Exercises 1–3 express in ordinary algebraic notation.

1. $\begin{pmatrix} 4 & 3 \\ 2 & -3 \end{pmatrix} \begin{pmatrix} x_1 \\ x_2 \end{pmatrix} = \begin{pmatrix} 11 \\ 7 \end{pmatrix}$

2. $\begin{pmatrix} 1 & 1 & 1 \\ 2 & 0 & 1 \\ 1 & -1 & 1 \end{pmatrix} \begin{pmatrix} x_1 \\ x_2 \\ x_3 \end{pmatrix} = \begin{pmatrix} 3 \\ 2 \\ 1 \end{pmatrix}$

3. $\begin{pmatrix} 3 & -1 & -2 \\ 5 & 3 & -1 \\ 2 & -7 & 3 \end{pmatrix} \begin{pmatrix} x_1 \\ x_2 \\ x_3 \end{pmatrix} = \begin{pmatrix} -13 \\ 4 \\ -30 \end{pmatrix}$

In Exercises 4–6 solve the matrix equation using A^{-1}.

4. $\begin{pmatrix} \frac{1}{3} & \frac{1}{3} \\ \frac{2}{3} & -\frac{1}{3} \end{pmatrix} \begin{pmatrix} x \\ y \end{pmatrix} = \begin{pmatrix} 6 \\ 9 \end{pmatrix}$; $A^{-1} = \begin{pmatrix} 1 & 1 \\ 2 & -1 \end{pmatrix}$

5. $\begin{pmatrix} -2 & 1 & 3 \\ 2 & -1 & -2 \\ 1 & 0 & -1 \end{pmatrix} \begin{pmatrix} x_1 \\ x_2 \\ x_3 \end{pmatrix} = \begin{pmatrix} -2 \\ 1 \\ 2 \end{pmatrix}$; $A^{-1} = \begin{pmatrix} 1 & 1 & 1 \\ 0 & -1 & 2 \\ 1 & 1 & 0 \end{pmatrix}$

6. $\begin{pmatrix} 2 & 1 & 1 \\ 1 & -1 & 5 \\ 0 & 1 & -1 \end{pmatrix} \begin{pmatrix} x_1 \\ x_2 \\ x_3 \end{pmatrix} = \begin{pmatrix} -6 \\ 12 \\ 18 \end{pmatrix}$; $A^{-1} = \begin{pmatrix} \frac{2}{3} & -\frac{1}{3} & -1 \\ -\frac{1}{6} & \frac{1}{3} & \frac{3}{2} \\ -\frac{1}{6} & \frac{1}{3} & \frac{1}{2} \end{pmatrix}$

In Exercises 7–18
(a) write the system as a matrix equation;
(b) compute A^{-1} (if it exists);
(c) use matrix multiplication to verify that A^{-1} is correct;
(d) use A^{-1} to solve the system;
(e) check your answers.

7. $x_1 + x_2 = 5$
    $2x_1 - x_2 = 1$

8. $x + y = 5$
    $2x + 2y = 16$

9. $x + y = 5$
    $3x + 3y = 15$

10. $2x_1 + x_2 + x_3 = 1$
     $x_1 - x_2 + 5x_3 = 2$
     $x_2 - x_3 = 3$

11. $x + y + z = 4$
     $-y + 2z = 1$
     $x + y = 2$

12. $x_1 + x_3 = 5$
     $x_2 - 2x_3 = 3$
     $2x_1 + x_2 + x_3 = -7$

13. $2x + y - z = 2$
     $x - z = 3$
     $x + y = 4$

14. $x + 2y + z = -2$
     $-x - y + z = 0$
     $y + 3z = 3$

15. $x + y = 0$
     $2x + y - z = 1$
     $x - z = 1$

16. $2x + y + z = 3$
     $x + y - z = 3$
     $x + z = 3$

17. $x + y = 4$
     $x - y = -2$

18. $x + z + w = 6$
     $-x + y + 2z + w = 12$
     $x - y + z + 2w = 24$
     $y + w = 6$

In Exercises 19–27 write the system of equations for the problems. Then follow the instructions given for Exercises 7–18.

19. Some students are working on a science fair project. They need 500 bolts, 300 nuts, and 300 washers. The hardware store has some odd packages of nuts, bolts, and washers that they want to sell at a reduced price. Type *A* packages contain 1 bolt, 2 nuts, and no washers. Type *B* packages contain 2 bolts, 1 nut, and 1 washer. Type *C* packages contain 1 bolt, no nuts, and 1 washer. How many of each type of package should the students purchase to have the required number of nuts, bolts, and washers? Let x = the number of *A*'s, y = the number of *B*'s, and z = the number of *C*'s.

20. In Exercise 19, if *A* costs 10 cents per package, *B* costs 8 cents per package, and *C* costs 3 cents per package, how much will the students have to spend?

21. One of the students in Exercise 19 who has worked on a similar project says that 10% of the nuts, bolts, and washers usually get lost or damaged. How many packages of each should be purchased to account for the expected loss?

22. Bill is taking 3 types of food in small packages on a hiking trip. He wants to take enough food to have 200 grams of protein, 300 grams of carbohydrates, and 80 grams of fat. Package *A* contains 1 gram of protein, 2 grams of carbohydrates, and 1 gram of fat. Package *B* contains 1 gram of protein, 1 gram of carbohydrates, and no fat. Package *C* contains 1 gram of protein, 3 grams of carbohydrates, and 1 gram of fat.
     (a) How many packages of each food should he take? Let x = the

number of packages of A, y = the number of packages of B, and z = the number of packages of C.

(b) How many packages of each should he take to have 100 grams of protein, 300 grams of carbohydrates, and 100 grams of fat?

C 23. A company has three warehouses in which it stores three types of items. The cost per day of storing each type of item in each warehouse is given in the following matrix.

	Warehouses		
	I	II	III
Refrigerators	$0.10	0.20	0.20
TV's	0.20	0.10	0.10
Dishwashers	0.10	0.10	0.20

The total cost per day of storing refrigerators is $300, the cost per day of storing TV's is $180, and the cost per day of storing dishwashers is $200.

(a) Find the number of refrigerators, TV's, and dishwashers presently in storage. Let x = the number of refrigerators, y = the number of TV's, and z = the number of dishwashers.

(b) Find the number of each type of appliance if the costs are $90, $90, and $60, respectively.

24. The perimeter of a triangle is 20. The sum of the lengths of the two shorter sides is equal to the length of the longest side. The longest side minus the shortest side is 8. What are the lengths of the sides? Let x = the length of the shortest side and z = the length of the longest side.

25. Laura is studying for final exams and wants to spend her day sleeping, eating, and studying. Time spent studying is two hours more than time spent sleeping. The total time spent sleeping and eating is 12 hours.

(a) How many hours are allotted to each occupation?

(b) If the total time spent sleeping and eating is 14 hours, how many hours are allotted to each occupation?

26. In a certain city the populations of the three major ethnic groups total 10,000. The largest ethnic group is 1000 more than twice as large as the other two groups combined. If the smallest group has 2000 less than the medium group, then how many people are in each group?

27. A test score does not reflect a student's true knowledge of the subject being tested. Part of the score is error due to the conditions under which the test is given. On a particular test, one cause of error is due to the number of items x on the test; another part of the error is due to the amount of light y (in arbitrary light units); and the rest of the error is due to the noise level z (in arbitrary noise units). The total error on a particular test is 6, the number of items minus the light is 2, and twice the light minus the noise is 4. How many questions are on the test, how much light is there, and what is the noise level?

7.8   Determinants and Cramer's Rule

Thus far we have seen two ways that a matrix can be used to solve a linear system. Now we will define the determinant of a square matrix and show how determinants can be used to solve linear systems. (Recall that a square matrix is one that has the same number of rows and columns.) Associated with each square matrix (M) of real numbers is a number called the determinant of M and denoted $\det(M)$.

The determinant "det" is a function whose *domain* is the set of all square matrices (containing real numbers) and whose *range* is the set of real numbers. We begin by defining the determinant of a square matrix that has two rows and two columns.

DEFINITION   The **determinant** of matrix

$$M = \begin{pmatrix} a & b \\ c & d \end{pmatrix} \text{ is } \det(M) = ad - bc$$

The determinant of a 2×2 matrix is called a determinant of order 2.

If $\det(M) = 0$, M is called a **singular matrix**.
If $\det(M) \neq 0$, M is called a **nonsingular matrix**.

Several other commonly used notations for $\det(M)$ are $|M|$, $\delta(M)$, $D(M)$, and simply D.

Example 1   Find the determinant of

(a) $\begin{pmatrix} 1 & -2 \\ 3 & 4 \end{pmatrix}$

(b) $\begin{pmatrix} x & \sqrt{2} \\ 3 & y \end{pmatrix}$

(c) $\begin{pmatrix} a_1 & b_1 \\ a_2 & b_2 \end{pmatrix}$

Solution   (a) $\det \begin{pmatrix} 1 & -2 \\ 3 & 4 \end{pmatrix} = (1)(4) - (-2)(3) = 4 + 6 = 10$

(b) $\det \begin{pmatrix} x & \sqrt{2} \\ 3 & y \end{pmatrix} = xy - 3\sqrt{2}$

(c) $\det \begin{pmatrix} a_1 & b_1 \\ a_2 & b_2 \end{pmatrix} = a_1 b_2 - a_2 b_1$ ∎

The relationship between determinants of 2×2 matrices and the solution of two linear equations in two variables can be shown by solving the system

$$A: \quad a_1 x + b_1 y = c_1$$
$$B: \quad a_2 x + b_2 y = c_2$$

by the addition method and applying the definition of a determinant of order 2.

$$
\begin{array}{lll}
\text{Multiply} & b_2 \times A: & a_1 b_2 x + b_1 b_2 y = b_2 c_1 \\
\text{and} & -b_1 \times B: & -a_2 b_1 x - b_1 b_2 y = -b_1 c_2
\end{array}
$$

We add the equations and solve for x:

$$a_1 b_2 x - a_2 b_1 x = b_2 c_1 - b_1 c_2$$

$$x(a_1 b_2 - a_2 b_1) = b_2 c_1 - b_1 c_2$$

$$x = \frac{b_2 c_1 - b_1 c_2}{a_1 b_2 - a_2 b_1}$$

In a similar manner, we find that

$$y = \frac{a_1 c_2 - a_2 c_1}{a_1 b_2 - a_2 b_1}$$

By the definition of a determinant of order 2,

$$
x = \frac{\det \begin{pmatrix} c_1 & b_1 \\ c_2 & b_2 \end{pmatrix}}{\det \begin{pmatrix} a_1 & b_1 \\ a_2 & b_2 \end{pmatrix}}
\quad \text{and} \quad
y = \frac{\det \begin{pmatrix} a_1 & c_1 \\ a_2 & c_2 \end{pmatrix}}{\det \begin{pmatrix} a_1 & b_1 \\ a_2 & b_2 \end{pmatrix}}
$$

The value of each unknown is a fraction consisting of two determinants. The denominator of both unknowns is the determinant of the coefficient matrix. The matrices in the numerators are obtained by replacing the coefficients of the unknown by the column of constants (assuming that both equations are in standard form).

Example 2   Solve the system A: $3x + 5y = 4$

$$\text{B:} \quad 2x - 3y = -10$$

by determinants.

Solution $x = \dfrac{\det \begin{pmatrix} 4 & 5 \\ -10 & -3 \end{pmatrix}}{\det \begin{pmatrix} 3 & 5 \\ 2 & -3 \end{pmatrix}} = \dfrac{38}{-19} = -2$

$$y = \frac{\det \begin{pmatrix} 3 & 4 \\ 2 & -10 \end{pmatrix}}{\det \begin{pmatrix} 3 & 5 \\ 2 & -3 \end{pmatrix}} = \frac{-38}{-19} = +2$$

Hence $x = -2$, $y = 2$ is the solution of the system. ■

The solution of a linear system by determinants as shown in Example 2 is called **Cramer's Rule.**

SYSTEM CLASSIFICATION

If the coefficient matrix is nonsingular (det $\neq$ 0), the system is consistent and has a unique solution (independent).

If the coefficient matrix is singular (det = 0), the system is either dependent or inconsistent. To decide which is the case, examine all the numerators of the unknowns. If the determinants in all numerators are 0, the system is dependent. If there is any nonzero numerator, the system is inconsistent.

Example 3 A chemist needs 250 milliliters of a 40% solution of sulphuric acid. Since there is no distilled water, the chemist decides to mix some 20% and 70% solution. How many milliliters of each are needed? Solve by determinants.

Solution Let x = the amount of 70% solution

y = the amount of 20% solution

The equations to be solved are

$$x + \quad y = 250 \qquad \text{or} \qquad x + \quad y = 250$$
$$0.70x + 0.20y = 0.40(250) \qquad \qquad 0.7x + 0.2y = 100$$

Then

$$x = \frac{\det \begin{pmatrix} 250 & 1 \\ 100 & 0.2 \end{pmatrix}}{\det \begin{pmatrix} 1 & 1 \\ 0.7 & 0.2 \end{pmatrix}} = \frac{-50}{-0.5} = 100 \text{ (milliliters of 70\% solution)}$$

$$y = \frac{\det \begin{pmatrix} 1 & 250 \\ 0.7 & 100 \end{pmatrix}}{\det \begin{pmatrix} 1 & 1 \\ 0.7 & 0.2 \end{pmatrix}} = \frac{-75}{-0.5} = 150 \text{ (milliliters of 20\% solution)} \quad \blacksquare$$

To solve systems of three equations in three unknowns by Cramer's Rule, we must define the determinant of a 3 × 3 matrix (a determinant of **order** 3). The determinant of a 3 × 3 matrix can be found by an *expansion* (or computation) into a series of second-order determinants (called minors).

DEFINITION   The **determinant** of the 3×3 matrix $M = \begin{pmatrix} m_{11} & m_{12} & m_{13} \\ m_{21} & m_{22} & m_{23} \\ m_{31} & m_{32} & m_{33} \end{pmatrix}$ by expansion along the first row is

$$\det(M) = m_{11} \det \begin{pmatrix} m_{22} & m_{23} \\ m_{32} & m_{33} \end{pmatrix} - m_{12} \det \begin{pmatrix} m_{21} & m_{23} \\ m_{31} & m_{33} \end{pmatrix}$$

$$+ \, m_{13} \det \begin{pmatrix} m_{21} & m_{22} \\ m_{31} & m_{32} \end{pmatrix}.$$

The determinant of the 2×2 matrix that remains when the row and column containing m_{ij} are deleted is called the **minor** of m_{ij}.

The **cofactor** of m_{ij} is the minor of m_{ij} if $i + j$ is even; it is the *negative* of the minor if $i + j$ is odd. This is why the signs in the expansion alternate. For m_{11}, $1 + 1 = 2$ is even, so the first term is $+$. For m_{12}, $1 + 2 = 3$ is odd, so the second term is $-$, and so on.

This method of finding $\det(M)$ is called "expansion by minors," but is essentially an expansion of cofactors.

The following example shows how to find the determinant of a 3×3 matrix by expansion along the first row.

Example 4   If $A = \begin{pmatrix} 1 & -1 & 1 \\ 2 & 0 & -1 \\ 0 & 1 & 1 \end{pmatrix}$, find $\det(A)$.

Solution $\det(A) = (1) \det \begin{pmatrix} 0 & -1 \\ 1 & 1 \end{pmatrix} - (-1) \det \begin{pmatrix} 2 & -1 \\ 0 & 1 \end{pmatrix} + (1) \det \begin{pmatrix} 2 & 0 \\ 0 & 1 \end{pmatrix}$

$\qquad = (1)[(0)(1) - (-1)(1)] - (-1)[(2)(1) - (-1)(0)]$
$\qquad \quad + (1)[(2)(1) - (0)(0)]$
$\qquad = (1)(1) + (1)(2) + (1)(2)$
$\qquad = 1 + 2 + 2$
$\qquad = 5$ ∎

The determinant of a matrix may be found by expansion along any single row (or column). Example 5 shows how to find the determinant of matrix A (given in Example 4) by expansion along the second column.

Example 5   Find the determinant of $A = \begin{pmatrix} 1 & -1 & 1 \\ 2 & 0 & -1 \\ 0 & 1 & 1 \end{pmatrix}$ by expanding along the second column.

Solution $\det(A) = -(-1)\det\begin{pmatrix}2 & -1\\0 & 1\end{pmatrix} + (0)\det\begin{pmatrix}1 & 1\\0 & 1\end{pmatrix} - (1)\det\begin{pmatrix}1 & 1\\2 & -1\end{pmatrix}$

$$= +1(2) + 0 - (1)(-3)$$
$$= 2 + 3$$
$$= 5 \quad \blacksquare$$

Of course, $\det(A) = 5$ no matter which row or column is chosen for the expansion, since "det" is a function.

The following example shows how Cramer's Rule can be used to solve a linear system containing three equations and three unknowns.

Example 6 Solve by Cramer's Rule:

$$x + z = y + 1$$
$$-z = 2 - 2x$$
$$z = 3 - y$$

Solution First put all equations in standard form:

$$x - y + z = 1$$
$$2x \qquad - z = 2$$
$$y + z = 3$$

Now form the appropriate matrices and fractions. (The determinant of the matrix in the denominator was computed in Example 4. Those in the numerator are left to you.)

$$x = \frac{\det\begin{pmatrix}1 & -1 & 1\\2 & 0 & -1\\3 & 1 & 1\end{pmatrix}}{\det\begin{pmatrix}1 & -1 & 1\\2 & 0 & -1\\0 & 1 & 1\end{pmatrix}} = \frac{8}{5}$$

$$y = \frac{\det\begin{pmatrix}1 & 1 & 1\\2 & 2 & -1\\0 & 3 & 1\end{pmatrix}}{\det\begin{pmatrix}1 & -1 & 1\\2 & 0 & -1\\0 & 1 & 1\end{pmatrix}} = \frac{9}{5}$$

$$z = \frac{\det\begin{pmatrix}1 & -1 & 1\\2 & 0 & 2\\0 & 1 & 3\end{pmatrix}}{\det\begin{pmatrix}1 & -1 & 1\\2 & 0 & -1\\0 & 1 & 1\end{pmatrix}} = \frac{6}{5} \quad \blacksquare$$

Computational Note: For 3 by 3 matrices *only*, there is a time-saving process for computing the determinant. To find the determinant of

$$\begin{pmatrix} 1 & -1 & 1 \\ 2 & 0 & -1 \\ 0 & 1 & 1 \end{pmatrix}$$

copy the first and second columns on the right side of the matrix. This gives the following:

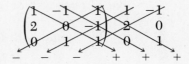

Now multiply the numbers along each arrow and put the sign at the point in front of the product; then combine the result. You have $-(0) - (-1) - (-2) + (0) + (0) + 2 = 1 + 2 + 2 = 5$. Hence, the determinant is 5 (as shown previously). Remember that this schematic approach does not work for larger matrices.

Determinants of Large Matrices

DEFINITION   The **determinant** of any square matrix

$$\begin{pmatrix} m_{11} & m_{12} & m_{13} & \ldots & m_{1n} \\ m_{21} & m_{22} & m_{23} & \ldots & m_{2n} \\ \cdot & \cdot & \cdot & & \cdot \\ \cdot & \cdot & \cdot & & \cdot \\ \cdot & \cdot & \cdot & & \cdot \\ m_{n1} & m_{n2} & m_{n3} & \ldots & m_{nn} \end{pmatrix}$$

is the sum of the products formed by multiplying each entry in any single row (or column) by its cofactor.

Example 7 shows how we can apply the definition to find a fourth-order determinant.

Computational Note: Expand the determinant along the row or column containing the greatest number of zeros. This will minimize the computation.

Example 7   Find the determinant of matrix

$$A = \begin{pmatrix} 1 & 0 & 2 & 1 \\ -1 & 0 & 3 & 4 \\ 0 & 3 & -1 & 1 \\ 1 & 0 & 0 & -1 \end{pmatrix}$$

Solution   We compute det(A) by expansion along the second column, which contains the most zeros.

$$\det(A) = -(0)\det\begin{pmatrix} -1 & 3 & 4 \\ 0 & -1 & 1 \\ 1 & 0 & -1 \end{pmatrix} + (0)\det\begin{pmatrix} 1 & 2 & 1 \\ 0 & -1 & 1 \\ 1 & 0 & -1 \end{pmatrix}$$

$$- (3)\det\begin{pmatrix} 1 & 2 & 1 \\ -1 & 3 & 4 \\ 1 & 0 & -1 \end{pmatrix} + (0)\det\begin{pmatrix} 1 & 2 & 1 \\ -1 & 3 & 4 \\ 0 & -1 & 1 \end{pmatrix}$$

Since the first, second, and fourth terms contain $a_{ij} = 0$, the product is 0. These determinants need not be computed. The third minor is expanded along the third row.

$$\det(A) = -3\left[+(1)\det\begin{pmatrix} 2 & 1 \\ 3 & 4 \end{pmatrix} - (0)\det\begin{pmatrix} 1 & 1 \\ -1 & 4 \end{pmatrix} \right.$$

$$\left. + (-1)\det\begin{pmatrix} 1 & 2 \\ -1 & 3 \end{pmatrix} \right]$$

$$= -3[(1)(5) - 0 + (-1)(5)]$$
$$= -3[5 - 0 - 5]$$
$$= 0$$

(Since det(A) = 0, A is singular.)   ■

Summary and Comments

Our "basic" determinant is of order 2. Third-order determinants are computed by expanding the 3×3 matrix (along any row or column) into a series of second-order determinants. Fourth-order determinants are computed by expanding into third-order determinants that are, in turn, expanded into second-order determinants.

In *each step*, expand along the row or column containing the most zeros to minimize computation.

Cramer's Rule can be used to solve (independent) systems of n equations in n unknowns for n greater than or equal to 2.

Exercises 7.8   In Exercises 1–12 find the determinant of the matrix. (See Examples 1 and 3.)

1. $\begin{pmatrix} 1 & 1 \\ 2 & 3 \end{pmatrix}$

2. $\begin{pmatrix} 1 & 5 \\ 2 & 12 \end{pmatrix}$

3. $\begin{pmatrix} 5 & 1 \\ 12 & 3 \end{pmatrix}$

4. $\begin{pmatrix} 2 & 1 \\ 1 & -2 \end{pmatrix}$

5. $\begin{pmatrix} 2 & -3 \\ 1 & 11 \end{pmatrix}$

6. $\begin{pmatrix} -3 & 1 \\ 11 & -2 \end{pmatrix}$

7. $\begin{pmatrix} x & 5 \\ \sqrt{3} & y \end{pmatrix}$

8. $\begin{pmatrix} a & b \\ c & d \end{pmatrix}$

9. $\begin{pmatrix} 1 & 1 & 1 \\ 0 & 1 & -1 \\ 1 & -1 & 0 \end{pmatrix}$

10. $\begin{pmatrix} 1 & 2 & 1 \\ 0 & 2 & -1 \\ 1 & 1 & 0 \end{pmatrix}$

11. $\begin{pmatrix} 2 & 1 & 1 \\ 2 & 1 & -1 \\ 1 & -1 & 0 \end{pmatrix}$

12. $\begin{pmatrix} 1 & 1 & 2 \\ 0 & 1 & 2 \\ 1 & -1 & 1 \end{pmatrix}$

In Exercises 13–22 solve the system using determinants. (See Examples 2 and 4.)

13. $\begin{aligned} x + y &= 5 \\ 2x + 3y &= 12 \end{aligned}$

14. $\begin{aligned} 2x + y &= -3 \\ x - 2y &= 11 \end{aligned}$

15. $\begin{aligned} x + 3y &= 1 \\ 4x + 2y &= 9 \end{aligned}$

16. $\begin{aligned} x - 2y &= 2 \\ x - 6y &= -\tfrac{2}{3} \end{aligned}$

17. $\begin{aligned} x + 2y &= 3 \\ x - 4y &= 0 \end{aligned}$

18. $\begin{aligned} 3x + 2y &= -3 \\ x - 2y &= -5 \end{aligned}$

19. $\begin{aligned} x + y + z &= 2 \\ y - z &= 2 \\ x - y &= 1 \end{aligned}$

20. $\begin{aligned} x + y &= 1 \\ y + z &= -1 \\ x + 2z &= 5 \end{aligned}$

21. $\begin{aligned} x - 3z &= 1 \\ 4x - y + 12z &= 12 \\ y + 12z &= -2 \end{aligned}$

22. $\begin{aligned} x + 2y &= -1 \\ 2x &= 2z - 7 \\ y - z &= -1 \end{aligned}$

23. In a certain two-digit number, the sum of the digits is 12. If the tens digit is doubled and then subtracted from the units digit, the result is 3. Find the number (solve for the digits).

24. The sum of the digits of a two-digit number is 6. If five times the units digit is subtracted from the number, the result is 4. Find the number. (*Hint:* If x is the tens digit and y is the units digit, then the number is $10x + y$.)

25. A parking meter that accepts pennies and nickels was found to contain 30 coins valued at 86¢. How many of each type of coin did the meter contain?

26. A landowner has 300 acres to sell. The acreage is in two tracts. One tract is placed on the market at $500 per acre, while the other tract is to be sold at $200 per acre. The expected revenue from the sale is $99,000. How many acres are there in each tract?

27. Towns A and B are 800 miles apart. An airplane made the trip in 4 hours flying with the wind and returned in 5 hours flying against the wind. Find the air speed of the airplane and the speed of the wind.

28. The equation $y = ax^2 + bx + c$ is satisfied by the (x, y) pairs $(1, 0)$, $(-1, -4)$, and $(2, 5)$. Find the equation (solve for a, b, c).

29. Find the determinant of each of the following matrices.

(a) $A = \begin{pmatrix} 1 & 1 & 0 & 1 \\ 0 & 3 & 2 & 1 \\ 0 & 0 & 3 & 1 \\ -2 & 1 & 1 & 1 \end{pmatrix}$
(b) $B = \begin{pmatrix} 1 & 0 & 0 & 0 \\ -1 & 1 & 2 & 1 \\ 1 & -1 & 1 & -1 \\ 1 & 1 & 3 & 1 \end{pmatrix}$

30. Solve the following system by the Gauss–Jordan method and Cramer's Rule.

$$
\begin{aligned}
x + 2y + 3z - w &= 6 \\
-2x + y + z + w &= -1 \\
x + 3y + 2z - 3w &= 4 \\
3x + y - z - 3w &= 3
\end{aligned}
$$

Key Terms and Formulas

Matrix
Vector
Order of a matrix
Conformable matrices
Additive identity
Additive inverse
Scalar
Matrix equation
Multiplicative identity

Row operations
Gauss–Jordan method
Augmented matrix
Standard form
Multiplicative inverse
Singular matrix
Determinant
Cramer's Rule

Review Exercises

In Exercises 1–19 perform the operation (if possible).

1. $\begin{pmatrix} 6 & 2 \\ -1 & 4 \end{pmatrix} + \begin{pmatrix} 3 & -5 \\ 0 & 0 \end{pmatrix}$

2. $\begin{pmatrix} x & 2 \\ -3 & -y \end{pmatrix} + \begin{pmatrix} -x & 2 \\ 3 & 2y \end{pmatrix}$

3. $\begin{pmatrix} 1 & 3 \\ -4 & 2 \end{pmatrix} + \begin{pmatrix} 0 \\ 0 \end{pmatrix}$

4. $(1 \quad 2 \quad 3) + \begin{pmatrix} 4 \\ 5 \\ 6 \end{pmatrix}$

5. $\begin{pmatrix} 10 \\ -3 \end{pmatrix} - \begin{pmatrix} 4 \\ 2 \end{pmatrix}$

6. $\begin{pmatrix} x + y \\ 2x - y \end{pmatrix} - \begin{pmatrix} 2x + 3y \\ x + y \end{pmatrix}$

7. $\begin{pmatrix} 7 & -6 \\ 4 & 2 \end{pmatrix} - \begin{pmatrix} 1 \\ 3 \end{pmatrix}$

8. $(a \quad b) \begin{pmatrix} c \\ d \end{pmatrix}$

9. $(1 \quad -2 \quad 4) \begin{pmatrix} 2 \\ 3 \end{pmatrix}$

10. $(4 \quad -1) \begin{pmatrix} 1 \\ 4 \end{pmatrix}$

11. $(1 \quad -2 \quad 4) \begin{pmatrix} 2 \\ 3 \\ 0 \end{pmatrix}$

12. $\begin{pmatrix} 1 & 2 \\ 3 & 4 \end{pmatrix} \begin{pmatrix} 1 \\ 2 \end{pmatrix}$

13. $\begin{pmatrix} 1 & 2 & -1 \\ 0 & 1 & 3 \\ 1 & 0 & 1 \end{pmatrix} \begin{pmatrix} 1 \\ 2 \\ 1 \end{pmatrix}$

14. $\begin{pmatrix} 1 & 2 & 3 \\ 4 & 5 & 6 \end{pmatrix} \begin{pmatrix} 1 \\ 2 \\ 3 \end{pmatrix}$

15. $\begin{pmatrix} 1 & 2 & 3 \\ 4 & 5 & 6 \end{pmatrix} (1 \quad 2 \quad 3)$

16. $x \begin{pmatrix} 1 & 2 \\ 3 & 4 \end{pmatrix}$

17. $\frac{1}{2} \begin{pmatrix} 1 & -3 \\ 4 & -2 \end{pmatrix}$

18. $2 \begin{pmatrix} x + y \\ a - b \end{pmatrix}$

C 19. $0.6 \begin{pmatrix} 0.8 & 0.2 \\ 1.3 & 5.0 \end{pmatrix}$

In Exercises 20–29 find the multiplicative inverse of the matrix if it exists.

20. $\begin{pmatrix} 1 & 2 \\ -1 & 0 \end{pmatrix}$

21. $\begin{pmatrix} 1 & 2 \\ 0 & 0 \end{pmatrix}$

22. $\begin{pmatrix} 3 & 4 \\ -6 & -8 \end{pmatrix}$

23. $\begin{pmatrix} 1 & 4 \\ 2 & 8 \end{pmatrix}$

24. $\begin{pmatrix} -1 & 1 & -1 \\ 2 & 3 & 0 \\ 0 & 2 & -1 \end{pmatrix}$

25. $\begin{pmatrix} 2 & 3 & 0 \\ -1 & 1 & -1 \\ 0 & 2 & -1 \end{pmatrix}$

26. $\begin{pmatrix} -1 & 1 & -1 \\ 2 & 3 & 0 \\ 2 & -2 & 2 \end{pmatrix}$

27. $\begin{pmatrix} -1 & 1 & -1 \\ 2 & 3 & 0 \\ 3 & -3 & 3 \end{pmatrix}$

28. $\begin{pmatrix} -1 & 1 & -1 \\ 2 & 3 & 0 \end{pmatrix}$

29. $\begin{pmatrix} a & 0 \\ 0 & a \end{pmatrix}$

In Exercises 30–38 solve the system, when possible, by each of the following methods:
(a) Gauss–Jordan method
(b) inverse method
(c) Cramer's Rule

30. $\begin{aligned} x - 3y &= 4 \\ 2x + y &= 1 \end{aligned}$

31. $\begin{aligned} 2x + y &= 1 \\ x - y &= 2 \end{aligned}$

32. $\begin{aligned} x_1 + x_2 + x_3 &= 6 \\ x_1 + x_2 &= 2 \\ -x_2 + 2x_3 &= 3 \end{aligned}$

33. $\begin{aligned} x_1 + x_2 + x_3 &= 3 \\ x_1 + x_2 &= 2 \\ -x_2 + 2x_3 &= 1 \end{aligned}$

34. $\begin{aligned} 2x + y + 2z &= 2 \\ x + y + z &= 1 \\ x + 2y + z &= 4 \end{aligned}$

35. $\begin{aligned} x + y &= 3 \\ x - z &= 2 \\ 2x + y - z &= 5 \end{aligned}$

C 36. $\begin{aligned} 0.1x + 0.2y &= 0.04 \\ 0.05x - 0.04y &= 0.09 \end{aligned}$

37. $\begin{aligned} x + y &= 5 \\ x - z &= 4 \\ 2x + y - z &= 9 \end{aligned}$

38. $\begin{aligned} x_1 + x_2 + x_3 &= 1 \\ 2x_1 + x_2 - 2x_3 &= 1 \\ 2x_1 + 3x_2 + 6x_3 &= 3 \end{aligned}$

39. A business has $4000 to spend on advertising an upcoming sale. The

money is to be divided between television (x), radio (y), and newspapers (z). The business manager has decided to spend three times as much money on television as on radio. The manager has also decided to spend $800 less on radio advertising than on newspapers.

(a) Write the system of equations defined by this problem.

(b) Put the system in standard form.

(c) Find the inverse of the coefficient matrix.

(d) Solve the system using the inverse.

40. If the business in Exercise 39 has $10,000 to spend and the manager decides to spend the same amount on radio and newspapers, how much money is spent on each type of advertising?

41. Solve Exercise 20 of Section 7.5 by using Cramer's Rule.

CHAPTER 8

Sequences and Sums

8.1 Arithmetic and Geometric Progressions

A **sequence** is a function whose domain is the set of positive integers. We may think of a sequence as a list of numbers $x_1, x_2, x_3, \ldots, x_n, \ldots$. Here x_1 represents the first term of the sequence or the value of the function at 1, and in general x_n represents the nth term of the sequence or the value of the function at n. A sequence is completely specified if we are given or can determine a formula that represents the nth term. A **finite sequence** is a sequence which has a finite number of terms.

Example 1 $1, 4, 9, 16, 25, \ldots, n^2, \ldots$ is a sequence. Here $x_1 = 1$, $x_2 = 4$, and $x_n = n^2$. The 100th term of this sequence is $100^2 = 10,000$, that is, $x_{100} = 10,000$. This is not a finite sequence, since it has no last term. ■

Example 2 $1, \frac{1}{2}, \frac{1}{3}, \frac{1}{4}, \frac{1}{5}, \ldots, \frac{1}{100}$ is a finite sequence, since it has exactly 100 terms. That is, 1 is the first term and $\frac{1}{100}$ is the last term.   ■

A sequence $x_1, x_2, \ldots, x_n, \ldots$ is usually denoted by writing $\{x_n\}_{n=1}^{\infty}$ or simply $\{x_n\}$. For example, the sequence

$$2, 4, 6, 8, \ldots, 2n, \ldots$$

may also be denoted by $\{2n\}$. The form

$$2, 4, 6, 8, \ldots, 2n, \ldots$$

is often simplified to

$$2, 4, 6, 8, \ldots$$

provided the established pattern is obvious.

Example 3   Compute the first three terms and the 10th term of the sequence $\{n^2 + 2n\}$.

Solution   Here $n^2 + 2n$ represents the nth term, x_n, of the sequence as a function of n. Hence

$$x_1 = \text{1st term} = 1^2 + 2(1) = 3$$
$$x_2 = \text{2nd term} = 2^2 + 2(2) = 8$$
$$x_3 = \text{3rd term} = 3^2 + 2(3) = 15$$

$$\vdots$$

$$x_{10} = \text{10th term} = 10^2 + 2(10) = 120 \quad ■$$

The first type of sequence we consider is called an arithmetic progression.

DEFINITION   An **arithmetic progression** is a sequence having the property that each term after the first can be obtained from the preceding one by addition of a fixed constant d. The fixed constant d is called the **common difference.**

Example 4 $3, 5, 7, 9, 11, \ldots$ is an arithmetic progression with first term 3 and common difference 2, since each term is obtained from the preceding term by adding the constant 2.   ■

Example 5 $2, -3, -8, -13, -18, \ldots$ is an arithmetic progression with first term 2 and common difference -5. Each term is obtained from the preceding term by adding -5. ∎

Suppose you have an arithmetic progression with first term a and common difference d. Then

1st term $= a$

2nd term $= $ (1st term) $+ d = a + d$

3rd term $= $ (2nd term) $+ d = (a + d) + d = a + 2d$

4th term $= $ (3rd term) $+ d = (a + 2d) + d = a + 3d$

5th term $= $ (4th term) $+ d = (a + 3d) + d = a + 4d$

The pattern is clear.

Note: In an arithmetic progression with first term a and common difference d, the formula for the nth term is given by

$$n\text{th term} = a + (n - 1)d$$

Example 6 List the first three terms and give the 100th term of an arithmetic progression with first term $a = 7$ and common difference $d = 5$.

Solution The formula for the nth term is $a + (n - 1)d = 7 + (n - 1)d$. Hence

1st term $= 7 + (1 - 1)5 = \boxed{7}$

2nd term $= 7 + (2 - 1)5 = \boxed{12}$

3rd term $= 7 + (3 - 1)5 = \boxed{17}$

$$\vdots$$

100th term $= 7 + (100 - 1)5 = 7 + 495 = \boxed{502}$ ∎

Example 7 Suppose $\{x_n\}$ is an arithmetic progression with $x_4 = 10$ and $x_8 = 26$. Find the 100th term of the progression.

Solution We must find a and d. Now the nth term is $a + (n - 1)d$. Hence,

$$x_4 = a + 3d = 10$$

$$x_8 = a + 7d = 26$$

Subtracting the first equation from the second, we get

$$4d = 16$$

$$d = 4$$

Substituting into the first equation, we get

$$a + 3(4) = 10$$

$$a + 12 = 10$$

$$a = -2$$

Hence

$$100\text{th term} = a + (100 - 1)d$$
$$= (-2) + 99(4)$$

$$= -2 + 396 = \boxed{394} \quad \blacksquare$$

Example 8 Mr. Smith borrows $2400 and agrees to repay it in 12 installments of $200 per month plus interest of 1.5% per month on the unpaid balance. What is the amount of Mr. Smith's 10th payment?

Solution 1st payment = $200 + (0.015)($2400) = $200 + $36 = $236

2nd payment = $200 + (0.015)($2200) = $200 + $33 = $233

3rd payment = $200 + (0.015)($2000) = $200 + $30 = $230

We see that the payment amounts form an arithmetic progression with first term $a = \$236$ and common difference $d = -\$3$. Hence

$$10\text{th payment} = 10\text{th term} = a + (10 - 1)d$$

$$= \$236 + 9(-\$3) = \boxed{\$209} \quad \blacksquare$$

Example 9 Mr. Jones begins work for Company ABC with a starting salary of $20,000 per year. Suppose he gets an automatic raise of $100 each six months. What is Mr. Jones's salary at the end of 10 years?

Solution Salary after 6 months = $20,100

Salary after 12 months = $20,200

Salary after 18 months = $20,300

.
.
.

We see that this is an arithmetic progression with first term $20,100 and common difference $100. Since 10 years consists of 20 intervals, each 6 months long, we see that Mr. Jones's salary after 10 years is the 20th term, or

$$20,100 + (19)(100) = \boxed{22,000 \text{ (dollars)}} \quad \blacksquare$$

Another important type of sequence is called a geometric progression.

DEFINITION

A **geometric progression** is a sequence having the property that each term after the first can be obtained from the preceding one by multiplying it by a fixed nonzero constant r. The constant r is called the **common ratio.**

Example 10

$3, 6, 12, 24, 48, \ldots$ is a geometric progression, since each term is obtained from the preceding one by multiplying by the fixed constant 2. Here $r = 2$ is the common ratio. $\blacksquare$

Example 11

$4, 2, 1, \frac{1}{2}, \frac{1}{4}, \ldots$ is a geometric progression, since each term is obtained from the preceding one by multiplying by $\frac{1}{2}$. Here $r = \frac{1}{2}$ is the common ratio. $\blacksquare$

In a geometric progression with first term a and common ratio r we notice the following pattern:

$$\text{1st term} \ = a$$
$$\text{2nd term} = (\text{1st term}) \cdot r = ar$$
$$\text{3rd term} \ = (\text{2nd term}) \cdot r = (ar)r = ar^2$$
$$\text{4th term} \ = (\text{3rd term}) \cdot r = (ar^2)r = ar^3$$
$$\text{5th term} \ = (\text{4th term}) \cdot r = (ar^3)r = ar^4$$

The pattern is clear.

Note: In a geometric progression with first term a and common ratio r, the formula for the nth term is given by

$$n\text{th term} = ar^{n-1}$$

Example 12 List the first three terms and give the 10th term of a geometric progression with first term $a = 5$ and common ratio $r = \frac{1}{2}$.

Solution The formula for the nth term is $ar^{n-1} = 5\left(\frac{1}{2}\right)^{n-1}$. Hence

$$\text{1st term} = \boxed{5}$$

$$\text{2nd term} = 5\left(\frac{1}{2}\right)^{2-1} = \boxed{\frac{5}{2}}$$

$$\text{3rd term} = 5\left(\frac{1}{2}\right)^{3-1} = \boxed{\frac{5}{4}}$$

$$\vdots$$

$$\text{10th term} = 5\left(\frac{1}{2}\right)^{10-1} = 5\left(\frac{1}{2}\right)^{9} = \boxed{\frac{5}{512}} \quad \blacksquare$$

Example 13 Mrs. Smith deposits \$1000 in a certain bank which pays interest at 10% per year compounded quarterly. How much money does she have after 5 years?

Solution The original principal is $P = \$1000$. An interest rate of 10% annually amounts to 2.5% per quarter. Therefore the amount at the end of the first quarter is

$$A_1 = 1000 + (1000)(0.025) = 1000(1 + 0.025) = 1000(1.025)$$

The amount at the end of the second quarter is

$$A_2 = \begin{bmatrix} \text{amount at the end} \\ \text{of first quarter} \end{bmatrix} + \begin{bmatrix} \text{amount at the end} \\ \text{of first quarter} \end{bmatrix}(0.025)$$

$$= A_1 + A_1(0.025) = A_1(1 + 0.025)$$
$$= A_1(1.025) = 1000(1.025)(1.025)$$
$$= 1000(1.025)^2$$

The amount at the end of the third quarter is

$$A_3 = \begin{bmatrix} \text{amount at end of} \\ \text{second quarter} \end{bmatrix} + \begin{bmatrix} \text{amount at end of} \\ \text{second quarter} \end{bmatrix}(0.025)$$

$$= A_2 + A_2(0.025) = A_2(1 + 0.025)$$
$$= A_2(1.025) = 1000(1.025)^2(1.025)$$
$$= 1000(1.025)^3$$

We see that we have a geometric progression

$$1000(1.025), \quad 1000(1.025)^2, \quad 1000(1.025)^3, \ldots$$

with first term $a = 1000(1.025)$ and common ratio $r = 1.025$. We want the amount at the end of the 20th quarter. This is simply the 20th term of the sequence. Hence

$$A_{20} = ar^{20-1} = [1000(1.025)] \cdot (1.025)^{19}$$

$$= 1000(1.025)^{20} \approx 1638.62 \text{ (dollars)} \quad \blacksquare$$

The ideas in this example can be used to establish the following result.

Note: If P dollars are invested at an interest rate of i per period, the compound amount at the end of n periods, A_n, is given by

$$A_n = P(1 + i)^n$$

Example 14 If \$200 is invested at an interest rate of 3% per quarter, what is the compound amount at the end of 4 quarters?

Solution $A_4 = \$200(1 + 0.03)^4 = \$225.10 \quad \blacksquare$

Exercises 8.1 In Exercises 1–12 list the first five terms of the sequence and determine the tenth term of the sequence.

1. $\{2n + 3\}$
2. $\{3n - 1\}$
3. $\{(-1)^n + 1\}$
4. $\left\{\left(\frac{1}{2}\right)^n\right\}$
5. $\left\{\frac{(-1)^n}{n + 1}\right\}$
6. $\{n^2 - 3n - 2\}$
7. $\left\{\frac{3n - 1}{n^2 + 1}\right\}$
8. $\{(-2)^n\}$
9. $\left\{\frac{n^2 - 1}{2n^2 + n}\right\}$
10. $\{n^n\}$
11. $\{2^{-n} + 1\}$
12. $\{(n + 1)^{-n}\}$

In Exercises 13–19 determine whether the sequence is an arithmetic progression. If the sequence is an arithmetic progression, find the common difference.

13. $5, 7, 9, 11, 13, \ldots$
14. $-5, -3, -1, 1, 3, \ldots$

15. $3, 0, -3, -6, -9, \ldots$
16. $1, 3, 6, 10, 15, \ldots$
17. $-4, -6, -8, -10, -12, \ldots$
18. $5, \frac{11}{2}, 6, \frac{13}{2}, 7, \ldots$
19. $\frac{1}{2}, \frac{5}{6}, \frac{7}{6}, \frac{3}{2}, \frac{11}{6}, \ldots$

In Exercises 20–28 list the first three terms and give the indicated nth term of the arithmetic progression. (Recall that a = first term and d = common difference.)

20. $a = 5$, $d = 2$, and $n = 10$
21. $a = 0$, $d = -3$, and $n = 50$
22. $a = -2$, $d = -4$, and $n = 70$
23. $x_4 = 20$, $x_{12} = 44$, and $n = 100$
24. $x_3 = -2$, $x_8 = -27$, and $n = 20$
25. $x_4 = 3$, $x_8 = 5$, and $n = 50$
26. $a = 7$, $x_4 = 22$, and $n = 10$
27. $a = 10$, $x_5 = 20$, and $n = 10$
28. $a = \frac{3}{4}$, $d = \frac{1}{2}$, and $n = 5$
29. Mr. Miller begins work for Company Z with a starting salary of $10,000 per year. Suppose he gets an automatic raise of $200 every 3 months. What is Mr. Miller's salary at the end of 10 years?
30. Suppose you own 130 shares of Company A stock. If you decide to buy 10 shares every month, how many shares will you have after 10 years?
31. Mrs. Faye borrows $2400 and agrees to repay it in 24 installments of $100.00 per month plus interest of 1% per month on the unpaid balance.
    (a) List the amount of each of the first three payments.
    (b) What will be the amount of the 10th payment?
32. Mr. Z borrows $3000 and agrees to repay $300 at the end of each 6 months plus interest of 2% on the unpaid balance.
    (a) How much has he paid at the end of two years?
    (b) What is the amount of the seventh payment?
33. Suppose you own 100 shares of Company B stock. If you decide to buy 20 shares every month, how many shares will you have after 5 years?
34. A free-falling body is dropped from an airplane. Suppose it falls 16 feet the first second and in each successive second it falls 32 feet farther than it did during the preceding second. How far does it fall
    (a) during the tenth second?
    (b) during the first five seconds?
35. Suppose you put a dollar in a piggybank on January 1, 1982. Suppose that on the first day of each month thereafter you save $0.50 more than you did the preceding month.
    (a) How much money will you save on December 1, 1982?
    (b) How much money will you save on December 1, 1983?

In Exercises 36–41 determine whether the sequence is a geometric progression. If so, what is the common ratio?

36. $2, 4, 8, 16, 32, \ldots$
37. $2, -6, 18, -54, 162, \ldots$
38. $3, 1, \frac{1}{3}, \frac{1}{9}, \frac{1}{27}, \ldots$
39. $1, 2, 6, 24, 120, \ldots$
40. $-2, 4, 12, 16, -32, \ldots$
41. $\frac{1}{2}, \frac{1}{3}, \frac{2}{9}, \frac{4}{27}, \frac{8}{81}, \ldots$

In Exercises 42–47 list the first three terms and give the indicated nth term of the geometric progression described.

42. $a = 5$, $r = 2$, and $n = 10$

43. $a = \frac{1}{2}$, $r = -1$, and $n = 100$

44. $a = \frac{2}{3}$, $r = \frac{3}{4}$, and $n = 5$

45. $a = 10$, 2nd term = 5, and $n = 10$

46.   3rd term = 9, 4th term = 27, and $n = 6$

47.   3rd term = 10, 4th term = -5, and $n = 8$

C 48.   Mr. X deposits $2000 at his bank, which pays interest at 8% per year compounded quarterly. How much money does Mr. X have after 6 years?

C 49.   Mrs. Z deposits $10,000 with her credit union, which pays interest at 12% per year compounded semiannually. How much money does Mrs. Z have after 10 years?

50.   Mrs. Brown owns 20 shares of Company A stock in January 1970. Suppose the stock splits (2 for 1) each December for n consecutive years. How many shares does Mrs. Brown have after the last split?

51.   Mr. Smith owns 30 shares of Company A stock in January 1970. Suppose the stock splits (3 for 1) each December for n consecutive years. How many shares does Mr. Smith have after the last split?

C 52.   A ball is dropped from a height of 20 feet. Each time it hits the ground it rebounds $\frac{4}{5}$ of the height it fell. How far does the ball rebound
(a) on the fifth bounce?
(b) on the seventh bounce?

53.   A new antibiotic dissolves in the stomach at the rate of 10% of the remaining weight each minute. How much remains after 5 minutes?

54.   For a satellite to be successfully recovered after reentry from orbit and splashdown in an ocean, at least one half of the heat shield material must remain until impact. A particular satellite loses one fourth of the remaining mass of its heat shield during each minute of reentry. Can this satellite be recovered if it takes 3 minutes for reentry?

8.2   Σ Notation and Sums

Associated with each finite sequence is a sum. For example, associated with the finite sequence 2, 4, 6, 8, 10 is the sum $2 + 4 + 6 + 8 + 10$. In general, associated with a finite sequence

$$x_1, x_2, x_3, \ldots, x_n$$

is the sum

$$x_1 + x_2 + x_3 + \cdots + x_n$$

The symbol Σ, which is the Greek letter sigma, is often used as a shorthand symbol indicating sum. For example,

$$\sum_{k=1}^{7} x_k$$

read "the sum of x_k as k goes from 1 to 7," means

$$\sum_{k=1}^{7} x_k = x_1 + x_2 + x_3 + x_4 + x_5 + x_6 + x_7$$

The following example illustrates other possibilities.

Example 1   (a) $\displaystyle\sum_{k=1}^{4} 2^k = 2^1 + 2^2 + 2^3 + 2^4 = 2 + 4 + 8 + 16$

(b) $\displaystyle\sum_{k=1}^{3} (2k + 1) = (2 \cdot 1 + 1) + (2 \cdot 2 + 1) + (2 \cdot 3 + 1) = 3 + 5 + 7$

(c) $\displaystyle\sum_{k=3}^{6} \frac{1}{k + 1} = \frac{1}{3 + 1} + \frac{1}{4 + 1} + \frac{1}{5 + 1} + \frac{1}{6 + 1} = \frac{1}{4} + \frac{1}{5} + \frac{1}{6} + \frac{1}{7}$

(d) $\displaystyle\sum_{k=2}^{8} x_k = x_2 + x_3 + x_4 + x_5 + x_6 + x_7 + x_8$ ∎

Example 2   Write $4 + 9 + 16 + 25$ in Σ notation.

Solution $4 + 9 + 16 + 25 = 2^2 + 3^2 + 4^2 + 5^2 = \displaystyle\sum_{k=2}^{5} k^2$ ∎

When using Σ notation, any letter may be used. For example,

$$\sum_{k=2}^{5} k^2 \quad \text{or} \quad \sum_{i=2}^{5} i^2 \quad \text{or} \quad \sum_{j=2}^{5} j^2$$

all indicate exactly the same sum, $2^2 + 3^2 + 4^2 + 5^2$. We now state a useful definition.

DEFINITION   If c is a fixed constant, we define

$$\sum_{k=1}^{n} c = \underbrace{c + c + \cdots + c}_{n \text{ times}} = nc$$

Example 3   (a) $\displaystyle\sum_{k=1}^{7} 5 = \underbrace{5 + 5 + 5 + 5 + 5 + 5 + 5}_{7 \text{ times}} = 7 \cdot 5 = \boxed{35}$

(b) $\displaystyle\sum_{i=1}^{4} \frac{2^{i+2}}{3+i} = \frac{2^{1+2}}{3+1} + \frac{2^{2+2}}{3+2} + \frac{2^{3+2}}{3+3} + \frac{2^{4+2}}{3+4}$

$$= 2 + \frac{16}{5} + \frac{32}{6} + \frac{64}{7} = \boxed{\frac{2066}{105}} \quad \blacksquare$$

The Σ notation has several useful properties, which we now state.

Note: Suppose that $x_1, x_2, \ldots, x_n$ and $y_1, y_2, \ldots, y_n$ are two finite sequences each with n terms, and suppose that C is a fixed constant. Then

(a) $\displaystyle\sum_{k=1}^{n} (x_k + y_k) = \sum_{k=1}^{n} x_k + \sum_{k=1}^{n} y_k$

(b) $\displaystyle\sum_{k=1}^{n} C \cdot x_k = C \cdot \sum_{k=1}^{n} x_k$

Proof of (a): $\displaystyle\sum_{k=1}^{n} (x_k + y_k) = (x_1 + y_1) + (x_2 + y_2) + (x_3 + y_3)$

$$+ \cdots + (x_n + y_n)$$

$$= (x_1 + x_2 + x_3 + \cdots + x_n)$$

$$+ (y_1 + y_2 + \cdots + y_n)$$

$$= \sum_{k=1}^{n} x_k + \sum_{k=1}^{n} y_k \quad \square$$

The proof of part (b) is left as an exercise.

Exercises 8.2

In Exercises 1–15 write out the sum term by term.

1. $\displaystyle\sum_{k=1}^{8} (4k - 1)$

2. $\displaystyle\sum_{k=3}^{7} (2k + 3)$

3. $\displaystyle\sum_{i=3}^{8} (i + 1)^2$

4. $\displaystyle\sum_{i=1}^{n} (2x_i + 1)$

5. $\displaystyle\sum_{j=4}^{7} \frac{1}{j^2}$

6. $\displaystyle\sum_{i=1}^{6} 3x^i$

7. $\displaystyle\sum_{k=1}^{10} 5$

8. $\displaystyle\sum_{i=1}^{n} (x_i + y_i)$

9. $\displaystyle\sum_{k=1}^{4} \left(\frac{1}{k+1}\right)^2$

10. $\displaystyle\sum_{j=1}^{5} \frac{(-1)^{j+1}}{j^2}$

11. $\displaystyle\sum_{i=3}^{10} 2^{i-1}$

12. $\displaystyle\sum_{k=1}^{5} (-1)^k 2^k$

13. $\displaystyle\sum_{k=1}^{5} 3\left(\frac{1}{2}\right)^{k-1}$

14. $\displaystyle\sum_{i=1}^{n} x_i z_i$

15. $\displaystyle\sum_{j=3}^{7} (2j^2 + 3j)$

In Exercises 16–26 write the expression in Σ notation.

16. $1 + 2 + 3 + 4 + \cdots + 20$

17. $2 + 4 + 8 + 16 + \cdots + 1024$

18. $2 \cdot 4 + 2 \cdot 5 + 2 \cdot 6 + 2 \cdot 7 + \cdots + 2 \cdot n$

19. $3 + 3 + 3 + 3 + 3 + 3 + 3$

20. $x_1^2 + x_2^2 + x_3^2 + x_4^2 + \cdots + x_{100}^2$

21. $1 + \dfrac{1}{2} + \dfrac{1}{3} + \dfrac{1}{4} + \cdots + \dfrac{1}{20}$

22. $\sqrt{3} + \sqrt{4} + \sqrt{5} + \sqrt{6} + \cdots + \sqrt{70}$

23. $\dfrac{1}{x_1} + \dfrac{1}{x_2} + \dfrac{1}{x_3} + \dfrac{1}{x_4} + \cdots + \dfrac{1}{x_n}$

24. $3 + 5 + 7 + 9 + \cdots + 21$

25. $\dfrac{3}{2} - \dfrac{6}{4} + \dfrac{9}{8} - \dfrac{12}{16} + \dfrac{15}{32} - \dfrac{18}{64}$

26. $\dfrac{3}{2} - \dfrac{4}{3} + \dfrac{5}{4} - \dfrac{6}{5} + \dfrac{7}{6} - \dfrac{8}{7} + \dfrac{9}{8} - \dfrac{10}{9}$

27. Prove that $\displaystyle\sum_{k=1}^{n} cx_k = c \sum_{k=1}^{n} x_k$ for any constant c.

28. Prove that $\displaystyle\sum_{k=1}^{n} (ax_k + by_k) = a \sum_{k=1}^{n} x_k + b \sum_{k=1}^{n} y_k$.

8.3   Arithmetic and Geometric Sums

Consider a sequence $x_1, x_2, x_3, \ldots, x_n, \ldots$. The sum of the first n terms, denoted by S_n, is called the **nth partial sum.** Thus,

$$S_1 = x_1$$

$$S_2 = x_1 + x_2 = \sum_{k=1}^{2} x_k$$

$$S_3 = x_1 + x_2 + x_3 = \sum_{k=1}^{3} x_k$$

$$S_4 = x_1 + x_2 + x_3 + x_4 = \sum_{k=1}^{4} x_k$$

$$\vdots$$

$$S_n = x_1 + x_2 + x_3 + \cdots + x_n = \sum_{k=1}^{n} x_k$$

$$\vdots$$

The associated sequence $S_1, S_2, S_3, \ldots, S_n, \ldots$ is called the **sequence of partial sums.**

Example 1 Give the first four partial sums (S_1, S_2, S_3, S_4) associated with the sequence $1, 4, 9, \ldots, n^2, \ldots$. Also indicate the formula for S_n using Σ notation.

Solution $S_1 = \boxed{1}$

$S_2 = 1 + 4 = \boxed{5}$

$S_3 = 1 + 4 + 9 = \boxed{14}$

$S_4 = 1 + 4 + 9 + 16 = \boxed{30}$

In general, $S_n = 1 + 4 + 9 + \cdots + n^2 = \displaystyle\sum_{k=1}^{n} k^2.$ ∎

The general formula for the partial sum S_n associated with an arithmetic progression may be obtained in the following manner. Consider an arithmetic progression

$$a, a + d, a + 2d, \ldots, a + (n - 1)d, \ldots$$

with first term a and common difference d. Then

(1) $S_n = a + (a + d) + (a + 2d) + \cdots + [a + (n - 2)d]$
$+ [a + (n - 1)d]$

Notice that if we list the terms of S_n in reverse order we get

(2) $S_n = [a + (n - 1)d] + [a + (n - 2)d] + \cdots + (a + d) + a$

Adding corresponding terms in (1) and (2), we get the following:

(1) $S_n = a + (a + d) + (a + 2d) + \cdots + [a + (n - 2)d]$
$[a + (n - 1)d]$

(2) $S_n = [a + (n - 1)d] + [a + (n - 2)d] + \cdots + (a + d) + a$

$2S_n = [2a + (n - 1)d] + [2a + (n - 1)d] + \cdots$
$+ [2a + (n - 1)d] + [2a + (n - 1)d]$

$\underbrace{}_{n \text{ identical terms}}$

So we see that $2S_n = n[2a + (n - 1)d]$, and hence

$$S_n = \frac{n[2a + (n - 1)d]}{2}$$

We have established the following important theorem:

THEOREM 1   In an arithmetic progression $a, a + d, a + 2d, \ldots, a + (n - 1)d$, $\ldots$ with first term a and common difference d, the sum of the first n terms, S_n, is given by

$$S_n = \frac{n[2a + (n - 1)d]}{2}$$

Example 2   Consider the sequence 5, 7, 9, 11, 13, ... Give the formula for the nth partial sum S_n and evaluate S_{50} and S_{100}.

Solution   Notice that the given sequence is an arithmetic progression with first term $a = 5$ and common difference $d = 2$. Therefore

$$S_n = \frac{n[2a + (n - 1)d]}{2}$$

$$= \frac{n[2 \cdot 5 + (n - 1)2]}{2} = \frac{n(10 + 2n - 2)}{2}$$

$$= \frac{n(2n + 8)}{2} = \frac{2n(n + 4)}{2} = \boxed{n(n + 4)}$$

Since $S_n = n(n + 4)$, we see that

$$S_{50} = 50(50 + 4) = 50(54) = \boxed{2700}$$

$$S_{100} = 100(100 + 4) = 100(104) = \boxed{10{,}400} \quad\blacksquare$$

Note: If r and s are real numbers, then

$$\sum_{k=1}^{n} (r + sk) = (r + s \cdot 1) + (r + s \cdot 2) + (r + s \cdot 3) + \cdots$$

$$+ (r + s \cdot n)$$

represents the sum of the first n terms of an arithmetic progression with first term $a = r + s$ and common difference $d = s$. Hence these sums may be evaluated using Theorem 1.

Example 3 Evaluate

(a) $\displaystyle\sum_{k=1}^{10} (3 + 2k)$ 	(b) $\displaystyle\sum_{k=1}^{20} (2 - 5k)$

Solution (a) Here the first term is $a = 3 + 2 = 5$ and the common difference is $d = 2$. Hence

$$\sum_{k=1}^{10} (3 + 2k) = S_{10} = \frac{10[2 \cdot 5 + (10 - 1)2]}{2}$$

$$= \frac{10[10 + 18]}{2} = 5(28) = \boxed{140}$$

(b) Here the first term is $a = 2 - 5 = -3$ and the common difference is -5. Hence

$$\sum_{k=1}^{20} (2 - 5k) = S_{20} = \frac{20[2 \cdot (-3) + (20 - 1)(-5)]}{2}$$

$$= 10[-101] = \boxed{-1010} \quad \blacksquare$$

Example 4 Mr. Perez borrows $1000 and will repay it in 10 monthly payments of $100 each plus interest of 1.5% per month on the unpaid balance. How much interest will he pay?

Solution The interest for the first month is 1.5% of $1000, or $15.
The interest for the second month is 1.5% of $900 or $13.50.
The interest for the third month is 1.5% of $800 or $12.00.
We see that the interest payments form an arithmetic progression with first term $a = \$15$ and common difference $d = -\$1.50$. Therefore the total interest he will pay is the sum of the first 10 terms of the sequence, or S_{10}. Now

$$S_{10} = \frac{10[2(15) + 9(-1.50)]}{2} = \frac{10(30 - 13.50)}{2}$$

$$= \boxed{82.50 \text{ (dollars)}} \quad \blacksquare$$

The general formula for the partial sums associated with a geometric progression may be obtained as follows. Consider a geometric progression

$$a, ar, ar^2, ar^3, \ldots, ar^{n-1}, \ldots$$

with first term a and common ratio r. The sum of the first n terms is

$$S_n = a + ar + ar^2 + \cdots + ar^{n-1}$$

Multiplying by r, we get

$$rS_n = ar + ar^2 + ar^3 + \cdots + ar^n$$

We see that

$$S_n - rS_n = S_n(1 - r) = (a + ar + ar^2 + \cdots + ar^{n-1})$$
$$- (ar + ar^2 + ar^3 + \cdots + ar^n)$$
$$= a - ar^n = a(1 - r^n)$$

Hence we have

$$S_n = \frac{a(1 - r^n)}{1 - r} \quad \text{provided that } r \neq 1$$

We have established the following important theorem.

THEOREM 2   In a geometric progression $a, ar, ar^2, \ldots, ar^{n-1}, \ldots$ with first term a and common ratio $r \neq 1$, the sum of the first n terms, S_n, is given by

$$S_n = \frac{a(1 - r^n)}{1 - r}$$

Example 5   Consider the sequence $3, \frac{3}{2}, \frac{3}{4}, \frac{3}{8}, \frac{3}{16}, \ldots$.
(a) Give the formula for the nth partial sum S_n, and evaluate the sum of the first 50 terms of the sequence.

(b) Evaluate $\displaystyle\sum_{k=1}^{20} 3\left(\frac{1}{2}\right)^{k-1}$.

Solution   (a) The given sequence is a geometric progression with first term $a = 3$ and common ratio $r = \frac{1}{2}$. Hence

$$n\text{th partial sum} = S_n = \frac{3\left[1 - \left(\frac{1}{2}\right)^n\right]}{1 - \frac{1}{2}} = \frac{3\left(1 - \frac{1}{2^n}\right)}{\frac{1}{2}}$$

$$= 6 - \frac{6}{2^n} = 6 - \frac{3}{2^{n-1}}$$

Using this formula for S_n, we see that $S_{50} = 6 - \dfrac{3}{2^{49}}$.

(b) $\displaystyle\sum_{k=1}^{20} 3\left(\frac{1}{2}\right)^{k-1} = 3 + \frac{3}{2} + \frac{3}{4} + \frac{3}{8} + \cdots + \frac{3}{2^{19}}$

$\qquad\qquad\qquad$ = sum of first 20 terms

$\qquad\qquad\qquad = S_{20} = \boxed{6 - \dfrac{3}{2^{19}}}$ ■

Example 6   Evaluate:

(a) $\displaystyle\sum_{k=1}^{10} 3(2)^k$
$\qquad\qquad\qquad\qquad\qquad$
(b) $\displaystyle\sum_{k=1}^{5} \left(\frac{3}{4}\right)^{k+1}$

Solution   (a) We see that $\displaystyle\sum_{k=1}^{10} 3(2)^k = 6 + 12 + 24 + \cdots + 3072$ is the sum of

the first 10 terms of a geometric progression with first term $a = 6$ and common ratio $r = 2$. Hence

$$\sum_{k=1}^{10} 3(2)^k = S_{10} = \frac{6(1 - 2^{10})}{1 - 2} = -6(-1023) = \boxed{6138}$$

(b) We see that $\displaystyle\sum_{k=1}^{5} \left(\frac{3}{4}\right)^{k+1} = \frac{9}{16} + \frac{27}{64} + \frac{81}{256} + \left(\frac{3}{4}\right)^5 + \left(\frac{3}{4}\right)^6$ is the sum of

the first 5 terms of a geometric progression with first term $a = \frac{9}{16}$ and common ratio $r = \frac{3}{4}$. Hence

$$\sum_{k=1}^{5} \left(\frac{3}{4}\right)^{k+1} = S_5 = \frac{\dfrac{9}{16}\left[1 - \left(\dfrac{3}{4}\right)^5\right]}{1 - \dfrac{3}{4}} = \frac{\dfrac{9}{16}\left(1 - \dfrac{243}{1024}\right)}{\dfrac{1}{4}}$$

$$= \frac{9}{4}\left(\frac{781}{1024}\right) = \boxed{\dfrac{7029}{4096}} \quad ■$$

Example 7   A ball is dropped from a height of 100 feet. Each time it hits the ground, it rebounds $\frac{1}{2}$ of the height it fell. How far has the ball traveled at the top of its 10th bounce?

Solution   The distance traveled during the first bounce is

$$100 + \frac{1}{2}(100) = 150 \text{ feet} \qquad 100 \Big\updownarrow \text{first bounce}$$
$$50$$

The distance traveled during the second bounce is

$$50 + \frac{1}{2}(50) = 75 \text{ feet}$$

$50 \left| \begin{array}{l} \uparrow \text{second bounce} \\ \downarrow | 25 \end{array} \right.$

The distance traveled during the third bounce is

$$25 + \frac{1}{2}(25) = 37.5 \text{ feet}$$

$25 \left| \begin{array}{l} \uparrow \text{third bounce} \\ \downarrow | 12.5 \end{array} \right.$

We now see that the pattern is a geometric progression 150, 75, 37.5, . . . , or

$$150, \ 150\left(\frac{1}{2}\right), \ 150\left(\frac{1}{2}\right)^2, \ . \ . \ .$$

with first term $a = 150$ and common ratio $r = \frac{1}{2}$. The distance traveled at the top of the 10th bounce is simply the sum of the first 10 terms, or S_{10}. Hence

$$S_{10} = \frac{150\left[1 - \left(\frac{1}{2}\right)^{10}\right]}{1 - \frac{1}{2}} = \frac{150\left(1 - \frac{1}{1024}\right)}{\frac{1}{2}}$$

$$= 300\left(\frac{1023}{1024}\right) \approx \boxed{299.71 \text{ (feet)}} \quad \blacksquare$$

Exercises 8.3

In Exercises 1–10, give the formula for S_n, and compute S_{20} for the arithmetic progression.

1. 1, 3, 5, 7, 9, . . .
2. 10, 13, 16, 19, 22, . . .
3. $-5, -3, -1, 1, 3, \ . \ . \ .$
4. 3, 0, $-3, -6, -9, \ . \ . \ .$
5. $\frac{1}{2}, \frac{3}{2}, \frac{5}{2}, \frac{7}{2}, \frac{9}{2}, \ . \ . \ .$
6. 1, $-1, -3, -5, -7, \ . \ . \ .$
7. $\frac{1}{2}, \frac{5}{6}, \frac{7}{6}, \frac{3}{2}, \frac{11}{6}, \ . \ . \ .$
8. 2, 4, 6, 8, 10, . . .
9. 5, $\frac{11}{2}$, 6, $\frac{13}{2}$, 7, . . .
10. $-4, -6, -8, -10, -12, \ . \ . \ .$

In Exercises 11–17 evaluate the indicated sum.

11. $\displaystyle\sum_{k=1}^{10} (2 + 3k)$
12. $\displaystyle\sum_{k=1}^{20} (3 - 2k)$
13. $\displaystyle\sum_{k=1}^{100} \left(2 - \frac{1}{2}k\right)$
14. $\displaystyle\sum_{k=1}^{20} (2k + 4)$
15. $\displaystyle\sum_{k=1}^{30} 3k$
16. $\displaystyle\sum_{k=1}^{100} (-2k + 10)$
17. $\displaystyle\sum_{k=1}^{99} \left(\frac{1}{2} + \frac{1}{3}k\right)$

18. Suppose n is a positive integer. Show that

$$\sum_{k=1}^{n} k = \frac{n(n + 1)}{2}$$

19. Show that

$$\sum_{k=1}^{n} (r + sk) = nr + \frac{sn(n + 1)}{2}$$

for any constants r and s. (*Hint:* Use Exercise 18 and the properties of Σ notation.)

20. In Example 3 we showed that $\sum_{k=1}^{10} (3 + 2k) = 140$ and $\sum_{k=1}^{20} (2 - 5k) = -1010$. Use the formula in Exercise 19 to check these answers.

21. Mr. Jones borrows $2400 and agrees to repay it in 12 monthly payments of $200 each plus interest of 1% per month on the unpaid balance. What is the total interest he will pay?

22. Mrs. Smith borrows $3600 and agrees to pay it in 36 monthly payments of $100 each plus interest of 1.5% per month on the unpaid balance. What is the total interest she will pay?

23. Consider the free-falling body described in Exercise 34 of Section 8.1. How far does the body fall during
    (a) the first 10 seconds?
    (b) the first 20 seconds?
    (c) the first 5 seconds?
    (d) Show that the distance the object falls in n seconds is $16n^2$ feet.

24. In Exercise 35 of Section 8.1, how much money will you have by
    (a) December 2, 1982?
    (b) December 2, 1983?

25. Suppose you save $1.00 the first day, $2.00 the second day, $3.00 the third day, $4.00 the fourth day, and so on. How much money will you have after
    (a) 10 days?
    (b) 20 days?
    (c) 1 year?

In Exercises 26–34 give the formula for S_n for the geometric progression.

26. 3, 6, 12, 24, 48, . . .   27. 1, 3, 9, 27, 81, . . .
28. 2, −2, 2, −2, 2, . . .   29. $\frac{1}{2}, \frac{1}{4}, \frac{1}{8}, \frac{1}{16}, \frac{1}{32}$, . . .
30. 1, −2, 4, −8, 16, . . .   31. 2, −1, $\frac{1}{2}$, −$\frac{1}{4}$, $\frac{1}{8}$, . . .
32. 27, 9, 3, 1, $\frac{1}{3}$, . . .   33. 6, 4, $\frac{8}{3}$, $\frac{16}{9}$, $\frac{32}{27}$, . . .
34. −3, −6, −12, −24, −48, . . .

C 35. Compute S_{10} in Exercises 27, 29, 30, and 31.

In Exercises 36–42 evaluate the sum.

36. $\sum_{k=1}^{10} 3(2)^{k-1}$ 37. $\sum_{k=1}^{7} 4\left(\frac{1}{2}\right)^{k-1}$

38. $\sum_{k=1}^{100} 2(-1)^{k+1}$ 39. $\sum_{k=1}^{75} 3(-1)^{k}$

40. $\displaystyle\sum_{k=1}^{8} 5\left(\frac{1}{2}\right)^{k+2}$ C 41. $\displaystyle\sum_{k=1}^{20} \left(\frac{1}{2}\right)^{k+2}$

C 42. $\displaystyle\sum_{k=1}^{10} 3(4)^{k+1}$

C 43. A ball is dropped from a height of 210 feet. Each time it hits the ground, it rebounds $\frac{1}{3}$ of the height it fell. How far has the ball traveled at the top of its 7th bounce?

44. A building cost \$100,000 and depreciates in value at a rate of 5% per year. What is the value of the building at the end of the 10th year?

C 45. Consider the ball described in Exercise 52 of Section 8.1. How far has the ball traveled at
(a) the top of its fifth bounce?
(b) the top of its seventh bounce?

C 46. Suppose the first day you save 1 penny, the second day you save 2 pennies, the third day you save 4 pennies, the fourth day you save 8 pennies, and so on. How much money will you have after
(a) 10 days?
(b) 100 days?

8.4 Mathematical Induction

The axiom of mathematical induction is one of the most important axioms of mathematics and can be stated as follows.

AXIOM OF MATHEMATICAL INDUCTION

Suppose that P_n is a proposition involving the positive integer n. Suppose also that

1. the proposition is true for $n = 1$ (that is, P_1 is true), and
2. under the assumption that the proposition is true for $n = k$, it can be proved true for $n = k + 1$ (that is, if P_k is true, then P_{k+1} is true).

Then the proposition is true for all positive integers n.

The axiom is intuitive if we consider the following example. Suppose there is an infinite staircase. You can climb the staircase to any step provided

1. you can climb the first step, and
2. the staircase is such that, if you are on any step, you can always climb to the next step.

In order to prove a statement by mathematical induction, two steps are required.

1. We must prove the statement is true for $n = 1$.
2. Assume the statement is true for $n = k$ and then prove that it is true for $n = k + 1$.

Example 1   Prove that $1 + 2 + 3 + \cdots + n = \dfrac{n(n + 1)}{2}$ for all positive integers n.

Solution   For $n = 1$ we have

$$1 = \frac{1(1 + 1)}{2}$$

which is clearly true. Thus the statement is true for $n = 1$ (P_1 is true). Now assume that the statement is true for $n = k$, that is, assume that P_k is true. This means that

$$(1) \quad 1 + 2 + 3 + \cdots + k = \frac{k(k + 1)}{2}$$

is true. By adding $(k + 1)$ to both sides of (1) we get

$$(2) \quad 1 + 2 + 3 + \cdots + k + (k + 1) = \frac{k(k + 1)}{2} + (k + 1)$$

Now, the right side of (2) can be simplified as follows:

$$\frac{k(k + 1)}{2} + (k + 1) = (k + 1)\left(\frac{k}{2} + 1\right) = (k + 1)\left(\frac{k + 2}{2}\right)$$

$$= \frac{(k + 1)(k + 2)}{2}$$

Hence we have proved that

$$(3) \quad 1 + 2 + 3 + \cdots + k + (k + 1) = \frac{(k + 1)(k + 2)}{2}$$

Equation (3) is precisely P_{k+1}, the statement we wished to establish with $n = k + 1$. Hence we have shown that if P_k is true then P_{k+1} is true. By the principle of mathematical induction we conclude that

$$1 + 2 + 3 + \cdots + n = \frac{n(n + 1)}{2}$$

is true for all positive integers n. ∎

Example 2   Prove $1 + 3 + 5 + \cdots + (2n - 1) = n^2$ for all positive integers n.

Solution For $n = 1$, we have $1 = 1^2$, which is true. Now assume that the statement is true for $n = k$, that is, assume that P_k is true. This means that

$$(1) \quad 1 + 3 + 5 + \cdots + (2k - 1) = k^2$$

is true. By adding $[2(k + 1) - 1]$ to both sides, we get

$$(2) \quad 1 + 3 + 5 + \cdots + (2k - 1) + [2(k + 1) - 1]$$
$$= k^2 + [2(k + 1) - 1]$$

But

$$k^2 + [2(k + 1) - 1] = k^2 + 2k + 2 - 1$$
$$= k^2 + 2k + 1 = (k + 1)^2$$

We have proved that

$$(3) \quad 1 + 3 + 5 + \cdots + (2k - 1) + [2(k + 1) - 1] = (k + 1)^2$$

which is the statement P_{k+1}. Thus, if P_k is true, then P_{k+1} is true. Thus

$$1 + 3 + 5 + \cdots + (2n - 1) = n^2 \quad \text{for all positive integers } n \quad \blacksquare$$

Exercises 8.4 In Exercises 1–14 prove that the statement is true for all positive integers n.

1. $5 + 10 + 15 + \cdots + 5n = \dfrac{5n(n + 1)}{2}$

2. $4 + 7 + 10 + \cdots + (3n + 1) = \dfrac{n(3n + 5)}{2}$

3. $1^2 + 2^2 + 3^2 + \cdots + n^2 = \dfrac{n(n + 1)(2n + 1)}{6}$

4. $3 + 5 + 7 + \cdots + (2n + 1) = n(n + 2)$

5. $1^3 + 2^3 + 3^3 + \cdots + n^3 = \dfrac{n^2(n + 1)^2}{4}$

6. $1 + 4 + 7 + \cdots + (3n - 2) = \dfrac{n(3n - 1)}{2}$

7. $\dfrac{1}{2} + \dfrac{1}{4} + \dfrac{1}{8} + \cdots + \dfrac{1}{2^n} = 1 - \dfrac{1}{2^n}$

8. $\dfrac{1}{(1)(2)} + \dfrac{1}{(2)(3)} + \dfrac{1}{(3)(4)} + \cdots + \dfrac{1}{(n)(n + 1)} = \dfrac{n}{n + 1}$

9. $1 \cdot 2 + 2 \cdot 3 + 3 \cdot 4 + \cdots + n(n + 1) = \dfrac{n(n + 1)(n + 2)}{3}$

10. $1 + 3^1 + 3^2 + 3^3 + \cdots + 3^n = \dfrac{3^{n+1} - 1}{2}$

11. $a + (a + d) + (a + 2d) + \cdots + [a + (n - 1)d] = \dfrac{n[2a + (n - 1)d]}{2}$

12. $a + ar + ar^2 + \cdots + ar^{n-1} = \dfrac{a(1 - r^n)}{1 - r}$

13. $(x - y)$ is a factor of $x^n - y^n$.
14. The number of subsets, counting the empty set, contained in a set with n elements is 2^n.

Key Terms and Formulas

Sequence
Arithmetic progression
  common difference of an
    arithmetic progression
  nth term of an arithmetic pro-
    gression: $a + (n - 1)d$
Geometric progression
  common ratio of a geometric
    progression
  nth term of a geometric pro-
    gression: ar^{n-1}
Compound amount at the end of
  n periods: $A_n = P(1 + i)^n$
Sum of a finite sequence

Σ notation
Sequence of partial sums
nth partial sum of an arithmetic
  progression:

$$S_n = \frac{n[2a + (n - 1)d]}{2}$$

nth partial sum of a geometric
  progression:

$$S_n = \frac{a(1 - r^n)}{1 - r}$$

Axiom of mathematical in-
  duction

Review Exercises

In Exercises 1–5 list the first five terms of the sequence and determine the tenth term.

1. $\{2^n\}$

2. $\left\{\dfrac{2 + (-1)^n}{n}\right\}$

3. $\left\{\dfrac{n + 1}{n + 2}\right\}$

4. $\left\{\dfrac{(-1)^n}{n^2}\right\}$

C 5. $\{(1.3)^n\}$

In Exercises 6–11 determine whether the sequence is an arithmetic progression. If so, find the common difference and evaluate S_{20}.

6. $7, 5, 3, 1, -1, -3, \ldots$
7. $3, 6, 9, 12, 15, \ldots$
8. $4, 5, 7, 10, 14, 19, \ldots$
9. $\sqrt{2}, 2\sqrt{2}, 3\sqrt{2}, 4\sqrt{2}, \ldots$
10. $\frac{1}{3}, \frac{11}{15}, \frac{17}{15}, \frac{23}{15}, \ldots$
11. $100, 101, 102, 103, \ldots$
12. What is the 21st term in an arithmetic progression with first term 3 and common difference 5?
13. What is the 51st term in an arithmetic progression with $x_4 = 10$ and $x_{10} = 34$?

14. A free-falling body is dropped from an airplane. Suppose that it falls 16 feet the first second and that in each successive second it falls 32 feet farther than it did during the preceding second. How far does it fall
    (a) during the 15th second?
    (b) during the first seven seconds?
    (c) during the first ten seconds?

In Exercises 15–20 determine whether the sequence is a geometric progression. If so, find the common ratio and compute S_{20}.

15. $\frac{1}{2}, -\frac{1}{2}, \frac{1}{2}, -\frac{1}{2}, \ldots$ 16. $4, 2, 1, \frac{1}{2}, \frac{1}{4}, \ldots$
17. $1, 2, 6, 24, 120, \ldots$ 18. $1, 2, 4, 8, 17, 34, 68, \ldots$
19. $4, -1, \frac{1}{4}, -\frac{1}{16}, \frac{1}{64}, \ldots$ 20. $5, 25, 125, 625, \ldots$
21. Compute the 10th term of a geometric progression with first term 3 and common ratio 2.

C 22. Compute the 10th term of a geometric progression with first term 1.5 and common ratio 1.73.

23. Compute the 10th term of a geometric progression with third term 32 and fourth term 64.

24. A ball is dropped from a height of 200 feet. Each time it hits the ground, it rebounds $\frac{1}{2}$ of the height that it fell.
    (a) How far does the ball rebound on the fifth bounce?
    (b) How far has the ball traveled at the top of the fifth bounce?

25. Write out $\displaystyle\sum_{k=3}^{10} \frac{2k-1}{k+1}$ term by term.

26. Write $1 + \frac{1}{2} + \frac{1}{3} + \frac{1}{4} + \frac{1}{5}$ in Σ notation.

27. Evaluate $\displaystyle\sum_{k=1}^{10} (3 + 2k)$.

28. Evaluate $\displaystyle\sum_{k=1}^{20} (5 - \frac{1}{2}k)$.

29. Evaluate $\displaystyle\sum_{k=1}^{5} 2(\frac{3}{4})^{k}$.

C 30. Evaluate $\displaystyle\sum_{k=1}^{5} 3(1.35)^{k+1}$.

31. Use the axiom of mathematical induction to prove that
$$2 + 4 + 6 + \cdots + 2n = n(n + 1)$$

32. Use the axiom of mathematical induction to prove that
$$1^2 + 3^2 + 5^2 + \cdots + (2n - 1)^2 = \frac{n(2n - 1)(2n + 1)}{3}$$

33. Mr. Polanco borrowed $3600 and will repay it in 12 monthly payments of $300 each plus interest of 2% per month on the unpaid balance. How much interest will he pay?

CHAPTER 9

Counting Techniques and Probability

This chapter is concerned with counting techniques and related formulas (Section 9.1) and with expanding certain types of algebraic expressions (Section 9.2). These topics are useful in themselves and are also useful in computing the likelihood or probability of many types of events (Section 9.3).

9.1   Counting, Permutations, and Combinations

In many practical situations we are interested in counting the number of possible ways different operations can occur. Most counting techniques follow from the following counting principle.

FUNDAMENTAL COUNTING PRINCIPLE

If one operation can be performed in N ways and a second operation can be performed in M ways, then both operations (the first followed by the second) can be performed in $N \cdot M$ ways. Furthermore, this principle can be extended to any number of operations.

This principle, as applied to k operations, can be thought of in the following manner: Suppose you have k blanks to fill

$$\underline{\hspace{2cm}} \quad \underline{\hspace{2cm}} \quad \underline{\hspace{2cm}} \quad \cdots \quad \underline{\hspace{2cm}}$$
$$1 \qquad\qquad 2 \qquad\qquad 3 \qquad\qquad\qquad k$$

and suppose there are only n_1 possible things which may be placed in the first blank, only n_2 things which may be placed in the second blank, only n_3 things which may be placed in the third blank, . . . , and only n_k things which may be placed in the kth blank. Then the total number of possible ways you can fill in the k blanks is

$$n_1 \cdot n_2 \cdot n_3 \cdots n_k$$

Example 1 How many three-letter initials (repetitions allowed) are possible (such as D.M.R., R.C.P., M.G.M., K.E.O.)?

Solution We can think of three blanks to fill:

$$\underline{\hspace{1.5cm}} \quad \underline{\hspace{1.5cm}} \quad \underline{\hspace{1.5cm}}$$

first   middle   last
initial   initial   initial

We have 26 possibilities for each blank. Hence, the number of possible three-letter initials is

$$26 \cdot 26 \cdot 26 = \boxed{17,576} \quad \blacksquare$$

Example 2 How many three-letter initials are possible if repetitions are not allowed?

Solution We have 26 possibilities for the first blank. Once we fill in the first blank, we have only 25 possibilities for the second blank, since we are not allowing repeated letters. Finally, after filling in the first two blanks, we have 24 possibilities for the third blank. Hence, the total number of three letter initials with no repeated letters is

$$26 \cdot 25 \cdot 24 = \boxed{15{,}600} \quad \blacksquare$$

Example 3 You are going to make license plates for the city of Houston. Suppose license plates are to have three letters followed by three digits, the first of which cannot be 0.
(a) How many possible license plates can you make?
(b) If repetitions are not allowed among the three letters, how many license plates can you make?

Solution (a) $\underline{-} \quad \underline{-}$
3 letters   3 digits

We have six blanks to fill with 26 possibilities for each of the first three blanks, 9 possibilities for the fourth blank (the digits 1, 2, 3, . . . , 9), and 10 possibilities for each of the last two blanks (the digits 0, 1, 2, 3, . . . , 9). Therefore the total number of possibilities is

$$26 \cdot 26 \cdot 26 \cdot 9 \cdot 10 \cdot 10 = \boxed{15{,}818{,}400}$$

(b) As in Example 2, we have 26 possibilities for the first blank, 25 possibilities for the second blank, and 24 possibilities for the third blank. There are 9 possibilities for the fourth blank and 10 possibilities for each of the last two blanks. Hence the number of possibilities is

$$26 \cdot 25 \cdot 24 \cdot 9 \cdot 10 \cdot 10 = \boxed{14{,}040{,}000} \quad \blacksquare$$

Example 4 Suppose you are going to the horse races and are going to buy a perfecta ticket. A perfecta wager is one in which you pick the correct first and second place horses. Suppose there are eight horses in the race.
(a) How many possible perfecta tickets can you buy?
(b) If you pick at *random*, what are your chances of winning the wager?

Solution (a) We have two blanks to fill:

$$\underline{} \quad \underline{}$$

first-place   second-place
   horse   horse

We have 8 possibilities for the first blank and 7 for the next blank.

Hence we have

$$8 \cdot 7 = \boxed{56}$$

possible combinations.

(b) There are 56 possible combinations (that is, 56 possible perfecta tickets). If you pick at random, your chances of winning are

$$\frac{1}{56} \approx \boxed{0.0179} \quad \blacksquare$$

We will often be considering an ordered arrangement of n objects. We introduce the following definition.

DEFINITION   An *ordered* arrangement of n distinct objects is called a **permutation** of the n objects.

Example 5   Consider the letters A, B, and C. We list *all* possible permutations of these three letters:

ABC   BCA
ACB   CAB
BAC   CBA

Thus there are six possible permutations.   ■

Note: When dealing with permutations, repetitions are not allowed. For example, BAC is a permutation of the letters A, B, and C, but AAB is *not*.

Example 6   How many permutations of the digits 1, 2, 3, 4, and 5 can be formed?

Solution   We can think of five positions to fill:

——— ——— ——— ——— ———

We have 5 possibilities for the first position, 4 possibilities for the next position, and so on. Hence the number of permutations is

$$5 \cdot 4 \cdot 3 \cdot 2 \cdot 1 = \boxed{120} \quad \blacksquare$$

For convenience we introduce the following definitions.

DEFINITION   We define

$$0! = 1$$
$$1! = 1$$
$$2! = 2 \cdot 1 = 2$$
$$3! = 3 \cdot 2 \cdot 1 = 6$$
$$4! = 4 \cdot 3 \cdot 2 \cdot 1 = 24$$

$$\cdot$$
$$\cdot$$
$$\cdot$$

In general,

$$n! = n(n - 1)(n - 2) \cdots (2)(1)$$

The symbol $n!$ is read "n factorial."

DEFINITION   Suppose we have n distinct objects. An ordered arrangement of r of the n objects is called a permutation of n distinct objects taken r at a time. The number of permutations of n distinct objects taken r at a time is denoted by $P(n, r)$. The symbol $_nP_r$ is sometimes used instead of $P(n, r)$.

Example 7   Consider the letters A, B, C, D, and E.
(a) List several (but not all) permutations of these 5 letters taken 3 at a time.
(b) Evaluate $P(5, 3)$.

Solution   (a) ADE
         EBD
         DAE
         CBD
         CDB
         $\cdot$
         $\cdot$
         $\cdot$

(b) $P(5, 3)$ denotes the number of permutations of 5 objects (say the letters A, B, C, D, and E) taken 3 at a time. To compute $P(5, 3)$, we can think of three blanks to fill:

$$\underline{\hspace{2cm}} \quad \underline{\hspace{2cm}} \quad \underline{\hspace{2cm}}$$
$$\text{1st} \qquad\quad \text{2nd} \qquad\quad \text{3rd}$$

We have 5 possibilities for the first blank (any of the letters A, B, C, D, E). Having filled the first position, we have 4 possibilities for the second position and then 3 possibilities for the third position. Hence

$$P(5, 3) = 5 \cdot 4 \cdot 3 = \boxed{60} \quad \blacksquare$$

Notice that in Example 7(a) above we listed 5 of the 60 total. However, arrangements such as ABA and BBC are *not* counted, since in dealing with $P(n, r)$ repetitions are not allowed.

The following theorem is proved using the same ideas demonstrated in Example 7.

THEOREM 1 The number of permutations of n distinct objects taken r at a time is given by

$$P(n, r) = n(n - 1)(n - 2) \cdots (n - r + 1)$$
$$= \frac{n(n - 1)(n - 2) \cdots (n - r + 1) \cdot (n - r)!}{(n - r)!}$$

$$\boxed{P(n, r) = \frac{n!}{(n - r)!}}$$

The number of permutations of n distinct objects is simply

$$P(n, n) = \frac{n!}{(n - n)!} = \frac{n!}{0!} = n!$$

A permutation is an ordered arrangement, and so each time you change the order you have a new permutation. In dealing with sets the order makes no difference. For example, ABC, ACB, BAC, BCA, CAB, CBA are 6 different permutations. However {A, B, C}, {A, C, B}, {B, A, C}, {B, C, A}, {C, A, B}, {C, B, A} all denote *exactly* the same set. With sets, changing the order does not change the set.

We often use the word **combination** instead of the word set. That is, a combination is simply a set.

Example 8 Consider the letters A, B, C, D. List *all* combinations of three of these four letters.

Solution {A, B, C}
{A, B, D}
{A, C, D}
{B, C, D}
Notice that {A, A, B} is *not* a combination of three letters, since {A, A, B} is exactly the same as the set {A, B} which is a combination of two letters. ∎

The following definition will be useful in connection with some of the concepts developed later.

DEFINITION The number of combinations of n distinct objects taken r at a time will be denoted by the symbol $\binom{n}{r}$. This symbol is sometimes read "n choose r" and sometimes written as $C(n, r)$ or $_nC_r$.

From Example 8, we see that

$$\binom{4}{3} = \begin{bmatrix} \text{number of combinations of 4 distinct} \\ \text{objects taken 3 at a time} \end{bmatrix} = 4$$

The following theorem shows us that the counting of combinations and the counting of permutations are closely related.

THEOREM 2 The number of combinations of n distinct objects taken r at a time is given by

$$\binom{n}{r} = \frac{n!}{r!(n-r)!}$$

Proof: Suppose we have n distinct objects (say 1, 2, 3, . . . , n). Suppose you single out r of these n objects and fix them in your mind. These particular r objects give you $r!$ different permutations but only *one* combination. Hence, we see that

$$(r!) \cdot \binom{\text{number of combinations of}}{n \text{ objects taken } r \text{ at a time}} = \binom{\text{number of permutations of } n}{\text{objects taken } r \text{ at a time}}$$

That is

$$(r!) \cdot \binom{n}{r} = P(n, r)$$

or

$$\binom{n}{r} = \frac{P(n, r)}{r!} = \frac{\dfrac{n!}{(n-r)!}}{r!} = \frac{n!}{r!(n-r)!}$$

This completes the proof of the theorem.   □

Example 9   How many committees of 5 people can be formed from a group of 15 people?

Solution   In this problem we are asking for the number of combinations of 15 objects taken 5 at a time. Hence the number is

$$\binom{15}{5} = \frac{15!}{5!(15-5)!} = \frac{15!}{5!10!} = \frac{15 \cdot 14 \cdot 13 \cdot 12 \cdot 11}{5 \cdot 4 \cdot 3 \cdot 2 \cdot 1}$$

$$= \boxed{3003} \quad ■$$

Example 10   In how many ways can a jury of 6 people reach a majority decision?

Solution   A majority decision can be reached if
(a) 4 vote yes and 2 vote no, or
(b) 5 vote yes and 1 votes no, or
(c) all 6 vote yes.

Now the number of ways exactly 4 can vote yes is $\binom{6}{4}$. The number of ways exactly 5 can vote yes is $\binom{6}{5}$. Finally, the number of ways exactly 6 can vote yes is $\binom{6}{6}$. Hence, the number of ways in which a majority decision can be reached is

$$\binom{6}{4} + \binom{6}{5} + \binom{6}{6} = \frac{6!}{4!2!} + \frac{6!}{5!1!} + \frac{6!}{6!0!} = 15 + 6 + 1$$

$$= \boxed{22} \quad \blacksquare$$

Example 11   A club consists of 8 boys and 7 girls. The club needs 10 players for a softball game. How many teams can be formed if the team is to have exactly 6 boys and 4 girls?

Solution   We can think of performing two separate operations, picking the 6 boys, and then picking the 4 girls. We can pick the 6 boys in

$$\binom{8}{6} = \frac{8!}{6!2!} = \frac{8 \cdot 7}{2 \cdot 1} = 28 \text{ ways}$$

We can pick the 4 girls in

$$\binom{7}{4} = \frac{7!}{4!3!} = \frac{7 \cdot 6 \cdot 5}{3 \cdot 2 \cdot 1} = 35 \text{ ways}$$

By the fundamental counting principle, both operations can be performed in $28 \cdot 35 = 980$ ways. Hence the number of teams with 6 boys and 4 girls is $\boxed{980}$. $\blacksquare$

Exercises 9.1

1.   In how many ways can the letters A, B, C, and D be arranged?

C 2.   Suppose license plates are to have two letters followed by four digits. How many license plates are possible if
   (a) the first digit cannot be 0?
   (b) no digit can be 0?
   (c) repetitions are not allowed among the two letters?
   (d) repetitions are not allowed among the two letters and among the four digits?

3.   How many three-letter initials are possible using only the letters D, M, R, S, T, and V, if
   (a) repetitions are allowed?
   (b) repetitions are not allowed?

4.   How many possible finishes exist in a five-horse race?

5.   Ten cars enter a race. In how many ways can trophies for first, second, and third place be awarded?

6.   How many four-digit numbers can be made if
   (a) repetitions are not allowed?
   (b) repetitions are allowed?

7.   How many three-letter "words" can be made using the letters A, B, C, D,

E, F if we allow no repeated letter? (A word is *any* arrangement of letters.)

8. How many committees of four people can be chosen from a group of ten?

C 9. A poker hand consists of five cards from a standard 52-card deck. How many poker hands are possible?

10. In poker, a flush consists of five cards all of the same suit. How many flushes exist?

11. How many permutations of the digits 2, 3, 5, 7, 8, and 9 can be formed?

12. How many permutations of the letters A, B, C, D, and E can be formed?

13. Consider the letters A, B, C, D, E, and F. List several (but not all) permutations of these six letters taken four at a time. How many possible *permutations* of six letters taken four at a time can be formed?

14. Consider the letters A, B, C, D, E, and F. List several (but not all) combinations of these six letters taken four at a time. How many possible *combinations* of six letters taken four at a time are possible?

15. In how many ways can a group of seven people reach a majority decision?

16. In how many ways can a group of eight people reach a majority decision?

17. In how many ways can a class of 30 people select a president, a vice president, and a secretary?

C 18. A class consists of 15 boys and 10 girls.
   (a) How many committees of seven students can be formed?
   (b) How many committees of seven students can be formed if the committee is to have exactly four boys?
   (c) How many committees of seven students can be formed which contain at least four boys?

19. Evaluate each of the following:

   (a) $P(7, 3)$ (b) $\binom{10}{4}$ (c) $P(10, 7)$ (d) $\binom{12}{7}$

20. How many ways can you put six people in six distinct seats?

21. How many ways can you put six people in ten seats?

22. A club consists of twelve boys and eight girls. The club needs nine players for a baseball team.
   (a) How many teams can be formed if the team is to have exactly four girls?
   (b) How many teams can be formed if the team is to have exactly four boys?

23. In how many ways can a five-man team be selected from a group of eight players?

24. In poker a full house consists of five cards, three of which are of one denomination and two of which are of a different denomination. How many full houses are possible?

25. How many five-member committees can be formed from a group of seven Republicans and ten Democrats if
   (a) no restriction is imposed on the composition of the committee?
   (b) the committee must contain exactly two Democrats?
   (c) the committee must contain at least two Republicans?

26. A statistics textbook committee contains the following members: Murphy, Oberhoff, Pierce, Rodriguez, Wood, and Rich.

> (a) How many three-member subcommittees can be formed containing Murphy?
> (b) How many three-member subcommittees can be formed if Murphy and Wood cannot serve on the same committee?

27. In how many possible orders can three girls and two boys be born to a family having five children?

28. How many possible distinct permutations can be formed with the letters of the word RESERVE? (*Hint:* The answer is not 7!.)

9.2   The Binomial Theorem

In Chapter 1 we dealt with exponents of a single term by using the laws of exponents. Exponents, or powers, of a sum or difference of two terms are not "multiplied out" or "expanded" quite as easily. This section will introduce you to the Binomial Theorem and will show you an application to an investment problem. Additional applications to computing the likelihood of certain success/failure situations are discussed in Section 9.3.

A binomial expression is an expression that contains exactly two terms separated by a plus or minus sign; for example, $x + y$, $2x - 5y$, $\frac{1}{2} - 3y$, and $x^2 + 5y^3$ are binomial expressions.

In this section we will develop a formula for the expansion of the nth power of a binomial. That is, we will develop a formula for $(a + b)^n$ where a and b are numbers and n is a positive integer. We first look at some examples.

$(a + b)^1 = a + b$ (here $n = 1$)

$(a + b)^2 = a^2 + 2ab + b^2$ (here $n = 2$)

$$\begin{aligned}
(a + b)^3 &= (a + b)(a + b)^2 \\
&= (a + b)(a^2 + 2ab + b^2) \\
&= a^3 + 2a^2b + ab^2 + a^2b + 2ab^2 + b^3 \\
&= a^3 + 3a^2b + 3ab^2 + b^3 \quad \text{(here } n = 3)
\end{aligned}$$

$$\begin{aligned}
(a + b)^4 &= (a + b)(a + b)^3 \\
&= (a + b)(a^3 + 3a^2b + 3ab^2 + b^3) \\
&= a^4 + 3a^3b + 3a^2b^2 + ab^3 + a^3b + 3a^2b^2 + 3ab^3 + b^4 \\
&= a^4 + 4a^3b + 6a^2b^2 + 4ab^3 + b^4 \quad \text{(here } n = 4)
\end{aligned}$$

$$\begin{aligned}
(a + b)^5 &= (a + b)(a + b)^4 \\
&= (a + b)(a^4 + 4a^3b + 6a^2b^2 + 4ab^3 + b^4) \\
&= a^5 + 4a^4b + 6a^3b^2 + 4a^2b^3 + ab^4 + a^4b + 4a^3b^2 + 6a^2b^3 \\
&\quad + 4ab^4 + b^5 \\
&= a^5 + 5a^4b + 10a^3b^2 + 10a^2b^3 + 5ab^4 + b^5 \quad \text{(here } n = 5)
\end{aligned}$$

From the above examples we notice the following properties:

PROPERTIES OF THE EXPANSION OF $(a + b)^n$

1. The first term is a^n, and the last term is b^n.
2. As we move from the first term to the last term, the exponent of a decreases by 1 and the exponent of b increases by 1.
3. The sum of the exponents of a and b in any term is n.
4. There are $n + 1$ terms in the expression.
5. The numerical coefficient of the first term is $\binom{n}{0}$, the numerical coefficient of the second term is $\binom{n}{1}$, the numerical coefficient of the third term is $\binom{n}{2}$, and so on. $\left(\text{Recall: For } 0 \le r \le n, \binom{n}{r} = \dfrac{n!}{r!(n-r)!}.\right)$

Using the above properties we now state the Binomial Theorem.

BINOMIAL THEOREM

For any positive integer n,

$$(a + b)^n = \binom{n}{0} a^n + \binom{n}{1} a^{n-1}b + \binom{n}{2} a^{n-2}b^2 + \binom{n}{3} a^{n-3}b^3 + \cdots$$
$$+ \binom{n}{n-1} ab^{n-1} + \binom{n}{n} b^n$$

Example 1 Expand $(x + 2y)^4$.

Solution Here $n = 4$, and the numerical coefficients are

$$\binom{4}{0} = \frac{4!}{0!4!} = 1, \qquad \binom{4}{1} = \frac{4!}{1!3!} = 4, \qquad \binom{4}{2} = \frac{4!}{2!2!} = 6,$$

$$\binom{4}{3} = \frac{4!}{3!1!} = 4, \qquad \binom{4}{4} = \frac{4!}{4!0!} = 1$$

We can write $(x + 2y)^4 = [x + (2y)]^4$. Here $a = x$ and $b = 2y$. Hence

$$(x + 2y)^4 = x^4 + 4x^3(2y) + 6x^2(2y)^2 + 4x(2y)^3 + (2y)^4$$
$$= x^4 + 8x^3y + 24x^2y^2 + 32xy^3 + 16y^4 \quad \blacksquare$$

Example 2 Expand $(2x^2 - 3y)^6$.

Solution Here $n = 6$, and the numerical coefficients are

$$\binom{6}{0} = \frac{6!}{0!6!} = 1, \qquad \binom{6}{1} = \frac{6!}{1!5!} = 6, \qquad \binom{6}{2} = \frac{6!}{2!4!} = 15,$$

$$\binom{6}{3} = \frac{6!}{3!3!} = 20, \qquad \binom{6}{4} = \frac{6!}{4!2!} = 15, \qquad \binom{6}{5} = \frac{6!}{5!1!} = 6,$$

$$\binom{6}{6} = \frac{6!}{6!0!} = 1$$

We write $(2x^2 - 3y)^6 = [(2x^2) + (-3y)]^6$. Here $a = 2x^2$ and $b = -3y$. Hence

$$(2x^2 - 3y)^6 = (2x^2)^6 + 6(2x^2)^5(-3y) + 15(2x^2)^4(-3y)^2 + 20(2x^2)^3(-3y)^3$$
$$+ 15(2x^2)^2(-3y)^4 + 6(2x^2)(-3y)^5 + (-3y)^6$$
$$= 64x^{12} - 576x^{10}y + 2160x^8y^2 - 4320x^6y^3 + 4860x^4y^4$$
$$- 2916x^2y^5 + 729y^6 \quad \blacksquare$$

Example 3 Find the first four terms of $(2x - y)^{10}$.

Solution Here $n = 10$, and the numerical coefficients of the first four terms are

$$\binom{10}{0} = \frac{10!}{0!10!} = 1, \qquad \binom{10}{1} = \frac{10!}{1!9!} = 10,$$

$$\binom{10}{2} = \frac{10!}{2!8!} = 45, \qquad \binom{10}{3} = \frac{10!}{3!7!} = 120$$

We write $(2x - y)^{10} = [(2x) + (-y)]^{10}$. Hence

$$\text{first term} = (2x)^{10} = \boxed{1024x^{10}}$$

$$\text{second term} = 10(2x)^9(-y) = \boxed{-5120x^9y}$$

$$\text{third term} = 45(2x)^8(-y)^2 = \boxed{11{,}520x^8y^2}$$

$$\text{fourth term} = 120(2x)^7(-y)^3 = \boxed{-15{,}360x^7y^3} \quad \blacksquare$$

From the Binomial Theorem,

$$(a + b)^n = \binom{n}{0} a^n + \binom{n}{1} a^{n-1}b + \binom{n}{2} a^{n-2}b^2 + \cdots + \binom{n}{n} b^n$$

$$= \frac{n!}{0!n!} a^n + \frac{n!}{1!(n-1)!} a^{n-1}b + \frac{n!}{2!(n-2)!} a^{n-2}b^2$$

$$+ \cdots + \frac{n!}{n!0!} b^n$$

$$= a^n + na^{n-1}b + \frac{n(n-1)}{2} a^{n-2}b^2 + \cdots + b^n$$

We notice the following pattern:

Note: In the expansion of $(a + b)^n$, if we multiply the coefficient of any term by the exponent of a in that term and then divide the product by the number of the term, we obtain the coefficient of the next term.

Example 4 Expand $(a + b)^5$.

Solution The first term is $a^5 = 1 \cdot a^5$. Hence

$$(a + b)^5 = 1 \cdot a^5 + 5a^4b + 10a^3b^2 + 10a^2b^3 + 5ab^4 + 1 \cdot b^5$$

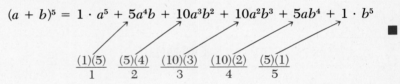

$$\frac{(1)(5)}{1} \quad \frac{(5)(4)}{2} \quad \frac{(10)(3)}{3} \quad \frac{(10)(2)}{4} \quad \frac{(5)(1)}{5}$$

Example 5 If you invest \$100 at 8% compounded annually, then the amount A after 4 years is given by

$$A = 100(1 + .08)^4 \quad \text{(see note at end of Section 8.1)}$$
$$= 100[(1)^4 + 4(1)^3(.08) + 6(1)^2(.08)^2 + 4(1)(.08)^3 + (.08)^4]$$
$$= 100[1 + 0.32 + 0.0384 + 0.002048 + 0.00004096]$$

$$= 100[1.36048896] = \boxed{136.05 \text{ (dollars)}} \quad \blacksquare$$

Exercises 9.2 In Exercises 1–20 expand using the Binomial Theorem.

1. $(x + y)^5$
2. $(2x + y)^4$
3. $(a + 3b)^4$
4. $(x - 2y)^6$
5. $(2x - y)^6$
6. $(3a + b)^4$
7. $(x^2 + 2y)^5$
8. $(2x^3 - 3y^2)^4$
9. $(3z^2 - 2w)^6$
10. $(2x^2 - 3y^2)^5$
11. $(a^3 - b)^7$
12. $(a + 2b)^7$

13. $(3x - 2y)^6$ 14. $(s - t)^5$
15. $(3s + 2t)^5$ 16. $(a - 4b)^5$
17. $(u - 2v)^6$ 18. $(u^2 + v^4)^4$
19. $(x + y)^{10}$ 20. $(x - y)^{10}$

In Exercises 21–26 compute the quantity using the Binomial Theorem. Round off the result to four decimal places.

21. $(1.02)^4 = (1 + 0.02)^4$ 22. $(1.01)^5$
23. $(2.03)^4$ 24. $(3.04)^4$
25. $(2.05)^4$ 26. $(2.01)^6$

C 27. Find the first four terms of $(2x - 3y)^{10}$.
   28. Find the first three terms of $(x + 2y)^8$.
C 29. Find the first five terms of $(2a^2 + 3b)^9$.
   30. Find the first four terms of $(x + y)^{20}$.
   31. Find the first five terms of $(x - y)^{20}$.
   32. Find the fifth term of $(x - 2y)^{20}$.
   33. Find the fourth term of $(2x - 3y)^{10}$.
   34. Find the seventh term of $(x^2 + y)^{15}$.
   35. Find the sixth term of $(x^2 + 2y^2)^{10}$.
   36. If you invest \$200 at 7% compounded annually, what is the amount A after 4 years? (See Example 5.)
C 37. If you invest \$100 at 12% compounded annually, what is the amount A after 5 years?

9.3   Elementary Probability

The word "probability" is often used synonymously with the word "chance" to indicate the likelihood that something will occur. We state some preliminary definitions and examples.

DEFINITION   An **experiment** is any operation whose outcome cannot be predicted beforehand. A sample space S of an experiment is the set of all possible outcomes for the experiment.

Example 1   Some examples of experiments are

(a) throwing a die
(b) selecting one card at random from a standard deck of 52
(c) throwing a pair of dice, one red and one green
(d) flipping three coins once (say, a nickel, a dime, and a quarter)
(e) selecting a student at random from the student body of the University of Houston

The corresponding sample spaces for the above experiments are:

(a) $S = \{1, 2, 3, 4, 5, 6\}$
(b) $S = \{$listing of all 52 cards$\}$
(c) $S = \{(1, 1), (1, 2), (1, 3), (1, 4), (1, 5), (1, 6)$
$\qquad (2, 1), (2, 2), (2, 3), (2, 4), (2, 5), (2, 6)$

$\qquad\qquad\qquad\quad \vdots$

$\qquad (6, 1), (6, 2), (6, 3), (6, 4), (6, 5), (6, 6)\}$

Here the first position corresponds to, say, the red die and the second position corresponds to the green die. For example, $(3, 5)$ represents the outcome that the red die came up 3 and the green die came up 5.

(d) $S = \{(H, H, H), (H, H, T), (H, T, H), (T, H, H), (T, T, H),$
$\qquad (T, H, T), (H, T, T), (T, T, T)\}$

Here, for example, (H, T, H) represents the outcome that the nickel came up heads, the dime came up tails, and the quarter came up heads.

(e) $S = \{$listing of all students at the University of Houston$\}$ ∎

DEFINITION

An **event** is a subset of a sample space.

Example 2

In Example 1(a) $S = \{1, 2, 3, 4, 5, 6\}$. Each of the following sets is an event:

$A = \{1, 2\}, B = \{2, 3, 6\}, C = \{2\}, D = \{1, 2, 3, 4\}, \varnothing$ ∎

We sometimes give a word description of an event, as is illustrated in the following example.

Example 3

In Example 1(b), let

$$A = \text{``a red card is drawn''}$$

$$B = \text{``a king is drawn''}$$

Then

$A = \{$ace of hearts, ace of diamonds, . . . ,

$\qquad\qquad\qquad\qquad\qquad$ king of hearts, king of diamonds$\}$

$\underbrace{\qquad\qquad\qquad\qquad\qquad\qquad\qquad\qquad\qquad}$
list of all 26 red cards

and

B = {king of hearts, king of diamonds, king of clubs, king of spades}

In Example 1(c), let

A = "the sum is seven"

B = "both dice show the same number"

Then

$$A = \{(1, 6), (6, 1), (2, 5), (5, 2), (3, 4), (4, 3)\}$$
$$B = \{(1, 1), (2, 2), (3, 3), (4, 4), (5, 5), (6, 6)\}$$

In Example 1(d), let

A = "exactly two heads"

B = "at least two heads"

Then

$$A = \{(H, H, T), (H, T, H), (T, H, H)\}$$
$$B = \{(H, H, T), (H, T, H), (T, H, H), (H, H, H)\}$$

Notice that "at least two heads" means "exactly two heads or exactly three heads." ∎

Throughout this section we will consider only experiments with a finite number of outcomes. The theory of probability is concerned with consistent ways of assigning numbers to events. The number assigned to an event will be called the probability of the event.

Most people have an intuitive feeling for what probability should be. For example, if we flip a fair coin, most people would quickly agree that the probability of a head is $\frac{1}{2}$, since if the coin is fair, then half of the flips should result in heads. If we pick a card at random from a standard deck of 52, then the probability of picking the king of hearts is $\frac{1}{52}$, since exactly one of the 52 cards is the king of hearts. The probability of picking a king is $\frac{4}{52}$ or $\frac{1}{13}$, since exactly four of the 52 cards are kings. Our intuitive notions of probability motivate certain axioms. Consider an experiment with sample space S. For *any* event A (recall that an event is simply a subset of S), the probability of A is denoted by $P(A)$. Then the assignment of probabilities must satisfy the rules given in the box at the top of page 343. These rules are known as the axioms of probability. Several important theorems can be proved from them.

AXIOMS OF PROBABILITY

1. $P(S) = 1$.
2. $0 \le P(A) \le 1$ for any event A.
3. If $A_1, A_2, \ldots, A_n$ are events such that $A_i \cap A_j = \varnothing$ if $i \ne j$ (that is, each pair have empty intersection), then

$$P(A_1 \cup A_2 \cup \cdots \cup A_n) = P(A_1) + P(A_2) + \cdots + P(A_n)$$

THEOREM 1 $P(\varnothing) = 0$.

Proof: $S \cup \varnothing = S$, and so $P(S \cup \varnothing) = P(S) = 1$ by axiom 1. But $S \cap \varnothing = \varnothing$, and so by axiom 3, $P(S \cup \varnothing) = P(S) + P(\varnothing) = 1 + P(\varnothing)$. Hence $1 + P(\varnothing) = 1$, and so $P(\varnothing) = 0$. $\square$

Before proving the next theorem we state the following preliminary definition.

DEFINITION   Let A be an event. The **complement** of A, denoted by $\overline{A}$, is the event consisting of all outcomes in S that are not in A. That is, $\overline{A} = \{x \mid x \in S$ and $x \notin A\}$.

For example, if we throw a die, and if

$$A = \text{``an even number occurs''}$$

then $A = \{2, 4, 6\}$ and

$$\overline{A} = \text{``an odd number occurs''} = \{1, 3, 5\}$$

If we flip three coins once, and if

$$A = \text{``exactly two heads''}$$

then $A = \{(H, H, T), (H, T, H), (T, H, H)\}$ and

$$\overline{A} = \{(T, T, H), (T, H, T), (H, T, T), (H, H, H), (T, T, T)\}$$

Notice that for any event A,

$$A \cup \overline{A} = S, \quad A \cap \overline{A} = \varnothing, \quad \overline{\varnothing} = S, \quad \text{and} \quad \overline{S} = \varnothing$$

THEOREM 2 For any event A, $P(\overline{A}) = 1 - P(A)$.

Proof: $A \cup \overline{A} = S$, and so $P(A \cup \overline{A}) = P(S) = 1$ by axiom 1. But $A \cap \overline{A} = \varnothing$, and so by axiom 3, $P(A \cup \overline{A}) = P(A) + P(\overline{A})$. Hence $P(A) + P(\overline{A}) = 1$, and so $P(\overline{A}) = 1 - P(A)$. This completes the proof of Theorem 2. $\square$

THEOREM 3 For any two events A and B, $P(A \cup B) = P(A) + P(B) - P(A \cap B)$

$A \cap B$

FIGURE 1

Intuitive Proof: The following is only an intuitive proof and not a rigorous proof. Let us think of probability as area. Then we have

$$P(A \cup B) = \text{area of } A \cup B$$
$$P(A) = \text{area of } A$$
$$P(B) = \text{area of } B$$
$$P(A \cap B) = \text{area of } A \cap B$$

From the diagram in Figure 1 it is clear that

$$\text{area of } A \cup B = \text{area of } A + \text{area of } B - \text{area of } A \cap B$$

Note that the area of $A \cap B$ must be subtracted, since it has been added twice. Hence

$$P(A \cup B) = P(A) + P(B) - P(A \cap B) \quad \square$$

THEOREM 4 If A is an event consisting of outcomes $x_1, x_2, \ldots, x_k$, that is, $A = \{x_1, x_2, \ldots, x_k\}$, then $P(A) = P(\{x_1\}) + P(\{x_2\}) + \cdots + P(\{x_k\})$.

Proof: We have

$$A = \{x_1\} \cup \{x_2\} \cup \cdots \cup \{x_k\} \quad \text{and} \quad \{x_i\} \cap \{x_j\} = \varnothing \text{ if } i \neq j$$

Hence, by axiom 3,

$$P(A) = P(\{x_1\} \cup \{x_2\} \cup \cdots \cup \{x_k\})$$
$$= P(\{x_1\}) + P(\{x_2\}) + \cdots + P(\{x_k\}) \quad \square$$

If α represents one outcome of an experiment, we write $P(\alpha)$ instead of $P(\{\alpha\})$. Thus, Theorem 4 would state that $P(A) = P(x_1) + P(x_2) + \cdots + P(x_k)$. Notice that the sample space S is an event, and so it follows from Theorem 4 that the sum of the probabilities of all the outcomes is 1. Furthermore, if we know the probabi'ity assigned to each outcome in S, then we can compute the probability of any event by applying Theorem 4. In solving a probability problem, first list the outcomes in S if possible, and then assign a probability to each outcome. Remember that the sum of the probabilities of all the outcomes must be 1. Carefully study the following examples.

Example 4   Suppose we roll a fair die. What is the probability of getting an even number?

Solution   Our sample space is $S = \{1, 2, 3, 4, 5, 6\}$. Since the die is fair, all six outcomes are equally likely to occur, and so each outcome has probability $\frac{1}{6}$. Define the event A = "an even number occurs." Then $A = \{2, 4, 6\}$, and we want $P(A)$. But by Theorem 4, $P(A) = P(2) + P(4) + P(6) = \frac{1}{6} + \frac{1}{6} + \frac{1}{6} = \boxed{\frac{1}{2}}$. ∎

Example 5   Mr. Jones goes to the grocery store. Suppose the probability that Mr. Jones buys bread is 0.4, that he buys meat is 0.7, and that he buys both meat and bread is 0.3. What is the probability that Mr. Jones buys meat or bread?

Solution   Define the events B and M by

$$B = \text{``Mr. Jones buys bread''}$$

$$M = \text{``Mr. Jones buys meat''}$$

Then $P(B) = 0.4$, $P(M) = 0.7$, and $P(B \cap M) = 0.3$. We want to compute $P(B \cup M)$. However, by Theorem 3,

$$P(B \cup M) = P(B) + P(M) - P(B \cap M)$$

$$= 0.4 + 0.7 - 0.3 = \boxed{0.8} \quad ∎$$

Example 6   We roll a pair of fair dice once. What is the probability that
(a) the sum of the two numbers is 7?
(b) it is 11?
(c) both dice show the same number?
(d) both dice show different numbers?

Solution   Here $S = \{(1, 1), (1, 2), (1, 3), (1, 4), (1, 5), (1, 6),$
$(2, 1), (2, 2), (2, 3), (2, 4), (2, 5), (2, 6),$
$(3, 1), (3, 2), (3, 3), (3, 4), (3, 5), (3, 6),$
$(4, 1), (4, 2), (4, 3), (4, 4), (4, 5), (4, 6),$
$(5, 1), (5, 2), (5, 3), (5, 4), (5, 5), (5, 6),$
$(6, 1), (6, 2), (6, 3), (6, 4), (6, 5), (6, 6)\}$

(S has a total of 36 outcomes.) Since both dice are fair, all 36 outcomes are equally likely to occur and so all 36 outcomes must have probability $\frac{1}{36}$. Define the events A and B by

$$A = \text{"the sum is seven"}$$
$$B = \text{"the sum is 11"}$$

Then

$$A = \{(1, 6), (6, 1), (2, 5), (5, 2), (3, 4), (4, 3)\}$$
$$B = \{(5, 6), (6, 5)\}$$

Hence using Theorem 4, we see that

(a) $P(A) = P(1, 6) + P(6, 1) + P(2, 5) + P(5, 2) + P(3, 4) + P(4, 3)$

$$= \frac{1}{36} + \frac{1}{36} + \frac{1}{36} + \frac{1}{36} + \frac{1}{36} + \frac{1}{36} = \frac{6}{36} = \boxed{\frac{1}{6}}$$

(b) $P(B) = P(5, 6) + P(6, 5) = \dfrac{1}{36} + \dfrac{1}{36} = \dfrac{2}{36} = \boxed{\dfrac{1}{18}}$

(c) Define the event $C = $ "both dice show the same number." Then

$$C = \{(1, 1), (2, 2), (3, 3), (4, 4), (5, 5), (6, 6)\}$$

and

$$P(C) = P(1, 1) + P(2, 2) + P(3, 3) + P(4, 4) + P(5, 5) + P(6, 6)$$

$$= \frac{1}{36} + \frac{1}{36} + \frac{1}{36} + \frac{1}{36} + \frac{1}{36} + \frac{1}{36} = \frac{6}{36} = \boxed{\frac{1}{6}}$$

(d) $\overline{C} = $ "both dice show different numbers," and so from Theorem 2,

$$P(\text{both dice show different numbers}) = P(\overline{C}) = 1 - P(C)$$

$$= 1 - \frac{1}{6} = \boxed{\frac{5}{6}} \quad \blacksquare$$

Many experiments have equally likely outcomes. As we have seen, in these cases the probability of each outcome is $1/n$, where n is the number of outcomes in S. Moreover, with equally likely outcomes,

$$P(A) = \frac{\text{number of outcomes in } A}{\text{number of outcomes in } S}$$

for each event A.

Independent Events

Events A and B are said to be **independent** if the occurrence of one of the events does not influence the probability of occurrence of the other. The following examples demonstrate this concept.

Example 7   An urn contains 7 red and 3 blue tags. Two tags are to be selected in succession with replacement (that is, after the first tag is selected it is replaced, and the second tag is selected from the original set of 10 tags). Suppose R_1 and R_2 are events given by

R_1 = "a red tag is selected on the first draw"

R_2 = "a red tag is selected on the second draw"

Then clearly R_1 and R_2 are independent events, since the occurrence of one of the events does not influence the probability of the other. For example, $P(R_2) = \frac{7}{10}$ and is independent of what tag you selected on the first draw.   ■

If we were to select two tags *without* replacement (that is, after the first tag is selected it is *not* replaced and the second tag is selected from the remaining 9 tags), then R_1 and R_2 would *not* be independent events, since $P(R_2)$ *is* influenced by whether or not R_1 occurred. If you selected a red tag on the first draw, then the probability of selecting a red tag on the second draw is $\frac{6}{9}$. However, if you selected a blue tag on the first draw, then the probability of selecting a red tag on the second draw is $\frac{7}{9}$.

If events A and B are independent, then we always assign a probability to the new event $A \cap B$ according to the following rule:

RULE OF MULTIPLICATION FOR INDEPENDENT EVENTS

If events A and B are independent, then $P(A \cap B) = P(A) \cdot P(B)$. The rule also holds for more than two independent events; for example, if A, B, C are independent events, then $P(A \cap B \cap C) = P(A) \cdot P(B) \cdot P(C)$.

Example 8 Two tags are selected with replacement from an urn containing 7 red and 3 blue tags.
(a) What is the probability that both tags are red?
(b) What is the probability that both tags are blue?
(c) What is the probability that the first tag is red and the second tag is blue?
(d) What is the probability that the first tag is blue and the second tag is red?
(e) What is the probability that the two tags selected are the same color?

Solution Since we are drawing with replacement, the outcome of the second draw is independent of the outcome of the first draw. Hence

(a) $P(R_1 \cap R_2) = P(R_1) \cdot P(R_2) = \dfrac{7}{10} \cdot \dfrac{7}{10} = \boxed{\dfrac{49}{100}}$

(b) $P(B_1 \cap B_2) = P(B_1) \cdot P(B_2) = \dfrac{3}{10} \cdot \dfrac{3}{10} = \boxed{\dfrac{9}{100}}$

(c) $P(R_1 \cap B_2) = P(R_1)P(B_2) = \dfrac{7}{10} \cdot \dfrac{3}{10} = \boxed{\dfrac{21}{100}}$

(d) $P(B_1 \cap R_2) = P(B_1)P(R_2) = \dfrac{3}{10} \cdot \dfrac{7}{10} = \boxed{\dfrac{21}{100}}$

(e) $P(\text{both tags are same color}) = P(R_1 \cap R_2) + P(B_1 \cap B_2)$

$$= \dfrac{49}{100} + \dfrac{9}{100} = \boxed{\dfrac{58}{100}} \quad \blacksquare$$

Example 9 A Tampa auto dealer has been troubled by new cars that are delivered with flawed paint and with tires that are one size too small. These two assembly line operations are clearly independent. If the probability of flawed paint is 0.05 and the probability of tires that are one size too small is 0.08, then what is the probability that the next car delivered will have flawed paint *and* small tires?

Solution $P(F \cap S) = P(F) \cdot P(S) = (0.05)(0.08) = \boxed{0.004} \quad \blacksquare$

Binomial Probabilities

Some types of experiments result in only two outcomes, success or failure. If the probability of success on any given trial is always the same, we can compute what is known as a **binomial probability.**

The probability of getting exactly r successes in n independent trials, if p = the probability of success on each trial and q = the probability of failure on each trial, is denoted by $B(n, r)$ and given by

$$B(n, r) = \binom{n}{r} p^r q^{n-r}$$

Note that $p + q = 1$, that the exponent of p is the number of successes, and that the exponent of q is the number of failures. Observe also that $B(n, r)$ is one of the terms of $(p + q)^n$; this is the reason for the name "binomial probability."

Example 10 Compute the probability of getting exactly 3 successful missile launchings on 5 attempts if the probability of success on any one launch is $\frac{2}{3}$.

Solution We have $n = 5$, $r = 3$, $p = \frac{2}{3}$, $q = \frac{1}{3}$ (note that $q = 1 - p$). The probability of exactly 3 successes in 5 trials is

$$B(5, 3) = \binom{5}{3}\left(\frac{2}{3}\right)^3\left(\frac{1}{3}\right)^2$$

$$= \frac{5!}{3!2!} \cdot \frac{8}{27} \cdot \frac{1}{9}$$

$$= \frac{80}{243}$$

$$B(5, 3) \approx 0.329 \quad \blacksquare$$

Example 11 A student takes a test with 10 questions and guesses on each question. Suppose the probability of guessing the correct answer is 0.2 on each question (that is, there are 5 choices on each: $\frac{1}{5} = 0.2$). Compute the probability that the student answers at least 9 questions correctly.

Solution Here $n = 10$, $p = 0.2$, and $q = 0.8$. The situation is satisfied if the student answers exactly 9 correctly ($r = 9$) or exactly 10 correctly ($r = 10$).

P(answers at least 9 questions correctly)
= P(answers exactly 9 questions correctly)
    + P(answers exactly 10 questions correctly)

$$= B(10, 9) + B(10, 10) = \binom{10}{9}(0.2)^9(0.8)^1 + \binom{10}{10}(0.2)^{10}(0.8)^0$$

$$= 0.000004096 + 0.0000001024 = 0.0000041984$$

The student should study next time! $\blacksquare$

Exercises 9.3 In Exercises 1–7 list the outcomes of a sample space S for the experiment.

1. Flipping two coins once
2. Flipping a coin and throwing a die at the same time
3. Firing a rifle at a target
4. Picking a tag at random and noting its color, from a box containing red, blue, green, and orange tags
5. Selecting three parts from a manufacturing process and noting whether or not the part is defective (let D stand for defective and let N stand for nondefective)
6. Selecting one of the digits 1, 2, 3 at random, then tossing a coin, and then throwing a die
7. Throwing three dice (determine the number of outcomes and list any 10)
8. In Exercise 2 list the outcomes in the event A = "even number on the die."
9. In Exercise 5 list the outcomes in the events A = "exactly two defective parts" and B = "at least two defective parts."
10. In Exercise 7 list the outcomes in the event
    (a) A = "all three dice show the same number"
    (b) B = "exactly two of the three dice show a 6"
    (c) C = "all three dice show a 6"
11. In Exercise 8 compute $P(A)$.
12. In Exercise 10 compute $P(A)$, $P(B)$, and $P(C)$.
13. Suppose one card is selected from an ordinary deck of 52 cards. Compute the probability of selecting
    (a) a king.
    (b) an ace.
    (c) either a king or an ace.
    (d) a card that is not a king.
14. Suppose we throw a pair of dice once. Compute the probability of the event A = "at least one of the dice comes up 6."
15. Suppose we throw three dice once. Compute the probability of the event A = "at least one of the dice comes up 6." [*Hint:* compute $P(\overline{A})$.]
16. An urn contains 8 red and 2 blue tags. Two tags are to be selected in succession with replacement.
    (a) List the four outcomes in the sample space S.
    (b) Compute the probability of each of the four outcomes in S.
    (c) Compute the probability the two tags selected are the same color.
17. Three cards are selected with replacement from an ordinary deck of 52 cards. Suppose we note only whether or not the card selected is a king.
    (a) List the eight outcomes of the sample space S (let K stand for king and N for not king.)
    (b) Compute the probability of each outcome in S.
18. In Exercise 5 suppose the probability that the manufacturing process will produce a defective part is 0.1. Compute the probability of the event
    (a) A = "exactly two defective parts are selected"
    (b) B = "at least two defective parts are selected"
19. An urn contains 6 gold and 2 silver coins. Two coins are selected in succession with replacement.
    (a) List the four outcomes in the sample space S.

(b) Compute the probability of each of the four outcomes in S.

(c) Compute the probability that the two coins selected are the same color.

20. A Houston auto dealer has been troubled by new cars that are delivered late, with flawed paint, and with tires one size too small. Assume that these three events are independent and suppose that the probability of late delivery is 0.5, the probability of flawed paint is 0.02, and the probability of tires that are one size too small is 0.01. Compute the probability that the next car will be delivered late, with flawed paint, and with tires one size too small.

21. Compute the probability of getting exactly 2 successful missile launchings on 4 attempts if the probability of success on any one launch is 0.7.

22. Compute the probability of getting exactly 3 heads in 6 flips of a balanced coin.

C 23. A sharpshooter fires 7 independent shots at a target. Suppose the probability of hitting the target on any particular shot is 0.6. Compute the probability of hitting the target

(a) exactly 5 times.

(b) exactly 6 times.

(c) exactly 7 times.

(d) at least 5 times.

Key Terms and Formulas

Fundamental Counting Principle

Permutation

n factorial:

$$n! = n(n - 1)(n - 2) \cdots 2 \cdot 1$$

Number of permutations of n objects taken r at a time:

$$P(n, r) = \frac{n!}{(n - r)!}$$

Combination

Number of combinations of n objects taken r at a time:

$$\binom{n}{r} = \frac{n!}{r!(n - r)!}$$

The Binomial Theorem

Experiment

Sample space

Event

Probability

Axioms of probability

Complement of an event: $\overline{A}$

$P(\varnothing) = 0$

$$P(A \cup B) = P(A) + P(B) - P(A \cap B)$$

$P(\overline{A}) = 1 - P(A)$

Independent events

Rule of multiplication for independent events:

$$P(A \cap B) = P(A) \cdot P(B)$$

Binomial probability:

$$B(n, r) = \binom{n}{r} \cdot p^r \cdot q^{n-r}$$

Review Exercises

1. How many possible orders of finish exist for a four-horse race?
2. How many three-digit numbers can be formed from the digits 1, 2, 3, 4, and 5 if
   (a) repeated digits are allowed?
   (b) repeated digits are not allowed?
3. How many committees of five people can be chosen from a group of nine?
4. In horse racing a quinella ticket is a wager indicating the two horses which will finish in first and second place (not necessarily in the correct order). If there are eight horses in the race, how many possible quinella tickets can you buy?
5. A club consists of 10 boys and 8 girls. The club needs 10 players for a softball game. How many teams are possible if the team is to have exactly 4 boys and 6 girls?
6. Evaluate $P(10, 6)$.
7. Evaluate $\binom{10}{5}$.
8. Evaluate $\binom{8}{8}$.
9. Evaluate $P(5, 5)$.
10. In how many ways can you put five people in five seats?
11. In how many ways can you put five people in seven seats?
12. Suppose that n and k are integers with $0 \le k \le n$. Show that $\binom{n}{k} = \binom{n}{n-k}$.
13. How many possible permutations can be formed with the letters in the word MATH?
14. How many possible distinct permutations can be formed with the letters in the word TREE?
15. A bridge hand consists of 13 cards from a standard 52-card deck. How many bridge hands are possible?

In Exercises 16–24 expand the expression using the Binomial Theorem.
16. $(2a + b)^4$
17. $(3x - 4y)^4$
18. $(x - y^2)^5$
19. $(3s + t^2)^5$
20. $(3s - t^2)^4$
21. $(2x^2 - 3y)^5$
22. $(2x^2 + y^2)^6$
23. $(3x^2 - 2)^6$
24. $(3 + x)^5$
C 25. Using the Binomial Theorem, find the first four terms of $(3x + 2y)^{10}$.
26. Using the Binomial Theorem, find the first four terms of $(3x^2 - 2y)^8$.
C 27. Using the Binomial Theorem, compute $(3.83)^5 = (3 + 0.83)^5$. Round off the result to four decimal places.

C 28. If you invest $200 at 10% compounded annually, what is the amount A after 4 years?

C 29. If you invest $300 at 12% compounded annually, what is the amount A after 4 years?

30. Find the sixth term of $(x^2 + y)^{10}$.

31. Find the seventh term of $(x^2 - 3y^2)^{10}$.

In Exercises 32–35 list the outcomes for a sample space S for the experiment.

32. Flipping two coins and then picking a number from 1, 2, or 3 at random

33. Throwing a die and then flipping a coin and then selecting one tag at random from an urn containing one red tag and one blue tag

34. Selecting two tags in succession without replacement from an urn containing a red, a blue, and a green tag

35. Selecting two tags in succession with replacement from an urn containing a red, a blue, and a green tag

36. In Exercise 32 list the outcomes in the event A = "exactly two heads and an even number come up" and then compute $P(A)$.

37. In Exercise 33 list the outcomes in the event B = "an even number comes up on the die" and then compute $P(B)$.

38. Suppose one card is selected from an ordinary deck of 52 cards. Compute the probability of selecting
(a) either a heart or a diamond.
(b) a card that is not a heart.
(c) a heart or a king.

39. An urn contains 6 blue and 3 green tags. Two tags are to be selected in succession with replacement.
(a) List the outcomes of the sample space S.
(b) Compute the probability of each of the four outcomes in S.
(c) Compute the probability that the two tags selected are the same color.

C 40. Compute the probability of getting exactly 3 successful missile launchings in 5 attempts if the probability of success on any one launch is 0.8.

41. Compute the probability of getting exactly 2 heads in 5 flips of a balanced coin.

42. Compute the probability of throwing a 6
(a) exactly once in five throws of a balanced die.
C (b) at least twice in five throws of a balanced die.

Table of Common Logarithms

	0	1	2	3	4	5	6	7	8	9
1.0	0.0000	0.0043	0.0086	0.0128	0.0170	0.0212	0.0253	0.0294	0.0334	0.0374
1.1	0.0414	0.0453	0.0492	0.0531	0.0569	0.0607	0.0645	0.0682	0.0719	0.0755
1.2	0.0792	0.0828	0.0864	0.0899	0.0934	0.0969	0.1004	0.1038	0.1072	0.1106
1.3	0.1139	0.1173	0.1206	0.1239	0.1271	0.1303	0.1335	0.1367	0.1399	0.1430
1.4	0.1461	0.1492	0.1523	0.1553	0.1584	0.1614	0.1644	0.1673	0.1703	0.1732
1.5	0.1761	0.1790	0.1818	0.1847	0.1875	0.1903	0.1931	0.1959	0.1987	0.2014
1.6	0.2041	0.2068	0.2095	0.2122	0.2148	0.2175	0.2201	0.2227	0.2253	0.2279
1.7	0.2304	0.2330	0.2355	0.2380	0.2405	0.2430	0.2455	0.2480	0.2504	0.2529
1.8	0.2553	0.2577	0.2601	0.2625	0.2648	0.2672	0.2695	0.2718	0.2742	0.2765
1.9	0.2788	0.2810	0.2833	0.2856	0.2878	0.2900	0.2923	0.2945	0.2967	0.2989
2.0	0.3010	0.3032	0.3054	0.3075	0.3096	0.3118	0.3139	0.3160	0.3181	0.3201
2.1	0.3222	0.3243	0.3263	0.3284	0.3304	0.3324	0.3345	0.3365	0.3385	0.3404
2.2	0.3424	0.3444	0.3464	0.3483	0.3502	0.3522	0.3541	0.3560	0.3579	0.3598
2.3	0.3617	0.3636	0.3655	0.3674	0.3692	0.3711	0.3729	0.3747	0.3766	0.3784
2.4	0.3802	0.3820	0.3838	0.3856	0.3874	0.3892	0.3909	0.3927	0.3945	0.3962
2.5	0.3979	0.3997	0.4014	0.4031	0.4048	0.4065	0.4082	0.4099	0.4116	0.4133
2.6	0.4150	0.4166	0.4183	0.4200	0.4216	0.4232	0.4249	0.4265	0.4281	0.4298
2.7	0.4314	0.4330	0.4346	0.4362	0.4377	0.4393	0.4409	0.4425	0.4440	0.4456
2.8	0.4472	0.4487	0.4502	0.4518	0.4533	0.4548	0.4564	0.4579	0.4594	0.4609
2.9	0.4624	0.4639	0.4654	0.4669	0.4683	0.4698	0.4713	0.4728	0.4742	0.4757
3.0	0.4771	0.4786	0.4800	0.4814	0.4829	0.4843	0.4857	0.4871	0.4885	0.4900
3.1	0.4914	0.4928	0.4942	0.4955	0.4969	0.4983	0.4997	0.5011	0.5024	0.5038
3.2	0.5051	0.5065	0.5079	0.5092	0.5105	0.5119	0.5132	0.5145	0.5159	0.5172
3.3	0.5185	0.5198	0.5211	0.5224	0.5237	0.5250	0.5263	0.5276	0.5289	0.5302
3.4	0.5315	0.5328	0.5340	0.5353	0.5366	0.5378	0.5391	0.5403	0.5416	0.5428
3.5	0.5441	0.5453	0.5465	0.5478	0.5490	0.5502	0.5514	0.5527	0.5539	0.5551
3.6	0.5563	0.5575	0.5587	0.5599	0.5611	0.5623	0.5635	0.5647	0.5658	0.5670
3.7	0.5682	0.5694	0.5705	0.5717	0.5729	0.5740	0.5752	0.5763	0.5775	0.5786
3.8	0.5798	0.5809	0.5821	0.5832	0.5843	0.5855	0.5866	0.5877	0.5888	0.5899
3.9	0.5911	0.5922	0.5933	0.5944	0.5955	0.5966	0.5977	0.5988	0.5999	0.6010
4.0	0.6021	0.6031	0.6042	0.6053	0.6064	0.6075	0.6085	0.6096	0.6107	0.6117
4.1	0.6128	0.6138	0.6149	0.6159	0.6170	0.6180	0.6191	0.6201	0.6212	0.6222
4.2	0.6232	0.6243	0.6253	0.6263	0.6274	0.6284	0.6294	0.6304	0.6314	0.6325
4.3	0.6335	0.6345	0.6355	0.6365	0.6375	0.6385	0.6395	0.6405	0.6415	0.6425
4.4	0.6435	0.6444	0.6454	0.6464	0.6474	0.6484	0.6493	0.6503	0.6513	0.6522
4.5	0.6532	0.6542	0.6551	0.6561	0.6571	0.6580	0.6590	0.6599	0.6609	0.6618
4.6	0.6628	0.6637	0.6646	0.6656	0.6665	0.6675	0.6684	0.6693	0.6702	0.6712
4.7	0.6721	0.6730	0.6739	0.6749	0.6758	0.6767	0.6776	0.6785	0.6794	0.6803
4.8	0.6812	0.6821	0.6830	0.6839	0.6848	0.6857	0.6866	0.6875	0.6884	0.6893
4.9	0.6902	0.6911	0.6920	0.6928	0.6937	0.6946	0.6955	0.6964	0.6972	0.6981
5.0	0.6990	0.6998	0.7007	0.7016	0.7024	0.7033	0.7041	0.7050	0.7059	0.7067
5.1	0.7076	0.7084	0.7093	0.7101	0.7110	0.7118	0.7126	0.7135	0.7143	0.7152
5.2	0.7160	0.7168	0.7177	0.7185	0.7193	0.7202	0.7210	0.7218	0.7226	0.7235
5.3	0.7243	0.7251	0.7259	0.7267	0.7275	0.7284	0.7292	0.7300	0.7308	0.7316
5.4	0.7324	0.7332	0.7340	0.7348	0.7356	0.7364	0.7372	0.7380	0.7388	0.7396

Table of Common Logarithms (continued)

	0	1	2	3	4	5	6	7	8	9
5.5	0.7404	0.7412	0.7419	0.7427	0.7435	0.7443	0.7451	0.7459	0.7466	0.7474
5.6	0.7482	0.7490	0.7497	0.7505	0.7513	0.7520	0.7528	0.7536	0.7543	0.7551
5.7	0.7559	0.7566	0.7574	0.7582	0.7589	0.7597	0.7604	0.7612	0.7619	0.7627
5.8	0.7634	0.7642	0.7649	0.7657	0.7664	0.7672	0.7679	0.7686	0.7694	0.7701
5.9	0.7709	0.7716	0.7723	0.7731	0.7738	0.7745	0.7752	0.7760	0.7767	0.7774
6.0	0.7782	0.7789	0.7796	0.7803	0.7810	0.7818	0.7825	0.7832	0.7839	0.7846
6.1	0.7853	0.7860	0.7868	0.7875	0.7882	0.7889	0.7896	0.7903	0.7910	0.7917
6.2	0.7924	0.7931	0.7938	0.7945	0.7952	0.7959	0.7966	0.7973	0.7980	0.7986
6.3	0.7993	0.8000	0.8007	0.8014	0.8021	0.8028	0.8035	0.8041	0.8048	0.8055
6.4	0.8062	0.8069	0.8075	0.8082	0.8089	0.8096	0.8102	0.8109	0.8116	0.8122
6.5	0.8129	0.8136	0.8142	0.8149	0.8156	0.8162	0.8169	0.8176	0.8182	0.8189
6.6	0.8195	0.8202	0.8209	0.8215	0.8222	0.8228	0.8235	0.8241	0.8248	0.8254
6.7	0.8261	0.8267	0.8274	0.8280	0.8287	0.8293	0.8299	0.8306	0.8312	0.8319
6.8	0.8325	0.8331	0.8338	0.8344	0.8351	0.8357	0.8363	0.8370	0.8376	0.8382
6.9	0.8388	0.8395	0.8401	0.8407	0.8414	0.8420	0.8426	0.8432	0.8439	0.8445
7.0	0.8451	0.8457	0.8463	0.8470	0.8476	0.8482	0.8488	0.8494	0.8500	0.8506
7.1	0.8513	0.8519	0.8525	0.8531	0.8537	0.8543	0.8549	0.8555	0.8561	0.8567
7.2	0.8573	0.8579	0.8585	0.8591	0.8597	0.8603	0.8609	0.8615	0.8621	0.8627
7.3	0.8633	0.8639	0.8645	0.8651	0.8657	0.8663	0.8669	0.8675	0.8681	0.8686
7.4	0.8692	0.8698	0.8704	0.8710	0.8716	0.8722	0.8727	0.8733	0.8739	0.8745
7.5	0.8751	0.8756	0.8762	0.8768	0.8774	0.8779	0.8785	0.8791	0.8797	0.8802
7.6	0.8808	0.8814	0.8820	0.8825	0.8831	0.8837	0.8842	0.8848	0.8854	0.8859
7.7	0.8865	0.8871	0.8876	0.8882	0.8887	0.8893	0.8899	0.8904	0.8910	0.8915
7.8	0.8921	0.8927	0.8932	0.8938	0.8943	0.8949	0.8954	0.8960	0.8965	0.8971
7.9	0.8976	0.8982	0.8987	0.8993	0.8998	0.9004	0.9009	0.9015	0.9020	0.9025
8.0	0.9031	0.9036	0.9042	0.9047	0.9053	0.9058	0.9063	0.9069	0.9074	0.9079
8.1	0.9085	0.9090	0.9096	0.9101	0.9106	0.9112	0.9117	0.9122	0.9128	0.9133
8.2	0.9138	0.9143	0.9149	0.9154	0.9159	0.9165	0.9170	0.9175	0.9180	0.9186
8.3	0.9191	0.9196	0.9201	0.9206	0.9212	0.9217	0.9222	0.9227	0.9232	0.9238
8.4	0.9243	0.9248	0.9253	0.9258	0.9263	0.9269	0.9274	0.9279	0.9284	0.9289
8.5	0.9294	0.9299	0.9304	0.9309	0.9315	0.9320	0.9325	0.9330	0.9335	0.9340
8.6	0.9345	0.9350	0.9355	0.9360	0.9365	0.9370	0.9375	0.9380	0.9385	0.9390
8.7	0.9395	0.9400	0.9405	0.9410	0.9415	0.9420	0.9425	0.9430	0.9435	0.9440
8.8	0.9445	0.9450	0.9455	0.9460	0.9465	0.9469	0.9474	0.9479	0.9484	0.9489
8.9	0.9494	0.9499	0.9504	0.9509	0.9513	0.9518	0.9523	0.9528	0.9533	0.9538
9.0	0.9542	0.9547	0.9552	0.9557	0.9562	0.9566	0.9571	0.9576	0.9581	0.9586
9.1	0.9590	0.9595	0.9600	0.9605	0.9609	0.9614	0.9619	0.9624	0.9628	0.9633
9.2	0.9638	0.9643	0.9647	0.9652	0.9657	0.9661	0.9666	0.9671	0.9675	0.9680
9.3	0.9685	0.9689	0.9694	0.9699	0.9703	0.9708	0.9713	0.9717	0.9722	0.9727
9.4	0.9731	0.9736	0.9740	0.9745	0.9750	0.9754	0.9759	0.9763	0.9768	0.9773
9.5	0.9777	0.9782	0.9786	0.9791	0.9795	0.9800	0.9805	0.9809	0.9814	0.9818
9.6	0.9823	0.9827	0.9832	0.9836	0.9841	0.9845	0.9850	0.9854	0.9859	0.9863
9.7	0.9868	0.9872	0.9877	0.9881	0.9886	0.9890	0.9894	0.9899	0.9903	0.9908
9.8	0.9912	0.9917	0.9921	0.9926	0.9930	0.9934	0.9939	0.9943	0.9948	0.9952
9.9	0.9956	0.9961	0.9965	0.9969	0.9974	0.9978	0.9983	0.9987	0.9991	0.9996

Answers to Odd-Numbered Exercises

Exercises 1.1 (page 5)

1. $3 \in A$ **3.** $3 \notin B$ **5.** $z \notin A$ **7.** $1 \notin A$ **9.** $A \notin B$
11. $\{1, 2, 3, 4, 5, 6\}$ **13.** $\{7, 8, 9, 10, \ldots\}$ **15.** $\{D, E, F, G\}$
17. $\{a, b, c\} = \{b, c, a\}$ **19.** $\{x, y, z\} \neq \{x, y, z, w\}$ **21.** $X \nsubseteq Y$
23. $X \subseteq X \cup Y$ **25.** $X \nsubseteq X \cap Y$ **27.** $\varnothing \subseteq Y$ **29.** $X \subseteq X$
31. $\{1, 2, 3, 4, 5, 6\}$ **33.** $\{2, 4, 6, 7, 8, 9\}$ **35.** $\{7, 8, 9\}$
37. $\{7, 8, 9\}$ **39.** $\{1, 3, 5\}$

41.

43.

$A = \{x \mid x$ is a natural number smaller than 5$\}$

45.

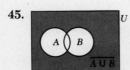

47. $\{(x, 4), (x, 5), (y, 4), (y, 5), (z, 4), (z, 5)\}$
49. $\{(2, 2), (2, 3), (3, 2), (3, 3)\}$ **51.** $A \times B = B \times A$ only when $A = B$.
53. $\{\varnothing, \{1\}, \{2\}, \{3\}, \{4\}, \{1, 2\}, \{1, 3\}, \{1, 4\}, \{2, 3\}, \{2, 4\}, \{3, 4\}, \{1, 2, 3\},$
$\{1, 2, 4\}, \{1, 3, 4\}, \{2, 3, 4\}, U\}$

Exercises 1.2 (page 12)

1. 8 (add)   **3.** -8 (add)   **5.** -5 (add)   **7.** $\frac{1}{6}$
2 (subtract) -2 (subtract)   5 (subtract)
15 (multiply)   15 (multiply)   0 (multiply)
9. $\frac{3}{2}$ **11.** $\frac{3}{8}$ **13.** 2   **15.** $\frac{1}{2}$ **17.** $-\frac{49}{10}$ **19.** $-\frac{1}{21}$
21. commutativity $(+)$ **23.** commutativity $(\cdot)$ **25.** identity $(+)$
27. distributivity   **29.** inverse $(+)$

31.   **33.** (a) $\frac{2}{3}$ (b) 3   (c) $\frac{1}{12}$ (d) $-\frac{1}{12}$

35. 2   **37.** 56   **39.** $\frac{3}{4}$

Exercises 1.3 (page 17)

1. (a) 3   (b) 8   (c) -3 **3.** (a) 4   (b) 3   (c) -1

5. (a) 5   (b) 4   (c) 3   **7.** $\frac{1}{z}$; variable in denominator not allowed.

9. $\frac{x}{y}$; variable in denominator not allowed.   **11.** $7x$

13. $17x^2y$ **15.** $\frac{17}{12}x$ **17.** $-x$ **19.** $6x^2y^3$ **21.** $8x - 5y$
23. $2x - y$ **25.** $3x^2y - xy^3 + x^3y - 2xy$ **27.** $-x^2y + \frac{5}{4}x - 8$

Exercises 1.4 (page 25)

1. $6x^3y$ **3.** $\frac{3}{4}xy^3z$ **5.** $30x^3y^5z$ **7.** $6x^2y^2 - 8xy^3$
9. $-15x^2y^2 - 10x^3y + 35x^4y^4$ **11.** $x^2 - y^2$ **13.** $x^2 - 2xy + y^2$
15. $x^2 + x - 6$ **17.** $10x^2 - 29x + 21$ **19.** $10x^2 - x - 21$
21. $x^2 + 2xy + y^2 - 5x - 5y + 6$ **23.** $x^3 - y^3$ **25.** $x^3 - 125$
27. $4x^2 - 12xy + 9y^2$ **29.** $x^4 - 16$ **31.** $(x - 3)(x + 3)$
33. $(x - 6)(x + 6)$ **35.** $(3 - 8x)(3 + 8x)$ **37.** $(x + 5)^2$
39. $(4x + y)^2$ **41.** $(x + 2)(x + 4)$ **43.** $(x - 2)(x - 5)$
45. $(x - 2)(x - 4)$ **47.** $(x - 1)(x + 7)$ **49.** $(x + 2)(x - 5)$
51. $(3x - 1)(x + 2)$ **53.** $(3x + 2)(x - 1)$ **55.** $(6x + 1)(x + 4)$
57. $(5x - 3y)(x + y)$ **59.** $(x - 1)(x + 1)(x - 2)(x + 2)$
61. $2a(x + 2y + 1)$ **63.** $3y(x + 1)^2$ **65.** $(x - 1)(x^2 + x + 1)$
67. $(x - 1)(x + 1)(x^2 + 1)(x^4 + 1)(x^8 + 1)$ **69.** $(3x - 200)(x + 50)$

Exercises 1.5 (page 30)

1. $\dfrac{1}{x + 2}$ 3. $\dfrac{1}{x}$ 5. $\dfrac{1}{x + y}$ 7. $\dfrac{x - 1}{(x + 1)(x^2 + 1)}$

9. $\dfrac{x + 5}{x}$ 11. $\dfrac{1}{x + 1}$ 13. $-\frac{1}{10}$ 15. $\dfrac{x(x - 3)}{-6y}$ 17. $\frac{1}{2}$

19. $\dfrac{x}{2}$ 21. $\dfrac{xy}{3z} - \dfrac{1}{3y}$ 23. $x + 2,\ r = -1$ 25. $-\dfrac{x}{3} + \dfrac{2}{3} - \dfrac{1}{x}$

27. $2x^3 + 6x^2 + 21x + 58,\ r = 171$ 29. $-\dfrac{2x}{y} + x - 1$

Exercises 1.6 (page 35)

1. 2 3. $x + 2$ 5. $\dfrac{2}{x}$ 7. $\dfrac{1}{2x}$ 9. $\dfrac{2x - 1}{(x - 3)(x + 2)}$

11. $\dfrac{-1}{x^2(x - 1)^2}$ 13. $\dfrac{4}{x - 1}$ 15. $\dfrac{2(x + 3)}{(x + 1)(3x - 2)}$

17. $\dfrac{2x - 3}{xy}$ 19. $\dfrac{2}{x^2}$ 21. $\dfrac{x}{x + 2}$ 23. $\dfrac{1}{y^3}$ 25. $\dfrac{x - 1}{x}$

27. $\dfrac{xy(3x + y)}{3(2x - y^2)}$ 29. $x + 2$

Exercises 1.7 (page 41)

1. y^9 3. z^4 5. x^3 7. $\dfrac{1}{y^3}$ 9. x^6 11. y^6

13. $27x^3$ 15. $27y^3$ 17. $\dfrac{9}{x^2}$ 19. x^4y^6 21. 1

23. $\dfrac{a}{2b^3}$ 25. 1 27. $\dfrac{1}{y^2}$ 29. $\dfrac{125}{x^3y^7}$ 31. $\dfrac{x^2y^3}{1 + x^2y^3}$

33. $\dfrac{x^2(1 + y^3)}{y}$ 35. $\dfrac{1}{a + 1}$ 37. 5 39. 2 41. -3

43. no real solution   45. -2 47. -2 49. 8

51. 125 53. $\frac{1}{9}$ 55. $\frac{1}{16}$ 57. no real solution

59. no real solution   61. 1 63. 1

Exercises 1.8 (page 48)

1. $\sqrt[5]{y}$ 3. $(\sqrt[4]{x + y})^3$ 5. $\sqrt[3]{x^2y^2}$ 7. $\sqrt[3]{ax + b}$

9. $(\sqrt{x + y})^5$ 11. $x^{1/2}$ 13. $x^{1/3}$ 15. $y^{4/3}$

17. $(2x + 1)^{3/4}$ **19.** $(x^2y)^{1/3}$ **21.** $10x$ **23.** $4x\sqrt{y}$

25. $5ab$ **27.** $3x\sqrt{y}$ **29.** $8x$ **31.** $-13 - 2\sqrt{10}$

33. $3x + 2\sqrt{xy} - 8y$ **35.** $x^2y + 2xy\sqrt{xy} + xy^2$ **37.** $\dfrac{\sqrt{10}}{5}$

39. $\dfrac{\sqrt{10} - 5\sqrt{5}}{-23}$ **41.** $-\frac{4}{3}$ **43.** $\dfrac{x - 2\sqrt{xy}}{x - 4y}$ **45.** $\frac{1}{3}$

Exercises 1.9 (page 53)

1. real part = 1   **3.** real part = 5
imaginary part = $1i$ imaginary part = 0

5. real part = $\frac{5}{3}$ **7.** real part = x
imaginary part = $\frac{2}{3}i$ imaginary part = yi

9. real part = $\sqrt{7}$ **11.** $7i$ **13.** $\frac{1}{2}i$ **15.** $2\sqrt{6}i$
imaginary part = wi

17. $3cd^2\sqrt{5di}$ **19.** $-2x^4i$ **21.** $7 + 2i$ **23.** $13 + 13i$

25. $-5 - 12i$ **27.** $5 + 8i$ **29.** $\frac{1}{2} - i$ **31.** $2 - 7i$

33. $1 - 4i$ **35.** $2 + x^2$ **37.** $i^3 = -i$ **39.** 0
                                       $i^4 = 1$
                                       $i^5 = i$

Review Exercises, Chapter 1 (page 55)

1. (a) $2 \in X$ (b) $c \notin X$ (c) $Y \subseteq X$ (d) $X \nsubseteq Y$

3. (a) $\{2, 4, 5\}$ (b) $\{4\}$ (c) $\{1, 3, 5\}$ (d) $\{1, 2, 3, 5\}$ (e) $\{4, 5\}$

5. $\{\{x, y\}, \{x\}, \{y\}, \varnothing\}$ **7.** 1   **9.** 0   **11.** $-\frac{17}{60}$ **13.** $\frac{1}{4}$

15. (a) 4   (b) 4   (c) $-\frac{2}{3}$ (d) 1   **17.** $3x^2y - 3x^3y$

19. $-6x^4y + 2x^2y^2 - 3x^3y^3 + xy^4$ **21.** $3x$ **23.** $-\frac{1}{12}xy^2 + \frac{3}{14}x^2y$

25. $(x + 2)(x + 5)$ **27.** $x(x + 15)$ **29.** $(3x - 1)(2x + 5)$

31. $x^2(x + 4)(x^2 - 4x + 16)$ **33.** $(9x - 4y)^2$

35. $3x(x - 1)(x + 1)(x - 3)(x + 3)$ **37.** $\dfrac{x + 4}{x}$ **39.** $\dfrac{x + 4}{2x + 1}$

41. $\dfrac{x^2 - x + 3}{(x - 1)(x - 2)(x + 1)}$ **43.** $\dfrac{-x + 12}{(x - 2)(x + 3)}$

45. $\dfrac{(-x - 3)(x + 2)}{(x - 1)(x^2 + 5x + 3)}$ **47.** $\dfrac{x^3}{y^5}$ **49.** $\dfrac{y^2}{x(1 + xy^2)}$ **51.** $\dfrac{1}{5y^3}$

53. $-6x$ **55.** $\dfrac{\sqrt{6x}}{3x}$ **57.** $\dfrac{\sqrt{(x + 1)(x - 1)}}{x - 1}$ **59.** $-5 - 2\sqrt{6}$

61. $9x^2$ **63.** $3xy\sqrt{3}$ **65.** $4x^2y\sqrt[3]{y}$

67. $7\sqrt{2} - \sqrt{14} + 2\sqrt{7} - 2$

69. $6x + 2\sqrt[3]{xy}$ **71.** $5x\sqrt{yi}$ **73.** $-4x^8i$ **75.** $\dfrac{\sqrt{2}}{3}x^2i$

77. $1 - i$ **79.** $8i$ **81.** $-7 + 6\sqrt{2}i$ **83.** $\frac{3}{2} + \frac{3}{4}i$ **85.** $x^2 + 7$

Exercises 2.1 (page 67)

3. $x = 4$ **5.** $x = -7$ **7.** $z = -3$ **9.** $x = -2$
11. $x = 4$ **13.** $x = 4$ **15.** $w = -2$ **17.** 0 **19.** $\frac{3}{2}$

21. $x = 2$ **23.** $y = -1$ **25.** $x = -1$ **27.** $x = \dfrac{7 - 3y}{2}$

29. $T = \dfrac{I}{PR}$ **31.** $r = \dfrac{d}{t}$ **33.** $x = \dfrac{S - 2yz}{2y + 2z}$

35. [number line: open circle at -2, $-3\,-2\,-1\,0\,1\,2$] x **37.** [number line: $-2\,-1\,0\,1\,2$] x

39. $x \le -7$; [number line shaded left from -7, $-9\,-8\,-7\,-6\,-5\,-4\,-3\,-2\,-1\,0\,1\,2$] x

41. $x \le -24$; [number line shaded left from -24, $-25\,-24\,-23\,-22\,-21$] x

43. $-4 < x < 2$; [number line open circles at -4 and 2, $-4\,-3\,-2\,-1\,0\,1\,2\,3$] x

45. $4 \ge x \ge -4$; [number line closed circles at -4 and 4, $-4\,-3\,-2\,-1\,0\,1\,2\,3\,4$] x **47.** $-2, 3$ **49.** 4

51. $x < 1$ or $x > 2$; [number line open circles at 1 and 2, $0\,1\,2\,3\,4$] x

53. $x \le -4$ or $x \ge 16$; [number line $-6\,-4\,-2\,0\,2\,4\,6\,8\,10\,12\,14\,16\,18$] x

55. $|x + 2| > 6$; [number line open circles, $-9\,-8\,-7\,-6\,-5\,-4\,-3\,-2\,-1\,0\,1\,2\,3\,4\,5\,6\,7$] x

Exercises 2.2 (page 76)

1. $x + 12$ **3.** $(x + 3)^2$ **5.** $4x$ **7.** $\dfrac{69 + 78 + x}{3}$ **9.** $25x$

11. 93   **13.** 212, 424, 848   **15.** 6, 8, 10 inches
17. 4 feet high, 6 feet wide, 10 feet long
19. The police car catches up with the truck just as it arrives at the department store.
21. 490 miles per hour, 1820 miles   **23.** 215 dimes, 174 quarters
25. 12 roses, 5 pieces of greenery   **27.** 114 milliliters
29. $4\frac{1}{2}$ hours, 9 hours   **31.** 12.7 centimeters   **33.** 26.4 pounds
35. 1.58 liters   **37.** 1.65 meters   **39.** 0.23 ounces

41. 1.61 pounds   **43.** $y = 3x$, $y = 15$ **45.** $y = \dfrac{800}{d}$, $y = 32$

47. $z = 16x^2$, $z = \frac{1}{4}$ **49.** $x = \dfrac{6s}{t}$, $x = 14$ **51.** $y = \dfrac{2m}{d^2}$, $y = \frac{2}{9}$

Exercises 2.3 (page 88)

1.

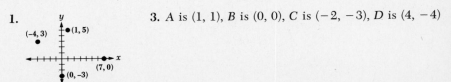

3. A is $(1, 1)$, B is $(0, 0)$, C is $(-2, -3)$, D is $(4, -4)$

5.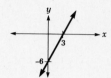

7.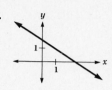

9.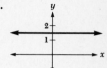

11.

13.

15. $m = 2$, $b = -6$; $m = -\frac{2}{3}$, $b = \frac{5}{3}$; $m = 0$, $b = \frac{3}{2}$; m is undefined, no b; $m = \frac{4}{3}$, $b = 4$

17. $y = 4x$

19. $y = -\frac{1}{2}$

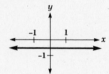

21. $y = \frac{2}{3}x + 3$

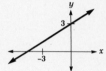

23. $y = 3x - 1$

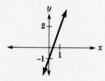

25. $y = -2x + 3$

27. $y = \frac{5}{9}x + 2$

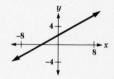

29. $y = -0.1x - 4.7$

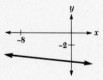

31. $y = -\frac{1}{2}x$

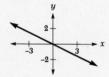

33. $y = -4x + 11$

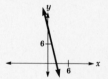

35. $y = \frac{7}{2}x + 5$

37. $y = \frac{1}{4}x + 6\frac{3}{4}$

39. $y = -x + 2$

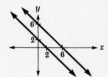

41. $y = -\frac{5}{2}x + 3$

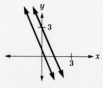

43. $y = x - 2$

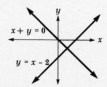

45. $y = -\frac{1}{3}x + 4$

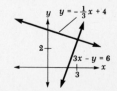

47. $6\frac{2}{3}$ tons   **49.** $y = -\frac{2}{5}x + 200$

Exercises 2.4 (page 95)

1. domain = $\{-7, -3, 5\}$,
range = $\{-1, 0, 2, 6\}$

3. domain = $\{-4, 1, 2\}$,
range = $\{0, 3, 4, 9\}$

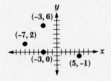

5. domain = $\{0, 1, 2, 3, 4\}$,
range = $\{0, 1, 4, 9, 16\}$

7. domain = $\{3\}$,
range = $\{0, 1, 2, 3, 4\}$

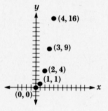

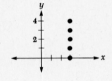

9. 2, 5, 6, 8
11. $f(-2) = 3, f(-1) = 4,$
$f(0) = 5, f(1) = 6$

13. $g(-2) = -1, g(1) = 0,$
$g(4) = 1, g(7) = 2$

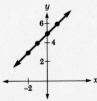

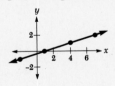

15. $h(-2) = -8, h(-1) = -1, h(0) = 0,$
$h(1) = 1, h(2) = 8$

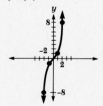

17. $g(-3) = -47, g(0) = 7, g(z) = 2z^3 + 7$
19. $f(-1) = -3, f(3) = 2\frac{1}{3}, f(x + 1) = \dfrac{5x + 4}{x + 4}$

21. function   **23.** function   **25.** not a function   **27.** function

29. $(f + g)(x) = x^2 + x + 3$, $(f - g)(x) = -x^2 + x + 1$,

$(fg)(x) = x^3 + 2x^2 + x + 2$, $\left(\dfrac{f}{g}\right)(x) = \dfrac{x + 2}{x^2 + 1}$, $(f \circ g)(x) = x^2 + 3$

31. $(f + g)(x) = (x - 1)^2 + \dfrac{1}{x}$, $(f - g)(x) = (x - 1)^2 - \dfrac{1}{x}$, $(fg)(x) = \dfrac{(x - 1)^2}{x}$,

$\left(\dfrac{f}{g}\right)(x) = x(x - 1)^2$, $(f \circ g)(x) = \left(\dfrac{1}{x} - 1\right)^2$

33. $(f + g)(x) = x^3 + 3x^2 + 4x - 8$, $(f - g)(x) = x^3 - 3x^2 + 6x - 6$,

$(fg)(x) = 3x^5 - x^4 + 14x^3 - 26x^2 + 2x + 7$, $\left(\dfrac{f}{g}\right)(x) = \dfrac{x^3 + 5x - 7}{3x^2 - x - 1}$,

$(f \circ g)(x) = 27x^6 - 27x^5 - 18x^4 + 17x^3 + 21x^2 - 8x - 13$

35. $(g + f)(x) = \dfrac{1}{x} + (x - 1)^2$, $(g - f)(x) = \dfrac{1}{x} - (x - 1)^2$, $(gf)(x) = \dfrac{(x - 1)^2}{x}$,

$\left(\dfrac{g}{f}\right)(x) = \dfrac{1}{x(x - 1)^2}$, $(g \circ f)(x) = \dfrac{1}{(x - 1)^2}$

37. $10x + 5h$ **39.** 4   **41.** 22   **43.** 20

45. $R(x) = 110x$ **47.** $\{x \mid x \geq 0$ and x is an integer$\}$

Exercises 2.5 (page 101)

1. 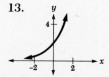 **3.** 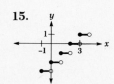 **5.**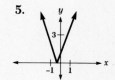

7. 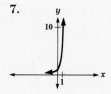 **9.** 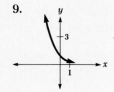 **11.**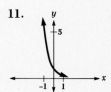

13. 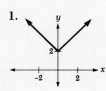 **15.** 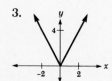 **17.**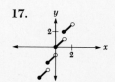

19. **21.** **23.** odd

25. even   **27.** neither   **29.** neither

Exercises 2.6 (page 106)

1. {(1, 0), (2, 1), (3, 2), (4, 3)}, inverse function
3. {(1, −1), (0, 0), (1, 1), (4, 2)}
5. {(3, −3), (1, −1), (1, 1), (3, 3), (5, 5)}
7. {(2, 4), (3, 9), (4, 16), (5, 25)}, inverse function
9. $f^{-1}(x) = x - 2$, inverse function   **11.** $f^{-1}(x) = \sqrt[3]{x}$, inverse function
13. $f^{-1}(x) = -\frac{1}{3}x$, inverse function
15. $f^{-1}(x) = x^2 + 2$, inverse function   **17.** $x = (y + 1)^2$
19. not one-to-one   **21.** one-to-one   **23.** one-to-one
25. not one-to-one
29. 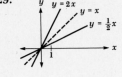   **31.** $x \le 0; x \ge 0$ **33.** $x \le 1; x \ge 1$

Review Exercises, Chapter 2 (page 109)

3. $y = 4$ **5.** no solution   **7.** $R = \dfrac{ST}{S + T}$

9. $x \le 9$;

11. $-6 < x < 3$;

13. Each side is 700 feet.   **15.** 50 gallons   **17.** 669 ounces
19. $A = 4s, A = 6$
21. 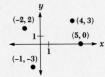   **23.**

25. $m = \frac{1}{3}, b = -2$ **27.** $y = 3x + 4$ **29.** $y = 3x - 17$
31. 675 tons   **33.**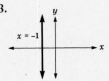

35. $g(-2) = -8, g(-1) = -1, g(0) = 0, g(1) = 1, g(2) = 8$

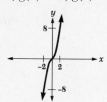

37. (a) function   (b) not a function

39. $(g + f)(x) = x^2 + 2x, (g - f)(x) = x^2 - 2x - 2,$

$(gf)(x) = 2x^3 + x^2 - 2x - 1, \dfrac{g}{f}(x) = \dfrac{x^2 - 1}{2x + 1}, (g \circ f)(x) = 4x^2 + 4x$

41. $\frac{11}{3}$ **43.**   **45.**

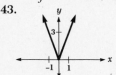

47. even   **49.** $f^{-1}(x) = \frac{1}{2}x + \frac{7}{2}$; yes   **51.** $-2 \le x \le 0, 0 \le x \le 3$

Exercises 3.1 (page 125)

1. independent   **3.** inconsistent   **5.** dependent
7. independent   **9.** inconsistent   **11.** dependent
13. (3, 1)   **15.** ($\frac{1}{2}$, 2)   **17.** (3, 0)   **19.** $(-\frac{1}{2}, \frac{5}{2})$

21. inconsistent   **23.** dependent   **25.** (3, 1)   **27.** $\left(\dfrac{1}{4}, \dfrac{1}{2}\right)$

29. (3, 2)   **31.** inconsistent   **33.** dependent   **35.** (1, −5)
37. $(\frac{3}{2}, \frac{3}{4})$ **39.** 15, 3   **41.** width = 3, length = 8
43. \$3000 at 6%, \$2000 at 8%   **45.** noise level = 3, test items = 5
47. 50 dimes, 100 quarters   **49.** $(\frac{14}{3}, -\frac{11}{3}, -\frac{4}{3})$ **51.** (3, 2, 1)
53. (7, −6, 5)   **55.** $(7 + c, 3 + 2c, c), (8, 5, 1), (7, 3, 0)$

Exercises 3.2 (page 136)

1.   **3.**   **5.**

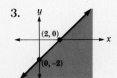

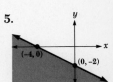

7.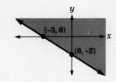

9.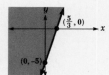

11.

13.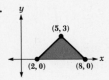

15.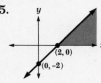

17.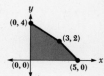

19. maximum $z = 6$ at $(2, 4)$ **21.** maximum $z = 20$ at $(\frac{10}{3}, 0)$

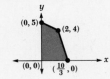

23. maximum $z = 20$ at $(0, 5)$ **25.** maximum $z = 26$ at $(6, 10)$
27. maximum $z = 420$ at $(6, 10)$ **29.** minimum $z = 8$ at $(0, 8)$
31. maximum revenue = \$60 for 30 A's and 0 B's
33. minimum cost = \$100 for 0 A's and 100 B's or 100 A's and 0 B's or any point on the line connecting $(0, 100)$ and $(100, 0)$
35. maximum profit = \$116 for $x = 17$ plain legs and $y = 13$ carved legs
37. minimum cost = \$3.50 for $x = 7$ units of vegetables and $y = 0$ units of meat
39. maximum revenue = \$52 for $x = 4$ A's and $y = 8$ B's
41. maximum revenue = \$1080 for $x = 280$ bags of N-Rich and $y = 120$ bags of P-Rich.

Review Exercises, Chapter 3 (page 138)

1. independent   **3.** inconsistent   **5.** independent
7. $(1, 0)$ **9.** inconsistent   **11.** $(-7, -3)$ **13.** $(1, 1)$
15. $(1, 0)$ **17.** inconsistent   **19.** $(20, 6)$ **21.** $(1, 1)$
23. $(10, 10)$ **25.** $(3, -1, 2)$ **27.** $(-3, 5, -2)$ **29.** $y = 2x + 3$
31. 8 cubic centimeters of 28% solution, 72 cubic centimeters of 8% solution
33. \$6000 at 15%, \$14,000 at 12%
35. maximum $z = 20$ at $(2, 2)$, $(4, 0)$, or any point on the boundary connecting $(2, 2)$ and $(4, 0)$
37. maximum $z = 42$ at $(2, 2)$ **39.** minimum $z = 13$ at $(2, 3)$
41. minimum $z = 6$ at $(0, 6)$

Exercises 4.1 (page 151)

1. $\{-3, -4\}$ **3.** $\{-1, -3\}$ **5.** $\{7, 8\}$ **7.** $\{-3, 6\}$
9. $\{3, -6\}$ **11.** $\{4\}$ **13.** $\{0, 4\}$ **15.** $\{0, -\frac{7}{2}\}$
17. $\{-3, -\frac{2}{3}\}$ **19.** $\{3, \frac{2}{3}\}$ **21.** $\{-2, \frac{1}{5}\}$ **23.** $\{\pm 5\}$
25. $\{\pm \frac{1}{2}\}$ **27.** $\{0, \frac{3}{2}\}$ **29.** $\{5, -1\}$ **31.** $\{1, -5\}$

33. $\{1, -\frac{3}{2}\}$ **35.** $\{2, -\frac{1}{2}\}$ **37.** $\left\{\dfrac{1}{2} \pm \dfrac{\sqrt{3}}{2} i\right\}$ **39.** $\{\pm \sqrt{7}\}$

41. real, not equal; $\{5, 3\}$ **43.** real, not equal; $\{-4, \frac{3}{2}\}$
45. real, not equal; $\{-\frac{2}{3}, -3\}$ **47.** real, not equal; $\{-1 \pm \sqrt{2}\}$

49. real, not equal; $\left\{\dfrac{-3 \pm \sqrt{17}}{4}\right\}$ **51.** imaginary; $\left\{-\dfrac{1}{2} \pm \dfrac{\sqrt{3}}{2} i\right\}$

53. x-intercepts $5, -1$;   **55.** x-intercepts $3, -2$;
y-intercept -5; y-intercept -6;
vertex $(2, -9)$ vertex $(\frac{1}{2}, -6\frac{1}{4})$

   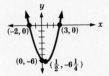

57. x-intercepts $-1, 5$;   **59.** no x-intercepts;
y-intercept 5; y-intercept 4;
vertex $(2, 9)$ vertex $(-\frac{1}{2}, 3\frac{3}{4})$

   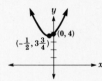

61. $\frac{1}{2}x(x + 4) = 6$; height $= x = 2$; base $= x + 4 = 6$
63. $x(x + 4) = 21$; width $= x = 3$; length $= x + 4 = 7$
65. $x^2 + (x + 1)^2 = 25$; $\{3, 4\}$ **67.** $x(10 - x) = 24$; $\{4, 6\}$
69. $x^2 + x^2 = (x + 1)^2$; $\{1 + \sqrt{2}\}$ **71.** $\frac{1}{2}x(x - 3) = 9$; $\{6\}$
73. $\frac{1}{2}x(x - 3) = 12$; no integer solution means no such polygon exists.
75. Minimum cost $= 4$ occurs at the vertex.

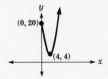

Exercises 4.2 (page 157)

1. $\{9\}$ **3.** $\{16\}$ **5.** $\{169\}$ **7.** $\{1\}$ **9.** $\{4\}$ **11.** $\{4\}$
13. $\{8, -1\}$ **15.** $\{5, 21\}$ **17.** $\{9\}$ **19.** $\{4, 1\}$ **21.** $\{9\}$

23. $\{\pm 1, \pm 2\}$ **25.** $\{\pm 1, \pm 2i\}$ **27.** $\{\pm i, \pm 2i\}$ **29.** $\{16, 1\}$
31. $\{2, 1\}$ **33.** $\{\frac{1}{2}, 1\}$ **35.** $\{\frac{1}{2}\}$ **37.** $\{2\}$ **39.** $\{\frac{9}{2}, 2\}$
41. $\{16\}$ **43.** $\{1, -\frac{10}{9}\}$ **45.** $\{\pm 3, \pm 3i\}$ **47.** $\{8, 12\}$

Exercises 4.3 (page 161)

1. $\{x \mid x \le -1 \text{ or } x \ge 1\}$ **3.** $\{x \mid x < -1 \text{ or } x > 1\}$
5. all real numbers   **7.** $\{x \mid x \ne 0\}$ **9.** all real numbers
11. all real numbers   **13.** $\{x \mid -2 \le x \le 2\}$
15. $\{x \mid x \le -3 \text{ or } x \ge 3\}$ **17.** $\{x \mid x < -4 \text{ or } x > 4\}$
19. $\{x \mid -5 < x < 5\}$ **21.** $\{x \mid -2 \le x \le 3\}$ **23.** $\{x \mid -5 \le x \le 4\}$
25. $\{x \mid -5 < x < 4\}$ **27.** $\{x \mid -5 \le x \le 2\}$
29. $\{x \mid x \le -5 \text{ or } x \ge 2\}$ **31.** all real numbers
33. no solution   **35.** $\{x \mid x < -1 \text{ or } x > \frac{3}{2}\}$ **37.** $\{x \mid x < -3 \text{ or } x > 2\}$
39. $\{x \mid x < 3 - \sqrt{3} \text{ or } x > 3 + \sqrt{3}\}$ **41.** $\{x \mid 200 < x < 800\}$
43. $\{x \mid 100 < x < 500\}$
45. (a) $y = 96t - 16t^2$ (b) $\{t \mid 0 \le t \le 6\}$ (c) $t = 3$ seconds

Exercises 4.4 (page 175)

1. $x^2 + y^2 = 25$ **3.** $x^2 + (y - 2)^2 = 16$

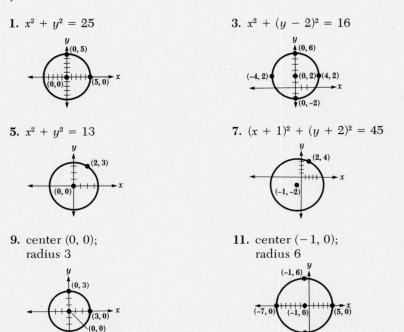

5. $x^2 + y^2 = 13$ **7.** $(x + 1)^2 + (y + 2)^2 = 45$

9. center $(0, 0)$;   **11.** center $(-1, 0)$;
    radius 3   radius 6

13. center $(-3, 7)$;
radius 2

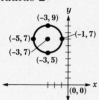

15. center $(1, \frac{5}{4})$;
radius $\frac{5}{4}$

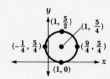

17. $(y - 2)^2 = 4(x + 1)$

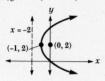

19. $(x - 2)^2 = -6(y - \frac{5}{2})$

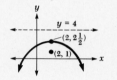

21. $(x + 2)^2 = -8(y - 3)$

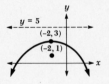

23. $x^2 = -8y$

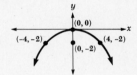

25. vertex $(0, 0)$;
focus $(0, 2)$

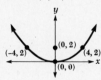

27. vertex $(-\frac{1}{2}, 0)$;
focus $(1\frac{1}{2}, 0)$

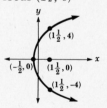

29. vertex $(-2, 2)$;
focus $(-\frac{3}{2}, 2)$

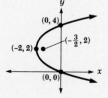

31. vertex $(\frac{1}{2}, \frac{1}{4})$;
focus $(\frac{1}{2}, \frac{3}{4})$

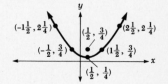

33. $\dfrac{x^2}{16} + \dfrac{y^2}{25} = 1$

35. $\dfrac{x^2}{36} + \dfrac{y^2}{20} = 1$

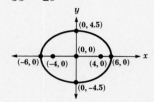

37. $\dfrac{x^2}{4} + \dfrac{y^2}{16} = 1$

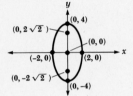

39. $\dfrac{x^2}{81} + \dfrac{y^2}{49} = 1$

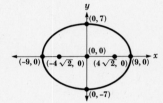

41. $\dfrac{(x-2)^2}{25} + \dfrac{(y-3)^2}{9} = 1$

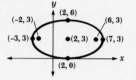

43. center $(0, 0)$;
vertices $(\pm 3, 0)$, $(0, \pm 2)$;
foci $(\pm\sqrt{5}, 0)$

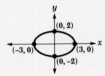

45. center $(0, 0)$;
vertices $(\pm 1, 0)$, $(0, \pm\frac{1}{2})$;
foci $\left(\pm\dfrac{\sqrt{3}}{2}, 0\right)$

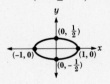

47. center $(0, 0)$;
vertices $(\pm 2, 0)$, $(0, \pm\sqrt{3})$;
foci $(\pm 1, 0)$

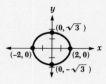

49. center $(-1, 0)$;
vertices $(-7, 0)$, $(5, 0)$, $(-1, 2)$, $(-1, -2)$;
foci $(-1 - 4\sqrt{2}, 0)$, $(-1 + 4\sqrt{2}, 0)$

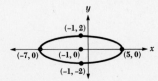

51. center $(1, -2)$;
vertices $(-1, -2), (3, -2), (1, 1), (1, -5)$;
foci $(1, -2 + \sqrt{5}), (1, -2 - \sqrt{5})$

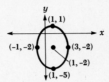

53. $\dfrac{x^2}{4} - \dfrac{y^2}{5} = 1$

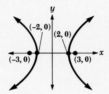

55. $\dfrac{y^2}{25} - \dfrac{x^2}{16} = 1$

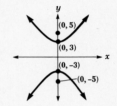

57. $\dfrac{(x - 1)^2}{16} - \dfrac{(y - 2)^2}{9} = 1$

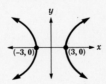

59. $\dfrac{(x - 2)^2}{16} - \dfrac{(y - 2)^2}{9} = 1$

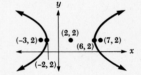

61. center $(0, 0)$;
foci $(\pm 3\sqrt{2}, 0)$;
vertices $(\pm 3, 0)$

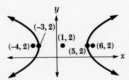

63. center $(0, 0)$;
foci $(0, \pm 5\sqrt{2})$;
vertices $(0, \pm 5)$

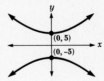

65. center $(0, 0)$;
foci $(\pm \sqrt{13}, 0)$;
vertices $(\pm 2, 0)$

67. center $(1, -2)$;
foci $(1, -2 + \sqrt{13})$, $(1, -2 - \sqrt{13})$;
vertices $(1, 0)$, $(1, -4)$

69.

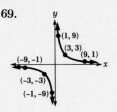

71. hyperbola   **73.** parabola

75. parabola; $y^2 = -(x - 4)$

77. hyperbola; $\dfrac{x^2}{4} - \dfrac{y^2}{4} = 1$

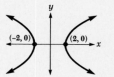

79. ellipse; $\dfrac{x^2}{4} + \dfrac{y^2}{1} = 1$

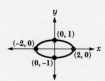

81. ellipse; $\dfrac{(x - 1)^2}{4} + \dfrac{(y - 2)^2}{36} = 1$

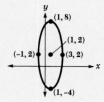

83. circle; $(x - 1)^2 + (y - 2)^2 = 6^2$

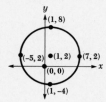

Exercises 4.5 (page 184)

1. $\{(\frac{5}{2}, -\frac{3}{2})\}$ **3.** $\{(2, 0), (\frac{6}{5}, \frac{4}{5})\}$ **5.** $\{(-1, 2), (\frac{1}{2}, \frac{5}{4})\}$
7. no real solutions   **9.** $\{(0, 3), (2, 5)\}$ **11.** $\{(2, 2)\}$
13. $\{(3, 3), (-3, -3)\}$ **15.** no real solutions
17. $\{(\pm 2, 0), (\pm\sqrt{3}, -1)\}$ **19.** $\{(\pm 2, 0)\}$ **21.** $\{(\pm\sqrt{2}, \pm\sqrt{3})\}$
23. $\{(0, \pm 2)\}$ **25.** no real solutions   **27.** $\{(\pm\frac{1}{2}, \frac{1}{3})\}$
29. $\{6, 8\}$ **31.** $\{9, 11\}$ **33.** $\{4, 8\}$

Review Exercises, Chapter 4 (page 186)

1. $\{3, -10\}$ **3.** $\{0, 9\}$ **5.** $\{\frac{3}{2}, -4\}$ **7.** $\{\pm\sqrt{5}\}$

9. $\left\{-\dfrac{1}{2} \pm \dfrac{\sqrt{5}}{2}\right\}$ **11.** $\{0, \frac{4}{3}\}$ **13.** $\left\{\pm \dfrac{\sqrt{15}}{3} i\right\}$ **15.** $\{-4 \pm \sqrt{2}i\}$

17. $\{0, \frac{7}{4}\}$

19. x-intercepts $-3, 5$;   **21.** x-intercepts $\pm\sqrt{2}$;
   y-intercept -15; y-intercept 8;
   vertex $(1, -16)$ vertex $(0, 8)$

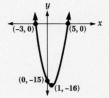   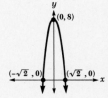

23. $\frac{1}{2}x(x + 5) = 18$; $\{4, 9\}$ **25.** $x^2 + (x + 1)^2 = 145$; $\{8, 9\}$

27. $\{\frac{25}{4}\}$ **29.** no solution   **31.** $\{3\}$ **33.** $\{0, -\frac{5}{2}\}$

35. $\{5\}$ **37.** $\{\pm\frac{1}{2}, \pm\frac{1}{2}i\}$ **39.** $\{x \mid x \leq -2 \text{ or } x \geq 2\}$

41. $\{x \mid x < -1 \text{ or } x > 8\}$ **43.** $\{x \mid -5 \leq x \leq \frac{1}{2}\}$

45. $\{x \mid 1.2 < x < 6.8\}$

47. $x^2 + y^2 = 10^2$ **49.** $(x - 1)^2 + (y - 1)^2 = (\sqrt{10})^2$

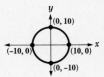   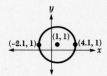

51. $x^2 = 8(y + 1)$ **53.** $(x - 1)^2 = 24(y + 3)$

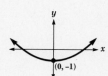   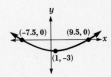

55. $\dfrac{x^2}{9} + \dfrac{y^2}{5} = 1$ **57.** $\dfrac{(x - 1)^2}{9} + \dfrac{(y - 3)^2}{25} = 1$

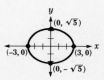   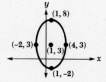

59. $\dfrac{y^2}{4} - \dfrac{x^2}{5} = 1$

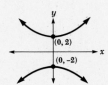

61. $\dfrac{(x + 1)^2}{4} - \dfrac{(y - 2)^2}{5} = 1$

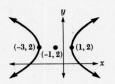

63. parabola; $x^2 = 16y$

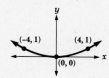

65. ellipse; $\dfrac{x^2}{25} + \dfrac{y^2}{\frac{25}{16}} = 1$

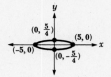

67. parabola; $(y + \frac{1}{2})^2 = 2(x + \frac{1}{2})$

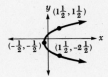

69. ellipse; $\dfrac{(x + 1)^2}{3} + \dfrac{(y - 2)^2}{2} =$

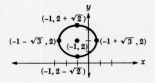

71. $\{(3, 4), (0, 5)\}$ **73.** $\{(3, 3), (-3, -3)\}$ **75.** $\{8, 15\}$

Exercises 5.1 (page 192)

1. -4 **3.** 4   **5.** 8   **7.** 0   **9.** 0

21. quotient $= x^3 + 4x^2 - x + 1$, remainder $= -4$

23. quotient $= 3x^2 + 4x + 15$, remainder $= 27$

25. quotient $= x^2 + 3x + 9$, remainder $= -3$

27. $x^4 - 5x^2 + 4$ **29.** $x^5 + 3x^4 - 6x^3 - 6x^2 + 8x$

Exercises 5.2 (page 196)

1. $3x^2 - x + 6$ with remainder -3

3. $x^3 - x^2 + 2x - 9$ with remainder 28

5. $2x^2 + 8x - 2$ with remainder 3

7. $x^2 - 2x + 4$ with remainder 0

9. $4x^6 + 8x^5 - 4x^4 - 8x^3 - 16x^2 + 2x + 4$ with remainder -12
11. $18x^4 - 6x^3 + 2x^2 - \frac{8}{3}x - \frac{1}{9}$ with remainder $\frac{28}{27}$
13. $\frac{33}{4}$, $\frac{13}{4}$ **15.** 4.9701, 9   **17.** 115, 318

Exercises 5.3 (page 202)

1. $x^4 - 2x^3 - 3x^2 + 4x + 4$ **3.** $x^6 + 5x^5 + 6x^4 - 4x^3 - 8x^2$
5. 2 is a zero of multiplicity 3, and -1 is a zero of multiplicity 5.
7. 0 is a zero of multiplicity 4, ± 2 are zeros each having multiplicity 2, and
1 is a zero of multiplicity 1.
9. $5i$ and $-5i$ are zeros of multiplicity 1, and $\frac{3}{2}$ is a zero of multiplicity 3.
11. $x^3 - 4x^2 + x + 6$ **13.** $x^4 - 4x^3 + 19x^2 - 30x + 50$
15. $x^4 + 6x^2 - 27$ **17.** $x^5 - 6x^4 + 38x^3 - 94x^2 + 221x$
19. $-2 \pm \sqrt{3}$ **21.** $1 \pm 2i$ **23.** $-1, \frac{1}{2}, -2$ **25.** $\pm\frac{1}{2}, \frac{1}{3}$
27. $\frac{1}{2}, -2, -2 \pm i$ **29.** $-2, -2, -1, \frac{1}{3}$

Exercises 5.4 (page 207)

7. 3 positive and 1 negative root; or
1 positive, 1 negative, and 2 complex roots
9. 1 positive and 2 negative roots; or
1 positive, no negative, and 2 complex roots
11. no positive, 3 negative, and 2 complex roots; or
no positive, 1 negative, and 4 complex roots
13. 2 positive and 3 negative roots; or
2 positive, 1 negative, and 2 complex roots; or
no positive, 3 negative, and 2 complex roots; or
no positive, 1 negative, and 4 complex roots
15. 3 positive and 2 negative roots; or
3 positive, no negative, and 2 complex roots; or
1 positive, 2 negative, and 2 complex roots; or
1 positive, no negative, and 4 complex roots
17. no positive, no negative, and 4 complex roots
19. 6 positive and no negative roots; or
4 positive, no negative, and 2 complex roots; or
2 positive, no negative, and 4 complex roots; or
no positive, no negative, and 6 complex roots

21.   **23.**   **25.**

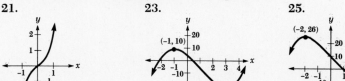

27.

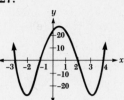

29.

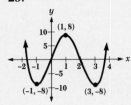

31.

Exercises 5.5 (page 213)

1.

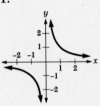

3.

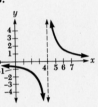

5.

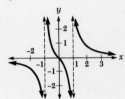

7.

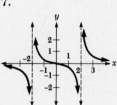

9.

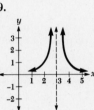

11.

13.

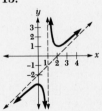

15.

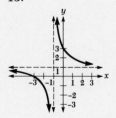

17.

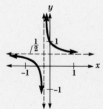

19.

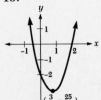

21.

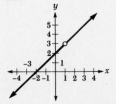

Review Exercises, Chapter 5 (page 214)

1. (a) -2 (b) -9 (c) -1 (d) 14 (e) 54.705 **3.** $x^3 - 2x^2 - x + 2$
5. quotient $= 2x^2 + 5x + 10$, remainder $= 40$
7. quotient $= 4x^6 + 4x^5 + 4x^4 + 4x^3 + 4x^2 + x + 3$, remainder $= -2$
9. quotient $= 2x^2 + 8x - 4$, remainder $= 4$
11. 72 **13.** $x^5 + 2x^4 + x^3$ **15.** $x^4 - 2x^3 + 6x^2 - 8x + 8$
17. $1, -1, 2, -2, 3$

21.   **23.**   **25.**

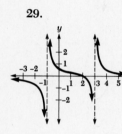

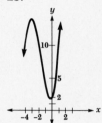

27.   **29.**   **31.**

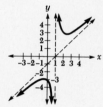

37. true   **39.** false

Exercises 6.1 (page 222)

1. $64 = 8^2$ **3.** $\frac{1}{2} = 2^{-1}$ **5.** $1,000,000 = 10^6$ **7.** $\frac{1}{8} = (\frac{1}{2})^3$
9. $5 = \log_2 32$ **11.** $\frac{1}{3} = \log_{27} 3$ **13.** $4 = \log_{10} 10,000$
15. $-4 = \log_{1/2} 16$
17.   **19.**   **21.**   **23.**

25. **27.** **29.** **31.**

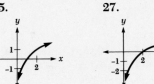

33. $y = 0$, $y = 0$, $y = 0$, $y = 0$, $x = 0$, $x = 0$, $x = -2$, $x = 1$
35. The number is initially 3, and it doubles every hour.

Exercises 6.2 (page 225)

1. $\log_{10} 2 + \log_{10} x$ **3.** $\log_3 a - \log_3 7$ **5.** $5 \log_2 x$ **7.** 4
9. $\log_2 x + 2 \log_2 y + \log_2 z$
11. $\frac{1}{2}(\log_3 11 + \log_3 a + 2 \log_3 b)$
13. 5   **15.** $\frac{1}{2} \log_{10} 2 + \log_{10} (x + y + z)$

17. $\log_2 9x$ **19.** $\log_3 \dfrac{2}{w}$ **21.** $\log_{10} x^2 z$ **23.** $\log_7 \sqrt{\dfrac{x}{y}}$

25. 2   **27.** $\log_3 \dfrac{9}{x}$ **29.** $\log_2 \dfrac{x}{5y^3}$ **31.** $\log_{10} \dfrac{10c^2}{x}$

Exercises 6.3 (page 227)

1. 1.92×10^3 **3.** 3.69×10^1 **5.** 2.1×10^{-1}
7. 5.29×10^0 **9.** 1.18×10^{-8} **11.** 72   **13.** 9980
15. 0.000050437   **17.** 655,000   **19.** 0.00000000000396
21. (a) 5.4×10^4 (b) 5.36×10^4 (c) 5.362×10^4
23. (a) 7.5×10^1 (b) 7.49×10^1 (c) 7.490×10^1
25. (a) 2.3×10^{-6} (b) 2.35×10^{-6} (c) 2.345×10^{-6} **27.** 3.8×10^6
29. 7.69×10^{-4} **31.** 8.873×10^3 **33.** 2.7×10^0

Exercises 6.4 (page 234)

1. 0.0000   **3.** $0.2122 + 1$ **5.** $0.3979 - 1$ **7.** $0.0569 - 4$
9. $0.2672 + 10$ **11.** 778   **13.** 2.65   **15.** 0.0217
17. 0.000438   **19.** 13,100   **21.** 42,000   **23.** 9.50
25. 117   **27.** 106,000   **29.** 10.3   **31.** 0.9550
33. $0.7449 - 2$ **35.** 546.6

Exercises 6.5 (page 238)

1. 1.431   **3.** 0.827   **5.** 5   **7.** 1.316   **9.** 5.773

11. $\dfrac{\log x}{\log 2} = 3.322 \log x$ **13.** $\dfrac{2 \log x}{\log 7} = 2.367 \log x$

15. $\dfrac{\log x}{2 \log 5} = 0.715 \log x$ **17.** 0.0296   **19.** 4.414

21. -2.025 **23.** -5.482 **25.** 1.112   **27.** 2.320   **29.** 5.140

Exercises 6.6 (page 242)

1. 3   **3.** 6   **5.** $\frac{9}{10}$ **7.** $\frac{5}{2}$ **9.** 3   **11.** 10

13. $\dfrac{1}{\log 5} = 1.431$ **15.** $\dfrac{\log 4}{2 \log 3} = 0.631$ **17.** $\frac{9}{2}$ **19.** 30.9

21. 1.09   **23.** $1 + \ln 3.14 = 2.144$ **25.** 1.74 hours

27. 11 years   **29.** 0.00214

Review Exercises, Chapter 6 (page 244)

1. $100 = 10^2$ **3.** $\frac{1}{3} = \log_{125} 5$ **5.** asymptote is $y = 0$

7. $2[\log (x + y) - \log z]$ **9.** $\log_3 9\sqrt{x}$ **11.** 3.7608×10^4

13. 0.00001992   **15.** 9.1×10^3 **17.** 4.1×10^2

19. $0.1367 + 2$ **21.** 0.0255   **23.** 1.41   **25.** $0.0178 + 3$

27. 3.322   **29.** $1.048 \log x$ **31.** 3.517   **33.** 8.19

35. 2   **37.** 1.09   **39.** 0.019 years

Exercises 7.2 (page 255)

1. $A: 2 \times 2$; $B: 1 \times 3$; $C: 3 \times 1$; $D: 3 \times 3$; $E: 3 \times 1$; $F: 2 \times 2$

3. $d_{21} = 1, d_{11} = 1, d_{33} = 5$ **5.** They do not have the same order.

7. $\begin{pmatrix} 1 \\ 2 & 3 \end{pmatrix}$; it is not rectangular.   **9.** C and E

11. $x = -1, y = 2$ **13.** $x = 4, y = 1$

15. $\begin{pmatrix} 1000 & 2000 & 5000 \\ 6000 & 3000 & 8000 \\ 0 & 3000 & 10{,}000 \end{pmatrix}$ **17.** $\begin{pmatrix} 2 & 2 \\ 3 & 6 \end{pmatrix}$

19. Not possible (different order)   **21.** $\begin{pmatrix} 0 \\ 0 \end{pmatrix}$ **23.** $\begin{pmatrix} 8 \\ 10 \end{pmatrix}$

25. $\begin{pmatrix} 5 & 3 \\ 3 & 11 \end{pmatrix}$ **27.** $(6 \quad 0)$ **29.** $\begin{pmatrix} -1 & -2 \\ -3 & -4 \end{pmatrix}, (-3 \quad 4)$

31. $\begin{pmatrix} 0 & 0 \\ 0 & 0 \end{pmatrix}$ **33.** (a) 75 (b) 150 (c) 40

Exercises 7.3 (page 260)

1. $\begin{pmatrix} -5 & 15 \\ 20 & 10 \\ 5 & 30 \end{pmatrix}$ **3.** $\begin{pmatrix} 0.4 & 1.4 \\ 1.6 & 1.2 \end{pmatrix}$ **5.** Not possible   **7.** $\begin{pmatrix} 5 \\ 19 \\ 14 \end{pmatrix}$

9. $\begin{pmatrix} 8 & 0 & 0 \\ 0 & -12 & 0 \\ 0 & 0 & 28 \end{pmatrix}$ **11.** $\begin{pmatrix} \$200 \\ \$300 \\ \$400 \\ \$600 \end{pmatrix}$ **13.** $\begin{pmatrix} \$160 \\ \$300 \\ \$430 \\ \$650 \end{pmatrix}$

Exercises 7.4 (page 268)

1. (11) **3.** $\begin{pmatrix} 3 & 6 \\ 4 & 8 \end{pmatrix}$ **5.** (0) **7.** $\begin{pmatrix} -4 & 28 \\ 4 & 2 \end{pmatrix}$

9. $\begin{pmatrix} x + 2y \\ -x + 3y \end{pmatrix}$ **11.** $\begin{pmatrix} 1 & 2 \\ 3 & 4 \end{pmatrix}$ **13.** $\begin{pmatrix} 1 & 0 & 0 \\ 0 & 1 & 0 \\ 0 & 0 & 1 \end{pmatrix}$ **15.** $\begin{pmatrix} 0 & 0 \\ 0 & 0 \end{pmatrix}$

17. $(\$890)$ **19.** $(\$1700)$

Exercises 7.5 (page 276)

1. $(2, -1)$ **3.** $(3, 2)$ **5.** $(\frac{3}{2}, -\frac{1}{2})$ **7.** $(1, -5)$
9. $(2, 1, -1)$ **11.** $(3, \frac{1}{2}, -2)$ **13.** inconsistent
15. dependent   **17.** 12 nickels, 6 dimes, 6 quarters
19. Inconsistent; the company cannot make the mix as stated.

Exercises 7.6 (page 283)

1. not inverses   **3.** not inverses   **5.** not inverses
7. $\begin{pmatrix} -2 & 1 & 3 \\ 2 & -1 & -2 \\ 1 & 0 & -1 \end{pmatrix}$ **9.** $\begin{pmatrix} \frac{1}{2} & -1 & -\frac{3}{2} \\ 0 & 1 & 2 \\ 0 & 1 & 1 \end{pmatrix}$ **11.** no inverse

13. $\begin{pmatrix} \frac{1}{3} & \frac{1}{3} \\ \frac{2}{3} & -\frac{1}{3} \end{pmatrix}$ **15.** $\begin{pmatrix} -\frac{6}{21} & \frac{9}{21} \\ \frac{5}{21} & -\frac{4}{21} \end{pmatrix}$ **17.** $\begin{pmatrix} 1 & 0 & 0 \\ -16 & -2 & 3 \\ 6 & 1 & -1 \end{pmatrix}$

19. $\begin{pmatrix} \frac{1}{2} & 0 & 0 \\ 0 & \frac{1}{3} & 0 \\ 0 & 0 & \frac{1}{4} \end{pmatrix}$

Exercises 7.7 (page 288)

1. $4x_1 + 3x_2 = 11$ **3.** $3x_1 - x_2 - 2x_3 = -13$
$\underline{2x_1 - 3x_2 = 7}$ $5x_1 + 3x_2 - x_3 = 4$
                              $2x_1 - 7x_2 + 3x_3 = -30$

5. $x_1 = 1, x_2 = 3, x_3 = -1$ **7.** $x_1 = 2, x_2 = 3$ **9.** dependent

11. $x = -1, y = 3, z = 2$ **13.** inconsistent   **15.** dependent

17. $x = 1, y = 3$

19. $x + 2y + z = 500$ $\begin{pmatrix} -1 & 1 & 1 \\ 2 & -1 & -2 \\ -2 & 1 & 3 \end{pmatrix}$; $\begin{aligned} x &= 100 \\ y &= 100 \\ z &= 200 \end{aligned}$
   $2x + y = 300$
   $y + z = 300$

21. 110, 110, 220

23. $0.1x + 0.2y + 0.2z = 300$ $\begin{pmatrix} -\frac{10}{3} & \frac{20}{3} & 0 \\ 10 & 0 & -10 \\ -\frac{10}{3} & -\frac{10}{3} & 10 \end{pmatrix}$ $\begin{aligned} x &= 200 \\ y &= 1000 \\ z &= 400 \end{aligned}$
   $0.2x + 0.1y + 0.1z = 180;$
   $0.1x + 0.1y + 0.2z = 200$

25. (a) $x + y + z = 24$ $\begin{pmatrix} 1 & -1 & -1 \\ -1 & 1 & 2 \\ 1 & 0 & -1 \end{pmatrix}$; $\begin{aligned} x &= 10 \\ y &= 2 \\ z &= 12 \end{aligned}$ (b) $\begin{aligned} x &= 8 \\ y &= 6 \\ z &= 10 \end{aligned}$
      $z = x + 2;$
   $x + y = 12$

27. $x + y + z = 6$ $x = 4$
   $x - y = 2;$ $y = 2$
   $2y - z = 4$ $z = 0$

Exercises 7.8 (page 297)

1. 1   **3.** 3   **5.** 25   **7.** $xy - 5\sqrt{3}$ **9.** -3

11. -6 **13.** $x = 3, y = 2$ **15.** $x = \frac{5}{2}, y = (-\frac{1}{2})$

17. $x = 2, y = \frac{1}{2}$ **19.** $x = 2, y = 1, z = -1$

21. $x = \frac{3}{2}, y = -4, z = \frac{1}{6}$ **23.** tens digit $= 3$, units digit $= 9$; 39

25. 14 nickels, 16 pennies

27. speed of plane $= 180$ miles per hour; speed of wind $= 20$ miles per hour

29. (a) $\det(A) = 21$ (b) $\det(B) = 0$

Review Exercises, Chapter 7 (page 299)

1. $\begin{pmatrix} 9 & -3 \\ -1 & 4 \end{pmatrix}$ **3.** not possible   **5.** $\begin{pmatrix} 6 \\ -5 \end{pmatrix}$

7. not possible   **9.** not possible   **11.** (-4) **13.** $\begin{pmatrix} 4 \\ 5 \\ 2 \end{pmatrix}$

15. not possible   **17.** $\begin{pmatrix} \frac{1}{2} & -\frac{3}{2} \\ 2 & -1 \end{pmatrix}$ **19.** $\begin{pmatrix} 0.48 & 0.12 \\ 0.78 & 3.0 \end{pmatrix}$

21. $\begin{pmatrix} 0 & -1 \\ \frac{1}{2} & \frac{1}{2} \end{pmatrix}$ **23.** does not exist   **25.** $\begin{pmatrix} -1 & -3 & 3 \\ 1 & 2 & -2 \\ 2 & 4 & -5 \end{pmatrix}$

27. does not exist   **29.** $\begin{pmatrix} \dfrac{1}{a} & 0 \\ 0 & \dfrac{1}{a} \end{pmatrix}$, $a \neq 0$ **31.** $x = 1, y = -1$

33. $(1, 1, 1)$ **35.** dependent   **37.** dependent
39. $x = \$1920$ (TV), $y = \$640$ (radio), $z = \$1440$ (newspapers)
41. $x = \$2700, y = \$900, z = \$1400$

Exercises 8.1 (page 308)

1. 5, 7, 9, 11, 13; 10th term = 23   **3.** 0, 2, 0, 2, 0; 10th term = 2
5. $-\frac{1}{2}, \frac{1}{3}, -\frac{1}{4}, \frac{1}{5}, -\frac{1}{6}$; 10th term = $\frac{1}{11}$
7. $1, 1, \frac{4}{5}, \frac{11}{17}, \frac{7}{13}$; 10th term = $\frac{29}{101}$ **9.** $0, \frac{3}{10}, \frac{8}{21}, \frac{15}{36}, \frac{24}{55}$; 10th term = $\frac{99}{210}$
11. $\frac{3}{2}, \frac{5}{4}, \frac{9}{8}, \frac{17}{16}, \frac{33}{32}$; 10th term = $\frac{1025}{1024}$ **13.** $d = 2$ **15.** $d = -3$
17. $d = -2$ **19.** $d = \frac{1}{3}$ **21.** $0, -3, -6$; 50th term = -147
23. 11, 14, 17; 100th term = 308   **25.** $\frac{3}{2}, 2, \frac{5}{2}$; 50th term = 26
27. $10, \frac{25}{2}, 15$; 10th term = $\frac{65}{2}$ **29.** $18,000
31. (a) $124, $123, $122   (b) $115   **33.** 1300
35. (a) $6.50   (b) $12.50   **37.** $r = -3$
39. not a geometric sequence   **41.** $r = \frac{2}{3}$
43. $\frac{1}{2}, -\frac{1}{2}, \frac{1}{2}$; 100th term = $-\frac{1}{2}$ **45.** $10, 5, \frac{5}{2}$; 10th term = $\frac{5}{256}$
47. $40, -20, 10$; 8th term = $-\frac{5}{16}$ **49.** $32,071.35   **51.** $10(3^{n+1})$
53. 59%

Exercises 8.2 (page 312)

1. $3 + 7 + 11 + 15 + 19 + 23 + 27 + 31$
3. $16 + 25 + 36 + 49 + 64 + 81$
5. $\frac{1}{16} + \frac{1}{25} + \frac{1}{36} + \frac{1}{49}$
7. $5 + 5 + 5 + 5 + 5 + 5 + 5 + 5 + 5 + 5$
9. $\frac{1}{4} + \frac{1}{9} + \frac{1}{16} + \frac{1}{25}$

11. $4 + 8 + 16 + 32 + 64 + 128 + 256 + 512$

13. $3 + \frac{3}{2} + \frac{3}{4} + \frac{3}{8} + \frac{3}{16}$

15. $27 + 44 + 65 + 90 + 119$

17. $\displaystyle\sum_{k=1}^{10} 2^k$ **19.** $\displaystyle\sum_{k=1}^{7} 3$ **21.** $\displaystyle\sum_{k=1}^{20} \frac{1}{k}$ **23.** $\displaystyle\sum_{k=1}^{n} \frac{1}{x_k}$

25. $\displaystyle\sum_{k=1}^{6} \left[(-1)^{k+1} \cdot \frac{3k}{2^k} \right]$

Exercises 8.3 (page 319)

1. $S_n = n^2,\ S_{20} = 400$ **3.** $S_n = n(n-6),\ S_{20} = 280$

5. $S_n = \dfrac{n^2}{2},\ S_{20} = 200$ **7.** $S_n = \dfrac{n(n+2)}{6},\ S_{20} = \dfrac{220}{3}$

9. $S_n = \dfrac{n(n+19)}{4},\ S_{20} = 195$ **11.** 185 **13.** -2325

15. 1395 **17.** $\frac{3399}{2}$ **21.** $\$156$

23. (a) 1600 feet   (b) 6400 feet   (c) 400 feet

25. (a) $\$55$ (b) $\$210$ (c) $\$66{,}795$ **27.** $\dfrac{3^n - 1}{2}$ **29.** $1 - \dfrac{1}{2^n}$

31. $\dfrac{4}{3}\left[1 - \left(\dfrac{-1}{2}\right)^n\right]$ **33.** $18\left[1 - \left(\dfrac{2}{3}\right)^n\right]$

35. $\dfrac{59{,}048}{3},\ \dfrac{1023}{1024},\ -341,\ \dfrac{341}{256}$ **37.** $\dfrac{127}{16}$ **39.** -3

41. $\dfrac{4{,}194{,}303}{16{,}777{,}216}$

43. $420[1 - (\frac{1}{3})^7] = \dfrac{30{,}604}{729} \approx 419.81$ feet

45. (a) $180[1 - (\frac{4}{5})^5] = \dfrac{75{,}636}{625} \approx 121.02$ feet

(b) $180[1 - (\frac{4}{5})^7] = \dfrac{2{,}222{,}676}{15{,}625} \approx 142.25$ feet

Review Exercises, Chapter 8 (page 324)

1. $2, 4, 8, 16, 32;\ 1024$ **3.** $\frac{2}{3}, \frac{3}{4}, \frac{4}{5}, \frac{5}{6}, \frac{6}{7};\ \frac{11}{12}$

5. $1.3,\ 1.69,\ 2.197,\ 2.8561,\ 3.71293;\ 13.78585$

7. $d = 3$ and $S_{20} = 630$ **9.** $d = \sqrt{2}$ and $S_{20} = 210\sqrt{2}$

11. $d = 1$ and $S_{20} = 2190$ **13.** 198 **15.** $r = -1$ and $S_{20} = 0$

17. not a geometric progression   **19.** $r = -\frac{1}{4}$ and $S_{20} = \dfrac{4^{20} - 1}{5(4^{18})}$

21. 1536 **23.** 4096 **25.** $\frac{5}{4} + \frac{7}{5} + \frac{9}{6} + \frac{11}{7} + \frac{13}{8} + \frac{15}{9} + \frac{17}{10} + \frac{19}{11}$

27. 140 **29.** $\frac{2343}{512}$ **33.** $\$468$

Exercises 9.1 (page 334)

1. 24   **3.** (a) 216 (b) 120   **5.** 720   **7.** 120

9. $\binom{52}{5} = 2{,}598{,}960$ **11.** 720   **13.** 360   **15.** 64

17. 24,360   **19.** (a) 210 (b) 210 (c) 604,800 (d) 792

21. $\binom{10}{6} \cdot 6! = 151{,}200$ **23.** 56

25. (a) 6188 (b) 1575 (c) 4466   **27.** 10

Exercises 9.2 (page 339)

1. $x^5 + 5x^4y + 10x^3y^2 + 10x^2y^3 + 5xy^4 + y^5$
3. $a^4 + 12a^3b + 54a^2b^2 + 108ab^3 + 81b^4$
5. $64x^6 - 192x^5y + 240x^4y^2 - 160x^3y^3 + 60x^2y^4 - 12xy^5 + y^6$
7. $x^{10} + 10x^8y + 40x^6y^2 + 80x^4y^3 + 80x^2y^4 + 32y^5$
9. $729z^{12} - 2916z^{10}w + 4860z^8w^2 - 4320z^6w^3 + 2160z^4w^4 - 576z^2w^5 + 64w^6$
11. $a^{21} - 7a^{18}b + 21a^{15}b^2 - 35a^{12}b^3 + 35a^9b^4 - 21a^6b^5 + 7a^3b^6 - b^7$
13. $729x^6 - 2916x^5y + 4860x^4y^2 - 4320x^3y^3 + 2160x^2y^4 - 576xy^5 + 64y^6$
15. $243s^5 + 810s^4t + 1080s^3t^2 + 720s^2t^3 + 240st^4 + 32t^5$
17. $u^6 - 12u^5v + 60u^4v^2 - 160u^3v^3 + 240u^2v^4 - 192uv^5 + 64v^6$
19. $x^{10} + 10x^9y + 45x^8y^2 + 120x^7y^3 + 210x^6y^4 + 252x^5y^5 + 210x^4y^6 +$
$\quad 120x^3y^7 + 45x^2y^8 + 10xy^9 + y^{10}$
21. 1.0824   **23.** 16.9818   **25.** 17.6610
27. $1024x^{10} - 15{,}360x^9y + 103{,}680x^8y^2 - 414{,}720x^7y^3 + \cdots$
29. $512a^{18} + 6912a^{16}b + 41{,}472a^{14}b^2 + 145{,}152a^{12}b^3 + 326{,}592a^{10}b^4 + \cdots$
31. $x^{20} - 20x^{19}y + 190x^{18}y^2 - 1140x^{17}y^3 + 4845x^{16}y^4 - \cdots$
33. $-414{,}720x^7y^3$ **35.** $8064x^{10}y^{10}$ **37.** \$176.23

Exercises 9.3 (page 350)

1. $\{(H, H), (H, T), (T, H), (T, T)\}$ **3.** {Hit, Miss}
5. $\{(D, D, D), (D, D, N), (D, N, D), (N, D, D), (N, N, D), (N, D, N), (D, N, N),$
$(N, N, N)\}$
7. $\{(1, 1, 1), (1, 1, 2), (1, 1, 3), (1, 1, 4), (1, 1, 5), (1, 1, 6), (1, 2, 1), (1, 2, 2),$
$(1, 2, 3), (1, 2, 4) \ldots\}$
(216 outcomes)
9. $A = \{(D, D, N), (D, N, D), (N, D, D)\}$
$B = \{(D, D, N), (D, N, D), (N, D, D), (D, D, D)\}$
11. $\frac{1}{2}$ **13.** (a) $\frac{1}{13}$ (b) $\frac{1}{13}$ (c) $\frac{2}{13}$ (d) $\frac{12}{13}$ **15.** $\frac{125}{216}$
17. (a) $\{(K, K, K), (K, K, N), (K, N, K), (N, K, K), (N, N, K), (N, K, N),$
$(K, N, N), (N, N, N)\}$
(b) $P(K, K, K) = \frac{1}{2197}$,
$P(K, K, N) = P(K, N, K) = P(N, K, K) = \frac{12}{2197}$,
$P(N, N, K) = P(N, K, N) = P(K, N, N) = \frac{144}{2197}$,
$P(N, N, N) = \frac{1728}{2197}$

19. (a) $\{(G, G), (G, S), (S, G), (S, S)\}$

(b) $P(G, G) = \frac{9}{16}$, $P(G, S) = \frac{3}{16}$, $P(S, G) = \frac{3}{16}$, $P(S, S) = \frac{1}{16}$ (c) $\frac{5}{8}$

21. 0.2646

23. (a) 0.2612736 (b) 0.1306368 (c) 0.0279936 (d) 0.419904

Review Exercises, Chapter 9 (page 352)

1. 24   **3.** 126   **5.** 5880   **7.** 252   **9.** 120

11. 2520   **13.** 24   **15.** $\binom{52}{13} = \dfrac{52!}{(13!)(39!)}$

17. $81x^4 - 432x^3y + 864x^2y^2 - 768xy^3 + 256y^4$

19. $243s^5 + 405s^4t^2 + 2700s^3t^4 + 90s^2t^6 + 15st^8 + t^{10}$

21. $32x^{10} - 240x^8y + 720x^6y^2 - 1080x^4y^3 + 810x^2y^4 - 243y^5$

23. $729x^{12} - 2916x^{10} + 4860x^8 - 4320x^6 + 2160x^4 - 576x^2 + 64$

25. $59{,}049x^{10} + 393{,}660x^9y + 1{,}180{,}980x^8y^2 + 2{,}099{,}520x^7y^3$

27. 824.1265   **29.** \$472.06   **31.** $153{,}090x^8y^{12}$

33. $\{(1, H, R), (1, H, B), (1, T, R), (1, T, B)$

$(2, H, R), (2, H, B), (2, T, R), (2, T, B)$

$(3, H, R), (3, H, B), (3, T, R), (3, T, B)$

$(4, H, R), (4, H, B), (4, T, R), (4, T, B)$

$(5, H, R), (5, H, B), (5, T, R), (5, T, B)$

$(6, H, R), (6, H, B), (6, T, R), (6, T, B)\}$

35. $\{(R, R), (R, B), (R, G), (B, B), (B, R), (B, G), (G, G), (G, R), (G, B)\}$

37. $B = \{(2, H, R), (2, H, B), (2, T, R), (2, T, B)$

$(4, H, R), (4, H, B), (4, T, R), (4, T, B)$

$(6, H, R), (6, H, B), (6, T, R), (6, T, B)\}$; $P(B) = \frac{1}{2}$

39. (a) $\{(B, B), (B, G), (G, B), (G, G)\}$

(b) $P(B, B) = \frac{4}{9}$, $P(B, G) = P(G, B) = \frac{2}{9}$, $P(G, G) = \frac{1}{9}$ (c) $\frac{5}{9}$

41. $\frac{5}{16}$

Index